LABORATORY METHODS IN DYNAMIC ELECTROANALYSIS

LABORATORY METHODS IN DYNAMIC ELECTROANALYSIS

Edited by

M. TERESA FERNÁNDEZ ABEDUL

Departamento de Química Física y Analítica, Facultad de Química,
Universidad de Oviedo, Oviedo, Spain

ELSEVIER

Elsevier
Radarweg 29, PO Box 211, 1000 AE Amsterdam, Netherlands
The Boulevard, Langford Lane, Kidlington, Oxford OX5 1GB, United Kingdom
50 Hampshire Street, 5th Floor, Cambridge, MA 02139, United States

Notices

Knowledge and best practice in this field are constantly changing. As new research and experience broaden our understanding, changes in research methods, professional practices, or medical treatment may become necessary.

Practitioners and researchers must always rely on their own experience and knowledge in evaluating and using any information, methods, compounds, or experiments described herein. In using such information or methods they should be mindful of their own safety and the safety of others, including parties for whom they have a professional responsibility.

To the fullest extent of the law, neither the Publisher nor the authors, contributors, or editors, assume any liability for any injury and/or damage to persons or property as a matter of products liability, negligence or otherwise, or from any use or operation of any methods, products, instructions, or ideas contained in the material herein.

Library of Congress Cataloging-in-Publication Data
A catalog record for this book is available from the Library of Congress

British Library Cataloguing-in-Publication Data
A catalogue record for this book is available from the British Library

ISBN: 978-0-12-815932-3

For information on all Elsevier publications visit our
website at https://www.elsevier.com/books-and-journals

Publisher: Susan Dennis
Acquisition Editor: Kathryn Eryilmaz
Editorial Project Manager: Ruby Smith
Production Project Manager: Swapna Srinivasan
Cover Designer: Greg Harris

Typeset by TNQ Technologies

To my daughters, Carla and Alejandra

Contents

II

Electroanalysis and microfluidics

13. Single- and dual-channel hybrid PDMS/glass microchip electrophoresis device with amperometric detection

Andrea González-López, Mario Castaño-Álvarez and
M. Teresa Fernández Abedul

14. Analysis of uric acid and related compounds in urine samples by electrophoresis in microfluidic chips

Diego F. Pozo-Ayuso, Mario Castaño-Álvarez and
Ana Fernández-la-Villa

15. Microchannel modifications in microchip reverse electrophoresis for ferrocene carboxylic acid determination

Rebeca Alonso-Bartolomé, Andrea González-López and
M. Teresa Fernández Abedul

16. Integrated microfluidic electrochemical sensors to enhance automated flow analysis systems

Mario Castaño-Álvarez, Diego F. Pozo-Ayuso and
Ana Fernández-la-Villa

III

Bioelectroanalysis

17. Bienzymatic amperometric glucose biosensor

María Carmen Blanco-López, M. Jesús Lobo-Castañón and
M. Teresa Fernández Abedul

VI
Multiplexed electroanalysis

VII
Spectroelectrochemical techniques

VIII
General considerations

36. Bibliographic resources in electroanalysis

M. Jesús Lobo-Castañón and M. Teresa Fernández Abedul

Contributors

Rebeca Alonso-Bartolomé Departamento de Química Física y Analítica, Universidad de Oviedo, Oviedo, Spain

Isabel Álvarez-Martos Interdisciplinary Nanoscience Center (iNANO). Aarhus University, Aarhus, Denmark

Olaya Amor-Gutiérrez Departamento de Química Física y Analítica, Universidad de Oviedo, Oviedo, Spain

Julien Biscay Department of Pure and Applied Chemistry, Technology and Innovation Centre, University of Strathclyde, Glasgow, United Kingdom

María Carmen Blanco-López Departamento de Química Física y Analítica, Universidad de Oviedo, Oviedo, Spain

Mario Castaño-Álvarez MicruX Technologies, Gijón, Asturias, Spain

Agustín Costa-García Departamento de Química Física y Analítica, Universidad de Oviedo, Oviedo, Spain

Estefanía Costa-Rama REQUIMTE/LAQV, Instituto Superior de Engenharia do Porto, Instituto Politécnico do Porto, Porto, Portugal; Departamento de Química Física y Analítica, Universidad de Oviedo, Oviedo, Spain

Noemí de los Santos Álvarez Departamento de Química Física y Analítica, Universidad de Oviedo, Oviedo, Spain

Pablo Fanjul-Bolado Metrohm Dropsens, Parque Tecnológico de Asturias, Edificio CEEI, Asturias, Spain

Ana Fernández-la-Villa MicruX Technologies, Gijón, Asturias, Spain

M. Teresa Fernández Abedul Departamento de Química Física y Analítica, Universidad de Oviedo, Oviedo, Spain

Raquel García-González Departamento de Química Física y Analítica, Universidad de Oviedo, Oviedo, Spain

Pablo García-Manrique Departamento de Química Física y Analítica, Universidad de Oviedo, Oviedo, Spain

María Begoña González-García Metrohm Dropsens, Parque Tecnológico de Asturias, Edificio CEEI, Asturias, Spain

Andrea González-López Departamento de Química Física y Analítica, Universidad de Oviedo, Oviedo, Spain

Charles S. Henry Department of Chemistry, Colorado State University, Fort Collins, CO, United States

David Hernández-Santos Metrohm Dropsens, Parque Tecnológico de Asturias, Edificio CEEI, Asturias, Spain

M. Jesús Lobo-Castañón Departamento de Química Física y Analítica, Universidad de Oviedo, Oviedo, Spain

Graciela Martínez-Paredes OSASEN Sensores S.L., Parque Científico y Tecnológico de Bizkaia, Derio, Spain

Rebeca Miranda-Castro Departamento de Química Física y Analítica, Universidad de Oviedo, Oviedo, Spain

Paula Inés Nanni Laboratorio de Medios e Interfases, Universidad Nacional de Tucumán, Tucumán, Argentina

Estefanía Núñez Bajo Department of Bio-engineering, Royal School of Mines, Imperial College London, London, United Kingdom

Alejandro Pérez-Junquera Metrohm Dropsens, Parque Tecnológico de Asturias, Edificio CEEI, Asturias, Spain

Diego F. Pozo-Ayuso MicruX Technologies, Gijón, Asturias, Spain

Preface

I started teaching graduate courses in Analytical Chemistry in 1994, one year before finishing my PhD on Electroanalysis. By that time, Polarography, and with the exciting renewable mercury drop, was very interesting from a didactic point of view but giving rise to an incredible amount of different solid electrodes, based on carbon and metals. By that time, I worked mainly with carbon paste electrodes. Technology advanced and conductive materials took the form of films, fibers, etc., to be employed in attractive applications. Combination with biochemistry started soon and was followed by the integration of nanotechnological approaches. The merging of flow systems with Electroanalysis was very successful and continued with the incorporation of detectors in microfluidic systems. Miniaturization was not yet a priority and analyses were centralized in most of the cases. However, innovative low-cost designs favored the decentralization of the analysis, relevant in both developed and developing countries. Some properties for analytical devices such as being stretchable, flexible, disposable, user-friendly, etc., were not in the electroanalytical vocabulary, even when Electroanalysis already had an intrinsic and enormous potential. It was also reinforced through the combination with other principles such as spectrometry or microscopy.

All along these years I could see a great advance, not only in Electroanalysis but also in Analytical Chemistry in general. It is time to place a new milestone in its evolution The arrival of sophisticated instruments and the democratization of computers generated excellent analytical equipment. However, the so many technological approaches that followed produced devices with new and unimaginable properties. Analytical Chemistry has entered a "bifid" era where the improvements of high-tech centralized instrumentation live together with the amazing advance of decentralized analytical platforms. It is here where Electroanalysis has an advantageous place. As most of the measurements are interfacial, only a conductive surface is required to act as electrode. It does not matter (or it really does!) if it is transparent, light, wearable, or paper-based; possibilities are infinite. The field is opened to creative and useful designs many different electrodes, electrochemical cells, and applications flourish in a very promising field. Topics such as bipolar electrochemistry, highly-multiplexed electroanalysis, single-cell analysis, application to genetics/epigenetics, development of tiny potentiostats, combination with energy devices, etc., are not covered in this book because this is a "to start with" treatise. It can be considered as a walk through different electroanalytical aspects to approach the field that have to be known by someone that approaches the field.

There are excellent books on Electroanalysis to learn fundamentals, principles, and applications, as well as scientific articles and review. This book aims to be of help to teachers, students, and researcher who want to enter the field, especially from the experimental side. The chapters are introduced with detailed instructions for experiments, not only in Electroanalysis (as a part of Analytical Chemistry), but also in other branches of Chemistry, Biochemistry, Environmental Sciences, and others that can benefit from it require understanding about in this specialty. During the years I have been teaching Analytical Chemistry (including Electroanalysis), I could see that this discipline provokes love and hate in a similar manner, and therefore, favoring a clear and deep understanding is the only way to take advantage of this fascinating field. Teachers and students, can find here a way to approach different aspects of Electroanalysis, not only in laboratories but also doing activities in classes that have to be converted into highly-motivating sessions. Chapters are commented as experiment guides with introduction, procedures, lab reporting, additional notes, diagrams, and schemes as well as questions and references. They can be distributed to groups of students to work with them. They can prepare topics on well-established techniques or on new trends, and a final (or initial) experiment can be performed to close (or start) a unit. Also researchers (juniors and seniors) who are starting in the area can benefit from this book because materials, designs, techniques, etc., are included in these 36 chapters. They are aimed to introduce dynamic electroanalysis, making easy what seems difficult (but it is not!).

Only dynamic electroanalytical techniques (with measurements taken while a current is flowing through the circuit), a part of interfacial electroanalysis, are considered here. Potentiometry is a static technique of paramount relevance that would need a specific treatise. After an introduction (Chapter 1), the book is divided into eight parts. The first one is devoted to the different techniques and the rest to some main trends: electroanalysis and microfluidics, bioelectrochemistry, nanomaterials and electroanalysis, low-cost

electroanalysis, multiplexed electroanalysis, and spectroelectrochemistry. A final section deals with the design of experiments (Chapter 35) and bibliographic resources in Electroanalysis (Chapter 36). Although bibliographic references are included in each of the chapters, the final one incorporates a general vision.

In the first part, devoted to different techniques, the electrochemical cell is seen as a system under study that can undergo an excitation signal (a potential in all the cases). The response signal (measurement of current or impedance) is the analytical signal that can be employed with qualitative or quantitative purposes. Coulometry is an absolute technique with important applications (e.g., Karl Fischer water determination) that will not be considered here. In the techniques related with the measurement of the current, voltammetry and amperometry are considered. Although amperometry is a generic name for designating the techniques in which the current is measured, the term is commonly employed for those where readout is taken at fixed potentials, especially under convection. This can be provided by stirring the solution (Chapter 7) or by injecting the samples in using flow (Chapters 9 and 28) or static (Chapter 10) systems. In the case where the diffusion controls mass transport, the measurement at a fixed time can be related to the concentration and the technique is commonly known as chronoamperometry (Chapter 8). Regarding voltammetry, when the current is measured during a potential scan, cyclic voltammetry is first considered (Chapter 2). Different waveforms can be employed for increasing the sensitivity, and hence differential pulse voltammetry and square wave voltammetry are considered later (Chapter 3). Other strategies aimed to increase the sensitivity are the incorporation of a preconcentration step (Chapters 4 and 5) or the use of specific electrode designs as occurs with interdigitated electrodes (Chapter 6). One special potential scan is this in which an alternating potential (frequency-dependent) is superimposed on a linear scan (alternating current voltammetry, Chapter 5). When different frequencies are scanned and the corresponding impedance is measured, the technique is named electrochemical impedance spectroscopy, very useful for label-free analysis (Chapter 11) or characterization of systems (Chapter 12).

Electroanalysis is evolving with the advances in technologies. As an example, it has been adapted not only to flow systems but also to microfluidics (part II). One of the possibilities is the integration with separation techniques, as in the case of microchip electrophoresis, working either in normal (Chapters 13 and 14) or reverse (Chapter 15) mode. This combination between microfluidics and electroanalysis has demonstrated to be very advantageous even for without separation purposes different from separation, as commented in Chapter 16.

The relevance of bioelectrochemistry is undoubted. The relationship between biochemistry and electroanalysis has proven to be very successful for a long time. The power of amplification of enzymes and the selectivity of the molecular recognition events are some of the causes of the enormous advance in the field of biosensors. Here (part III), examples of enzymatic (Chapters 17–19), immune (Chapter 20), or DNA (Chapters 21 and 22, the last one with aptamers) assays are considered. The cost and activity of Miniaturization and low-cost approaches can be very advantageous to the field. On the other hand, if biomaterials have demonstrated their relevance, nanomaterials run in parallel. This is a field that would deserve a specific treatise (as does biosensing); here (part IV) only some examples with metal nanoparticles (Chapter 23) and carbon nanotubes (Chapter 24) are discussed.

Society demands information, and analysis is a way to obtain it. This is a golden age for Analytical Chemistry and also for Electroanalysis as provider of powerful tools for decentralization. Low-cost assays (part V) are an example. Here, the use of paper (Chapters 25 and 26) or the employ of elements in "out-of-box" applications, such as pins (Chapters 27 and 28) or staples (Chapter 29) are some examples. Another requirement is multiplexing (part VI), a term that comes from telecommunications and refers to the strategy to obtain several signals with the same medium. In this case, the possibility of performing two (Chapter 31) or eight (Chapter 30) simultaneous measurements is discussed. The part VII is devoted to the combination of electroanalysis with optical principles to form hybrid approaches, as in the case of electrochemiluminescence (Chapters 32 and 33) or surface-enhanced Raman scattering (Chapter 34).

Most of the experiments described in the book have been adapted from research articles, made by the groups working in Electroanalysis at the University of Oviedo or in companies that emerged from results of PhD students, such as MicruX or DropSens. Chapter 4 was started at Colorado State University under the kind advice of Prof. Henry, but research was resumed and finished in Oviedo. Many more experiments could have been included, and also the field extended to all the outstanding work is being made in the area, but limits have to be established.

In conclusion, this book is a starting point. I encourage readers to submit comments, suggestions, and ideas to improve it. I really hope you find it informative and helpful for your research. It is, for sure, a fascinating field. Good luck!

M. Teresa Fernández Abedul
June 1, 2019, Oviedo

Acknowledgments

Science is not only about experiments but also about people and environments. Science is a place to be. Then, I would like to thank all the people who shared it with me and contributed to increase the knowledge I acquired in this field, as well as the motivation to continue learning. I have started my research in the group of Electroanalysis lead by Prof. Paulino Tuñón Blanco. Later on, I continued in the group of Immuno-electroanalysis lead by Prof. Agustín Costa García, my PhD advisor. I am very grateful to him for all the conversations and moments shared around electrodes, cells, and techniques. I am also thankful to all the colleagues and students who shared with me those electroanalytical moments, especially Begoña González García, excellent colleague and friend. Also, I would like to thank MicruX people: Mario Castaño Álvarez, Ana Fernández la Villa, and Diego Pozo Ayuso, outstanding students, entrepreneurs and friends.

I am also very grateful to Prof. George S. Wilson (Kansas University) for hosting me in his group when I was PhD student and wanted to learn about Immunoanalysis. Also to Prof. William R. Heineman (University of Cincinnati) who welcomed me, very kindly, in a brief postdoctoral research stay. I keep with affection a dedicated second edition of his book "Laboratory Techniques in Electroanalytical Chemistry." In a third stage of my research life, I want to thank Prof. George M. Whitesides (Harvard University) for hosting me during four fascinating and creative summers I spent in his lab. I include in my thanks all the wonderful people in the groups. I keep nice memories as a precious treasure.

Also, I want to thank colleagues and students (Andrea González López and Olaya Amor Gutiérrez as current excellent PhD students, but I could name many more) from my Department and others as well as in other Universities, with whom I enjoy commenting and discussing scientific and other issues. More related to this book, I would like to thank PhD student Pablo García Manrique for showing me a book of similar structure. This gave me the idea of putting together, in the form of experiments, some of the research done. I also want to thank all the authors for their excellent contributions. I would like to give special thanks to Prof. M. Jesús Lobo Castañón, Dr. Estefanía Costa Rama, and Dr. Arturo.J. Miranda Ordieres for their useful suggestions. Working on this project was easier with the patience and kind reminders of Ruby Smith, Indhumati Mani and Swapna Srinivasan (Elsevier). Finally, needless to say, my last but warmest thanks are given to my family and friends.

M. Teresa Fernández Abedul

Dynamic electroanalysis: an overview

M. Teresa Fernández Abedul

Departamento de Química Física y Analítica,
Universidad de Oviedo, Oviedo, Spain

1.1 Dynamic electroanalysis

In the era in which we require more and more information and this has to be obtained by everyone, everywhere, and at any time, Electroanalytical Chemistry is becoming of tremendous relevance. The trend toward decentralization (that can benefit from several others: miniaturization [1], low cost, multiplexing, etc.) is becoming very strong in Analytical Chemistry. Then, traditional laboratories are being replaced for places where autonomous and portable devices can provide this information. Therefore, "flying laboratories" that refer to devices mounted on drones to analyze environmental samples [2], "edible sensors" concerning the manufacturing of ingestible (pills that can monitor events inside the body [3]) or digestible (sensors fabricated using real food [4]) monitoring components, and "lab-on-paper" devices, related to paper-based platforms that include different steps of the analytical process [5], are some of the examples related to the current implementation of in situ analysis. Unstoppable decentralization will surely extend the applications of Electroanalysis, a field with huge possibilities.

Electroanalysis comes from the combination of two "chemistries": Electrochemistry and Analytical Chemistry and then it is also referred to as Electroanalytical Chemistry and also as Analytical Electrochemistry. Electrochemistry developed from the single contributions of famous researchers and scientists in the 150 years spanning 1776 and 1925. Then, discoveries of Galvani, Volta, Faraday, Coulomb, and Ohm are very familiar, and most of the instruments and computers operate with electrical current [6]. In the past century, Nobel prizes to Arrhenius, Ostwald, Nernst, Tiselius, Heyrovsky, Taube, and Marcus were related also to Electrochemistry and that of Heyrovsky (1959) directly to Electroanalysis "for his discovery and development of the polarographic methods of analysis", which are based on the use of mercury electrodes. Electroanalysis deals with the analysis of electroactive species, but also non-electroactive (through indirect methodologies or derivatization procedures) employing electrochemical methods for a vast range of applications. Electrical entities (mainly charge, potential, current and impedance) are measured and correlated with the concentration of the analyte. Advances in the last decades of the 20th century, including the development of ultramicroelectrodes, the design of tailored interfaces and molecular monolayers, the integration of biological components and electrochemical transducers, the coupling with microscopes and spectroscopes, the microfabrication of devices, or the development of efficient flow detectors have lead to a substantial increase in popularity of Electroanalysis [7]. Evolution has continued during the first decades of the millennium, especially with the inclusion of nanotechnological approaches, the use of new conductive surfaces, the exponential miniaturization of cells and equipment, the incorporation of low-cost approaches, or the integration with smartphones. Certainly, this is the golden age of Electrochemistry. Never before has this discipline found itself at the nexus of so many developing technologies, not only in which refers to analytical applications (biomedical, food, environmental) but also industrial applications, material science, or theoretical chemistry. In this context, advances on electrochemical (bio)(nano)sensors and energy-related applications are notorious [8].

One of the main advantages of Electroanalysis is the variety of techniques that can be employed for extracting information from systems. In this book, electrodic (also named interfacial, related to processes happening in the electrode–electrolyte interface) techniques are considered (Fig. 1.1).

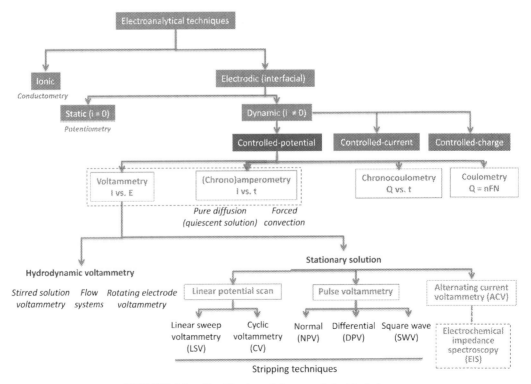

FIGURE 1.1   Classification of electroanalytical techniques.

Among them, we can distinguish static techniques (i.e., potentiometric) where the information about the concentration of the analyte is obtained from the measurement of a potential under equilibrium (zero current) conditions. The relevance and current advances of this technique [9] require a specific treatise for itself. Here, only dynamic (nonzero current) electroanalytical techniques, those based on measurements taken when current flows through the cell, are considered. There are many criteria to classify the electro-analytical techniques. These techniques rely on the active observation by the experimenter of the response signal that a system under study produces after an excitation signal (Fig. 1.2) [10]. Then, a simple criterion is the electrical entity that is measured (response signal): e.g.,

FIGURE 1.2   Block diagram illustrating experimental design with feedback from previous experiments (active observation of a system). *From P.T. Kissinger, W.R. Heineman, Laboratory Techniques in Electroanalytical Chemistry, second ed., Marcel Dekker, New York, 1996 with modifications.*

current is measured in amperometry, charge in coulometry, or impedance in electro-chemical impedance spectroscopy (EIS). On the other hand, according to the excitation signal here only those techniques in which potential is controlled, either performing a potential step or a potential scan, or both, are considered.

Misunderstandings may arise also due to the names employed for the different techniques. Thus, in general, the term amperometry relates to methodologies based on the measurement of the current (hence "amp"). However, this is also the term employed for a specific technique in which a potential is applied to the working electrode (WE) of an electrochemical cell and the current is measured under a steady-state mass transport regime. On the other hand, vol-tammetry refers to the measurement of the current when a potential is scanned (hence "volt"). Here, many different subclasses are included depending on the excitation signal that is applied to the WE: if a linear scan is made from an initial potential E_i to a final one E_f, the technique is linear sweep voltammetry; if the scan reverses either to E_i or to a different potential, then we have cyclic voltammetry (CV). This cycle (forward and reverse scans) can be repeated once and again to obtain information on a specific system. Therefore, it can be summarized that the term "amperometric" comprehends all electrochemical techniques measuring the current as a function of an independent variable that is, typically, time or electrode potential. In this book, however, we have considered separately amperometric (operating at fixed potential) and voltammetric (performing a potential scan) techniques. Both can be recorded under different mass transport regimes.

To increase the sensitivity of voltammetric techniques, potential pulses can be applied. They increase the ratio between faradaic current (due to electron transfer processes, i.e., oxidation and reduction processes, governed by Faraday's law) and nonfaradaic current (due to processes that do not involve electron transfer, e.g., adsorption and desorption of species at the interface electrode—solution that produces flow of external currents). Both faradaic and nonfaradaic processes occur when electrochemical reactions happen, but the nonfaradaic (capacitive) component decreases faster when a potential pulse is applied. With an adequate current sampling, it could be discriminated. Pulses can be of increased amplitude (normal pulse voltammetry) or same amplitude (differential pulse voltammetry). A train of pulses can be also applied as in square wave voltammetry (SWV). They all increase faradaic to nonfaradaic current ratio. Moreover, in alternating current voltammetry (ACV), an alternating potential is superimposed on the ramp of a linear potential, and the current (with a phase shift compared to the potential) is measured. A technique that is related to ACV is EIS where the impedance of the system, after excitation with this alternating signal, is measured. Therefore, this is not a voltammetric but an impedimetric technique. In this case, instead of a potential scan, a frequency scan (with an alternating potential of constant amplitude) is made. The small perturbation can provoke the transfer of electrons in electronic conductors or the transport of charged species from electrode to electrolyte and vice versa. Then, two different approaches, faradaic and nonfaradaic EIS, are possible. In the first, a redox probe is required and there are processes of reduction/oxidation of electroactive species at the electrode. In the nonfaradaic EIS, a redox probe is not required because processes of charging and discharging of the double-layer capacitance are studied. The parameters that correlate to the concentration of the analyte are the resistance to the charge transfer in the first case and the capacitance in the second one. This technique is gaining enormous interest because it can be performed in situ and is adequate for label-free applications.

On the other hand, voltammetric subclasses arise depending on different criteria. For example, when voltammograms are recorded under convective conditions, the name of the technique is hydrodynamic voltammetry (e.g., flow systems, stirred solutions). WEs in voltammetry are commonly solid electrodes (carbon-based, metals). However, when mercury electrodes (such as the dropping mercury electrode) are employed, the correct name is polarography (voltammetry at the dropping mercury electrode). Moreover, some of the experimental conditions, such as the use of a specific electrode, can give the name to the technique, as in "rotating disc electrode voltammetry."

Besides, current can be simply measured at a fixed potential. In conventional amperometry the measurement is performed under convective conditions: e.g., flow, rotating disc electrode, stirred solution. When it is measured at a fixed time after applying a single or double potential step, the technique is known as chronoamperometry. In this case, a quiescent solution is employed (no convection) and the time becomes an important variable (as reflected in the name), since the current is measured at a fixed time. In this way, amperometry (measurement of current) can be just amperometry (convective control, fixed potential), chronoamperometry (diffusive control, potential step) or voltammetry when current is recorded against potential. The last two follow an operational nomenclature where an independent variable part (potential in "volt" ammetry or time in "chrono" amperometry) is followed by a dependent variable part (current).

Alternatively to the application of pulses, to increase the sensitivity of voltammetric techniques, a previous preconcentration step can also be performed. The species of interest is accumulated on the electrode by, for example, reduction (anodic stripping voltammetry, ASV), oxidation (cathodic stripping voltammetry), or simple adsorption processes (adsorptive stripping voltammetry) and, once preconcentrated, is stripped off giving increased signals. Among the stripping techniques, ASV has demonstrated to be extremely sensitive especially for metal species because of the favored accumulation on metallic surfaces (e.g., on mercury electrodes, or on carbon electrodes modified with mercury or bismuth films). The sensitivity enhancement can be even higher if it is combined with an appropriate potential waveform (e.g., SWV). Moreover, apart from voltammetric techniques, chronoamperometry can be used also for the stripping step.

Charge is also an important entity that relates directly (through Faraday's law) with the number of moles electrolyzed. As current (faradaic) is the flow of electrons per unit of time, the charge can be calculated by measuring the area under the chronoamperometric curve (coulometric readout). If the charge is measured with time, the technique becomes chronocoulometry. When total electrolysis is produced, the charge (Q) can be related to the number of moles (N) using the Faraday's law ($Q = nFN$, where n is the number of electrons and F the Faraday's constant). Coulometry is an absolute technique; therefore, calibration is not required because the slope of the relationship between the magnitude measured (Q) and the number of moles (N) is the product nF, that is constant. Total electrolysis must be assured as well as 100% efficiency in the current. According to this, techniques can be separated in categories depending on the degree of electrolysis. In all the techniques considered here apart from coulometry, microelectrolysis is occurring and then, if curves are recorded under diffusion control, stirring the solution between measurements will restore the initial conditions. Special cases are e.g., paper-based methodologies where the number of moles that are electrolysed is very small, and others where stirring is not possible. However, in most of these cases single-use devices are employed.

In Fig. 1.3, some of the criteria that can be followed for classification of the electroanalytical techniques, most of them considered above, are reported.

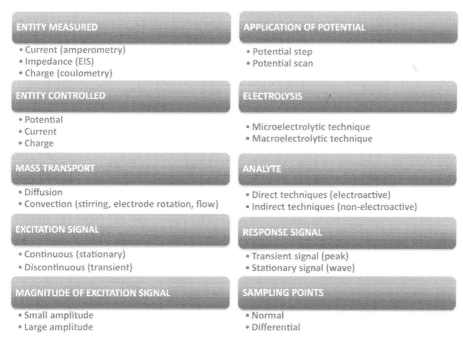

FIGURE 1.3 Different criteria employed for classifying the electroanalytical techniques.

Electroanalytical techniques are very adaptable and can be combined with many other principles. The integration with separation techniques such as liquid chromatography or capillary electrophoresis is well known. They also fit perfectly as detectors in flow systems (e.g., flow injection analysis). A convective mass transport regime is attained with the flow of the solutions and detection can be performed at maximum concentration gradient and constant diffusion layer (distance from the electrode where diffusion is the main mass transport phenomenon) thickness. In a different dimension, an interesting integration is possible in the field of microfluidics, where fluids are manipulated in channels with dimensions of tens of micrometers, not only in association with separation techniques, e.g., capillary electrophoresis [11], but also with other low-cost devices (e.g., paper-based platforms [12]).

Similarly, combination with several microscopes and spectroscopes can be made. Spectroelectrochemical techniques have attracted great interest in the last years because they allow obtaining simultaneous information of both electrochemical and optical character. An example is Raman spectroelectrochemistry that provides information about the vibrational states of molecules and, therefore, about their functional groups and structure, so that it is extremely useful. The use of adequate electrodes as substrates to enhance the optical signal (e.g., with activated silver screen-printed electrodes in Raman spectroelectrochemical measurements) or appropriate electrochemical processes to produce ultrasensitive methodologies, as is in the case of the electrochemiluminescence (ECL), justify the integration. ECL is based on a process in which electrochemically generated species combine to undergo electron transfer reactions and form excited, light-emitting species, giving place to the development of highly sensitive and selective assays. Dual detection (optical and electrochemical) could be performed, expanding the information about the system under study.

As can be seen in previous paragraphs, the toolbox of electroanalytical techniques is full of possibilities. The choice of one or another will depend on the main requirement, as for example:

- Simple: chronoamperometry,
- Fast: square wave voltammetry,
- Sensitive: stripping voltammetry,
- Absolute: coulometry,
- Informative: cyclic voltammetry,
- Label-free: electrochemical impedance spectroscopy,
- Combinable with flow techniques: amperometric detection, and
- Hybrid: electrochemiluminescence.

Apart from the different techniques that are available, Electroanalysis can be combined with different disciplines and approaches to produce advantageous methodologies, as indicated in Fig. 1.4.

Thus, the combination of electroanalytical and biochemical methodologies produced very interesting devices for the electroanalytical detection of bioassays (e.g., enzymatic, immuno or DNA assays). Electroanalytical biosensors are gaining in popularity and actually,and biochemical methodologies produced very interesting devices for the electroanalytical detection of bioassays (e.g., enzymatic, immuno or DNA assays). Electroanalytical biosensors are gaining in popularity and actually, the number of entries for "electrochemical biosensors" from 2015 in Google Scholar (May 16, 2019) is ca. 31,500, in many cases

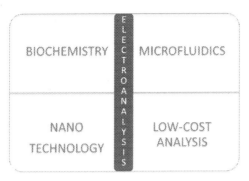

FIGURE 1.4 Diagram showing the current interplay between Electroanalysis and other disciplines and approaches.

incorporating nanotechnological approaches (nanobiosensors). In the journal *Biosensors and Bioelectronics* [13], the first in the category of *Chemistry, Analytical*, of the JCR (2018), the 10 most downloaded articles by May 16, 2019 were related to (1) a needle-shaped microelectrode for electrochemical detection of a sepsis biomarker, (2) advantages and challenges of microfluidic cell culture in polydimethylsiloxane devices, (3) detection principles of biological and chemical field-effect transistor sensors, (4) a wearable multisensing patch for continuous sweat monitoring, (5) a screen-printed paper microbial fuel cell biosensor for detection of toxic compounds in water, (6) graphene-based electrochemical biosensors for monitoring noncommunicable disease biomarkers, (7) two-dimensional oriented growth of a ZnZr bimetallic metal organic framework as a highly sensitive and selective platform for detecting cancer markers, (8) aptasensors for pesticide detection, (9) electrochemical sensor and biosensor platforms based on advanced nanomaterials for biological and biomedical applications, and (10) a review on various electrochemical techniques for heavy metal ions detection with different sensing platforms. Eight of them are directly related to electrochemical techniques and the other two review aptasensing and microfluidics, which can also involve electrochemical detection. Therefore, the interest in the integration of biochemical approaches in electroanalytical methodologies is clear and with a promising future.

The same happens with nanotechonological strategies that interact with Electroanalysis in three main ways: (1) through the modification of electrode surfaces to improve the characteristics of the methodologies, or (2) the use of nanomaterials as electroactive labels, and (3) the use of nanostructures as substrates for bioreagent immobilization. "Conventional" nanoparticles and nanotubes gave rise to the explosion of well-characterized nanostructures such as nanoonions, nanocubes, nanoflakes, nanourchins, etc., together with nanocatalysts, nanocomposites, or nanonetworks, etc., in single or hybrid configurations and in combination with different conductive electrodes to produce advantageous developments.

All these possibilities can be performed using conventional setups or simpler and low-cost devices. In any case, traditional or innovative designs are based on two elements: electrochemical cells and instrumentation. Dynamic techniques commonly use a three-electrode

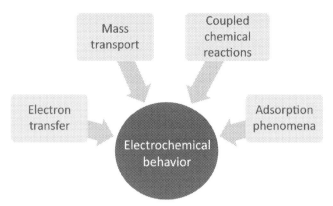

FIGURE 1.5   Diagram showing the main different phenomena that influence the electrochemical behavior of an electroactive species.

configuration with a potentiostat (that also includes an energy-feeding element as well as control and communication components). In these dynamic techniques, although nonfaradaic processes also exist, those faradaic are of main interest. Oxidations and/or reductions occur on the surface of a WE, and other homogeneous chemical (C) reactions might be coupled. Then, for example, EC (electrochemical/chemical), CE (chemical/electrochemical), ECE (electrochemical/chemical/electrochemical), or CEC (chemical/electrochemical/chemical) mechanisms are possible. Besides, adsorption phenomena can be present (Fig. 1.5). Information on all these can be obtained, and similarly, they have to be taken into account and controlled to perform quantitative analysis.

In the common three-electrode setup, the potential is monitored in relation to the reference electrode (RE, stable potential) and the current flows between the CE (counter electrode) and the WE (or indicator). Although the electrochemical processes allowing to do so are not well understood, a pseudoreference electrode could fulfill the role of an RE. On the other hand, if the passage of current does not affect the potential of the RE, a two-electrode cell could be used instead. In the case the WE is an ultramicroelectrode and the current flowing is small, highly resistive solutions could be used. In the rest of the cases, an inert electrolyte (supporting electrolyte) has to be used to decrease migration phenomena (e.g., cations attracted by a negatively charged electrode and repelled by the positively charged). Regarding the CE, it must have an area higher than the WE not to limit the flow of the current between CE and WE. Although products of the CE reaction could arrive to the WE and interfere the process, as microelectrolysis occur, this is not very probable. In case it is (long experiments), CE should be isolated from the solution (the same for the RE if material leaking is probable).

In any case, the possibilities of design are enormous, and along this book different configurations are presented. As commented in the first paragraph, the urge to decentralization implies that not only the electrochemical cells have to be inexpensive, miniaturized, disposable, etc., but also the instrumentation has to be autonomous and portable. Among all the components required, potentiostat represents the core unit. Small potentiostats that can be included in a ring [14], held on the wrist [15], incorporated in eyeglasses [16], or adhered as a patch to the skin [17] fit perfectly with this trend/need. In any case, whether a conventional or innovative setup is used, electroanalysis opens up an enormous field of research and applications, with a new generation of inexpensive and disposable electrochemical instruments advancing fast [18], especially when combination with either first-generation mobile phones [19] or smartphones [20] is possible. In this way, analysis could be performed in several locations at the point-of-need, either nonhospital settings (when clinical applications are involved), farms, industrial processes, ...or remote locations, i.e., everywhere, by everyone, and at any time. Additionally, with low-cost disposable sensors integrated in tiny "use-and-throw" instruments.

1.2 Additional notes

The IUPAC convention for i-E curves has a normal *x*-axis, i.e., with the potential moving from negative (left side) to positive (right side). Then, oxidation occurs moving the potential to the positive way and reduction to the negative way. Anodic currents are positive and cathodic currents are negative. This is the convention employed in this book. The other possibility comes from polarography that records *i-E* curves employing mercury as WE. As the potential window of this electrode is mainly cathodic, polarograms were represented starting the potential scan at about 0 V and moving to negative potentials (here at the right side). The currents (cathodic) were drawn as positive. As solid electrodes are replacing mercury, voltammetric techniques have almost displaced polarography and the IUPAC convention is gaining adepts. Most of the journals use this one, although most of the American books employ the so-called polarography convention (reverse of IUPAC).

References

[1] A. Ríos, A. Escarpa, B. Simonet, Miniaturization of Analytical Systems: Principles, Designs and Applications, Wiley, New York, 2009.
[2] Scentroid Flying Laboratory DR1000, http://scentroid.com/scentroid-dr1000/.
[3] K. Kalantar-zadeh, N. Ha, J.Z. Ou, K.J. Berean, Ingestible sensors, ACS Sens. 2 (2017) 468−483.
[4] J. Kim, I. Jeerapan, B. Ciui, M.C. Hartel, A. Martin, J. Wang, Edible electrochemistry: food materials based electrochemical sensors, Adv. Healthc. Mater. 6 (2017) 1700770.
[5] Y. Xu, M. Liu, N. Kong, J. Liu, Lab-on-paper micro- and nano-analytical devices: fabrication, modification, detection and emerging applications, Microchim. Acta 183 (2016) 1521−1542.
[6] C. Breitkopf, K. Sweider-Lyons (Eds.), Handbook of Electrochemical Energy, Springer, Berlin, 2017.
[7] J. Wang (Ed.), Analytical Electrochemistry, second ed., Wiley-VCH, New York, 2000.
[8] R.M. Penner, Y. Gogotsi, The rising and receding fortunes of electrochemists, ACS Nano. 10 (2016) 3875−3876.
[9] E. Bakker, E. Pretsch, Advances in potentiometry, in: Electroanalytical Chemistry, CRC Press, 2016, pp. 16−89.
[10] P.T. Kissinger, W.R. Heineman, Laboratory Techniques in Electroanalytical Chemistry, second ed., Marcel Dekker, New York, 1996.
[11] https://www.micruxfluidic.com.
[12] W. Dungchai, O. Chailapakul, C.S. Henry, Electrochemical detection for paper-based microfluidics, Anal. Chem. 81 (2009) 5821−5826.
[13] https://www.journals.elsevier.com/biosensors-and-bioelectronics.
[14] J.R. Sempionatto, R.K. Mishra, A. Martín, G. Tang, T. Nakagawa, X. Lu, A.S. Campbell, K.M. Lyu, J. Wang, Wearable ring-based sensing platform for detecting chemical threats, ACS Sens. 2 (2017) 1531−1538.
[15] R.K. Mishra, L.J. Hubble, A. Martín, R. Kumar, A. Barfidokht, J. Kim, M.M. Musameh, I.L. Kyratzis, J. Wang, Wearable flexible and stretchable glove biosensor for on-site detection of organophosphorus chemical threats, ACS Sens. 2 (2017) 553−561.

[16] J.R. Sempionatto, L.C. Brazaca, L. García-Carmona, G. Bolat, A.S. Campbell, A. Martín, G. Tang, R. Shah, R.K. Mishra, J. Kim, V. Zucolotto, A. Escarpa, J. Wang, Eyeglasses-based tear biosensing system: non-invasive detection of alcohol, vitamins and glucose, Biosens. Bioelectron. 137 (2019) 161−170.

[17] A. Martín, J. Kim, J.F. Kurniawan, J.R. Sempionatto, J.R. Moreto, G. Tang, A.S. Campbell, A. Shin, M.Y. Lee, X. Liu, J. Wang, Epidermal microfluidic electrochemical detection system: enhanced sweat sampling and metabolite detection, ACS Sens. 2 (2017) 1860−1868.

[18] V. Beni, D. Nilsson, P. Arven, P. Norberg, G. Gustafsson, A.P.F. Turner, Printed electrochemical instruments for biosensors, ECS J. Solid State Sci. Technol. 4 (2015) S3001−S3005.

[19] A. Nemiroski, D.C. Christodouleas, J.W. Hennek, A.A. Kumar, E.J. Maxwell, M.T. Fernández-Abedul, G.M. Whitesides, A universal mobile electrochemical detector designed for use in resource-limited applications, Proc. Natl. Acad. Sci. USA 111 (2014) 11984−11989.

[20] A. Ainla, M.P.S. Mousavi, M.-N. Tsaloglou, J. Redston, J.G. Bell, M.T. Fernández-Abedul, G.M. Whitesides, Open-source potentiostat for wireless electrochemical detection with smartphones, Anal. Chem. 90 (2018) 6240−6246.

PART I

Dynamic electroanalytical techniques

CHAPTER

2

Determination of ascorbic acid in dietary supplements by cyclic voltammetry

Noemí de los Santos Álvarez, M. Teresa Fernández Abedul

Departamento de Química Física y Analítica, Universidad de Oviedo, Oviedo, Spain

2.1 Background

As current flows through the circuit during measurements, voltammetry is considered a dynamic electrochemical technique. The potential (E) imposed to a working electrode (WE) is scanned and the current intensity (i) circulating is measured (i-E curve, voltammogram). An electrochemical cell with three electrodes is commonly employed for voltammetric experiments. On the surface of the WE, the electrochemical processes of interest, which can be related to the analyte concentration, take place. Several WEs can be employed, mainly carbon-based or metallic. There are many possibilities and the only condition is to present a conductive surface that allows precise measurements. In this case, an easy-to-prepare carbon paste electrode (CPE) is employed. Graphite powder is very conductive but, to handle it in an easy way, a binder (paraffin or silicon oil) has to be employed. After mixing thoroughly both components, the carbon paste can be inserted in an electrode body (e.g., Teflon cylinder with an inner stainless steel connector). The as-prepared electrode is then compacted, polished, and introduced in the solution for measurement.

The reference electrode (RE) has a stable potential so that a fixed difference can be established between RE and WE. In this case, a common silver/silver chloride/saturated potassium chloride electrode is employed as a RE. It is important to locate both electrodes as close as possible to decrease the resistance between them (ohmic drop). Although a two-electrode configuration can be employed, when currents are high, a third electrode is introduced, the counter electrode (CE). Then, the current (flow of charges per unit of time or rate of charge flow) passes between WE and CE. Because the current does not flow between WE and RE, the potential remains stable. A platinum wire is commonly used as

Laboratory Methods in Dynamic Electroanalysis
https://doi.org/10.1016/B978-0-12-815932-3.00002-4

CE, but many different conductive materials of lower cost can be used (e.g., stainless steel). The CE should have an area high enough to not limit the current flow.

The entire electrochemical cell comprises not only the electrodes but also the solution where they are immersed. Electrons flow through solid or liquid conductors (electrodes), but in solution the charges are carried by ions. Therefore, to close the circuit, both electrical and ionic conduction are needed. This is one of the reasons why a background electrolyte is employed in voltammetric experiments (e.g., 0.1 M phosphate buffer pH 7.4, 0.5 M KCl, 0.1 M H_2SO_4, etc.), in this case in a glass electrochemical cell containing 20 mL. An additional reason is related to minimization of ion migration, a mode of undesirable mass transfer that causes the movement of ions in solution under the influence of an electric field. In voltammetry, mass transfer is limited to diffusion by carrying the experiments under conditions where migration and convection are negligible. Migration of the electroactive species is significantly reduced using a high concentration of an inert electrolyte (background or supporting electrolyte) and convection is prevented in the absence of stirring and vibrations.

Among the voltammetric techniques, cyclic voltammetry (CV) is considered a diagnostic technique that provides knowledge about the electrochemical behavior of electroactive analytes on electrode surfaces. Ascertaining the mechanism, the number of electrons, the presence of coupled reactions, and adsorption processes are some of the information that can be obtained with this technique. Therefore, recording a cyclic voltammogram is the first step in most of electroanalytical methodologies. In CV, the potential (excitation signal, Fig. 2.1A) is swept from an initial potential, E_i, (V, volts) at a specific scan rate (mV/s) until a switching or reverse potential (E_λ), at which the potential is reversed (scanned in the opposite direction) back to the initial potential at the same scan rate. Several cycles can be performed. The current in A, amperes (analytical signal), is recorded as a function of the potential yielding a graph known as voltammogram (see Fig. 2.1B). If the scan starts at a potential well below (negative) the formal potential ($E^{o'}$) of the redox process, only nonfaradaic currents flow.

Although charge does not cross the interface, external currents can flow when the potential, electrode area, or solution composition changes [1]. When the potential (E) is changes, the electrode—solution interface acts as a capacitor and a charging process occurs. Then, charge (q) will accumulate on its plates, with excess of electrons in one plate (electrode or solution) and deficiency in the other (solution or electrode). This happens until the charge

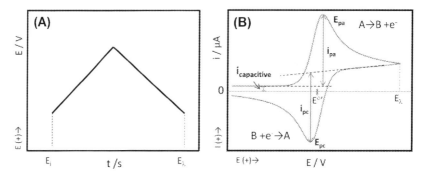

FIGURE 2.1   (A) Excitation and (B) response signals in cyclic voltammetry. The voltammogram is recorded in a solution of the electroactive species A.

satisfies the equation $q/E = C$, where C is the capacitance. The magnitude of this current depends also on the resistance in the circuit. In any case, both types of processes occur at electrodes and although the faradaic processes are usually of primary interest, nonfaradaic (capacitive) processes must be taken into account.

When the electrode potential reaches the proximity of the $E^{\circ\prime}$, the oxidation starts and an anodic current begins to flow. The surface concentration of the redox compound starts decreasing creating a concentration gradient, so the flux of the compound to the surface (and the current) increases. As the potential moves past $E^{\circ\prime}$, the surface concentration drops nearly to zero because all species are oxidized immediately on the surface and the compound diffusion to the electrode cannot maintain the current anymore. Consequently, a decline in the current is observed and a peak-shaped i versus E curve is recorded. When the potential is switched, the potential is swept in the negative direction. In the vicinity of the electrode, there is a large concentration of oxidized species. As the potential approaches $E^{\circ\prime}$ again, the oxidized species is reduced and a cathodic current flows. The shape of the curve is similar to the forward scan for the same reasons [1].

The most important parameters that can be derived from a CV are the peak potential (characteristic of the species in a specific medium) and the peak current (faradaic current related to the concentration of analyte), which are very useful for identification and determination of analytes. The way these parameters are extracted from the CV is indicated in Fig. 2.1B. From their values, the reversibility can be studied. An electrochemically reversible process follows the Nernst equation and this occurs when the system can attain rapidly the equilibrium after a perturbation (compared with the measurement time scale). So, in general, a fast electron transfer process is reversible [1]. Then, in these processes, i_{pa}/i_{pc} is equal to 1 and the difference between anodic (E_{pa}) and cathodic peak (E_{pc}) potentials ($\Delta E_p = E_{pa} - E_{pc}$) is $2.3RT/nF$ or lower, i.e., $(0.059/n)$ V at 25 °C independently of the scan rate.

CV is also very useful to ascertain whether a process is controlled by diffusional mass transport. In a reversible, diffusion-controlled process, the peak current is defined by the Randles—Sevcik which indicates that i_p is proportional to the concentration of the electroactive species in solution and to the square root of the scan rate:

$$i_p = 2.69 \times 10^5 n^{1.5} A\, C_0 D_0^{1/2} v^{1/2}$$

where n is the number of exchanged electrons, A is the electrode area in cm^2, D_0 is the diffusion coefficient of the electroactive species in cm^2/s, C_0 is the bulk concentration of the electroactive species in mol/cm^3, and v is the scan rate in V/s.

An adsorption-controlled process, however, exhibits a peak current dependent on the surface concentration of the adsorbed redox species and on the scan rate. The peak separation tends to zero and the faradaic current drops quickly to the nonfaradaic current at potentials positive or negative the peak potentials. In this way, these processes are easily identified by visual inspection of the CV since electrochemical response can be affected quite significantly by the adsorption, either strongly or weakly, of the species A (reactant) or B (product). Similarly, the presence of other chemical (or electrochemical) processes influences the response.

In this experiment, two antioxidant compounds will be examined by CV: ascorbic acid (AA) and gallic acid (GA). AA or vitamin C is one of the most important vitamins in human diet as it participates in many biochemical pathways and acts as a powerful antioxidant. Then, it is important to monitor AA levels in dietary sources (fruits and vegetables, food supplements, and vitamin formulations) and in biological liquids, such as urine and serum. As an electron donor, AA serves as one of the most important small molecular weight antioxidants in food industry to prevent unwanted changes in color or flavor. It contributes to the total antioxidant capacity, an important quality indicator of foods and drinks. Owing to the crucial role of vitamin C in biochemistry and in industrial applications, its determination presents high interest. Rapid monitoring of vitamin C levels during production and quality control stages is also important. On the other hand, GA, a naturally occurring plant phenol, was also found to be a strong antioxidant in emulsion or lipid systems, in some cases more effective than several water-soluble antioxidants, such as AA [2].

Determination can be carried out employing different techniques but electrochemical detection is low-cost, easily miniaturizable, simple, and fast, which are characteristics very convenient for decentralized analysis. The easy oxidation of AA is the basis of many electrochemical methods for its determination [3] with carbon electrodes widely used, either conventional such as CPEs (modified in many cases to increase the sensitivity [4] or the selectivity [5]) or new approaches as, e.g., pencil drawn electrodes [6]. Both species are oxidized in processes involving two electrons each (see reactions in Fig. 2.2). These redox processes will be studied under different experimental conditions by CV.

This experiment has two objectives: on one hand, the characterization of redox processes by CV in terms of reversibility and diffusion/adsorption control and on the other hand, the determination of the content of AA in a vitamin complex.

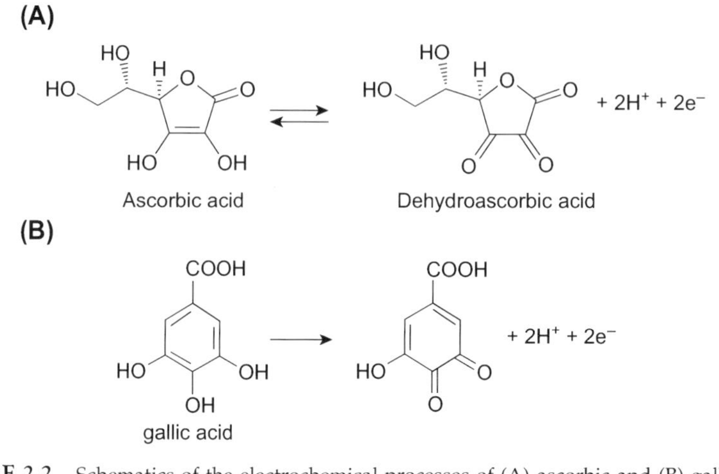

FIGURE 2.2   Schematics of the electrochemical processes of (A) ascorbic and (B) gallic acids.

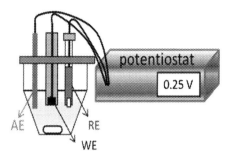

FIGURE 2.3 Conventional electrochemical cell setup. *RE*, reference electrode; *WE*, working electrode; *AE*, auxiliary electrode.

2.2 Electrochemical cell

A potentiostatic system with a three-electrode setup will be used. Specifically, in this experiment, a carbon paste WE, a Pt CE, and a Ag|AgCl|KCl(sat) RE will be employed. Electrodes are immersed in a 10- or 20-mL cuvette, located in a cell stand, as shown in Fig. 2.3. The cell is connected to a potentiostat.

2.3 Chemicals and supplies

- *Working electrode:* Carbon paste is prepared by thorough mixing of 1 g of spectroscopy-grade graphite powder and 0.323 g of silicone high-vacuum grease in CPE holder.
- *Background electrolyte*: Concentrated sulfuric acid, concentrated acetic acid, and sodium hydroxide.
- *Analytes*: Ascorbic acid and gallic acid.
- *Samples*: Dietary supplements containing AA.
- *General materials and apparatus*: Volumetric (amber) flasks, 1.5-mL microcentrifuge tubes, 100 and 1000-μL micropipettes and corresponding tips, magnetic bars, and a stirrer. Analytical balance, pH meter, and potentiostat.
- Milli-Q or distilled water is employed for preparing solutions and washing.

2.4 Hazards

Concentrated acids are corrosive and cause serious burns. Students are required to wear a lab coat, appropriate gloves, and safety glasses.

2.5 Experimental procedure

2.5.1 Preparation of solutions

- A volume of 250 mL of acetic acid/acetate buffer solution of pH 3.5, 4.5, and 5.5 prepared using the appropriate volume of concentrated acetic acid for 0.1 M concentration and adding small portions of concentrated sodium hydroxide until the desired pH is reached (measured with the pH meter).
- A volume of 250 mL of 0.01 M sulfuric acid solution.
- Stock solutions of AA (1.0 g/L) in 50 mL of acetic/acetate buffer solutions (different pH values) and a stock solution of GA (1.0 g/L) in 0.01 M sulfuric acid.

2.5.2 Characterization of the redox processes

2.5.2.1 Redox process of ascorbic acid

1. Prepare the WE by filling the hole of the Teflon holder (approximately 2 mm deep) with the carbon paste.
2. After compacting and polishing on a clean piece of paper, place it together with the RE and the AE in the cell stand.
3. Add 20 mL of buffer solution of pH 4.5 to the electrochemical cell and a magnetic bar.
4. Attach the cell to the stand and connect the three electrodes to the potentiostat.
5. Record the CV by scanning the potential between −0.1 and +0.6 V at a scan rate of 50 mV/s in the background electrolyte (without electroactive species) to check the cleanliness of the solution and/or electrode.
6. Add the appropriate volume of AA to the solution under stirring to obtain a final concentration in the cell of 1×10^{-4} M.
7. Stop the stirring and start recording the CV in the same potential window at identical scan rate.
8. Obtain the peak potentials (E_p) and peak currents (i_p) to ascertain the nature of the process (oxidation/reduction and reversible/irreversible).

2.5.2.2 Scan rate study

1. Follow the steps 1−6 in the previous section to prepare the electrochemical setup and confirm that there is no any impurity in the system.
2. Record the CVs (without stirring) between −0.1 and +0.6 V varying the scan rate between 10 and 500 mV/s (stir the solution between measurements) without changing the solution or the WE (see Fig. 2.4A as orientation).
3. Plot the E_p versus the square root of scan rate ($v^{1/2}$) and versus v. Decide if it is an adsorption or diffusion-controlled process.

2.5.2.3 pH study

1. Prepare the WE by filling the hole of the Teflon holder (approximately 2 mm deep) with the carbon paste.
2. After compacting and polishing on a clean piece of paper, place it together with the RE and the AE in the cell stand.

(A)   **(B)**

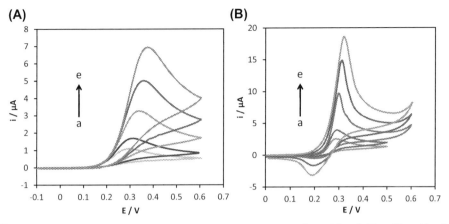

FIGURE 2.4   Example of cyclic voltammograms recorded in (A) ascorbic acid and (B) gallic acid solutions at different scan rates increasing from a to e.

3. Add 20 mL of buffer solution of pH 2 to the electrochemical cell and a magnetic bar.
4. Attach the cell to the stand and connect the three electrodes to the potentiostat.
5. Record the CV by scanning the potential between 0 and +0.65 V at a scan rate of 50 mV/s to check the cleanliness of the solution and/or electrode.
6. Add the appropriate volume of AA to the solution under stirring to obtain a final concentration in the cell of 1×10^{-4} M.
7. Stop the stirring and record the corresponding CV from 0 to +0.65 V at 50 mV/s.
8. Repeat the experiment in the buffer solution of pH 3.5 scanning the potential from 0 to +0.6 V.
9. Repeat the experiment in the buffer solution of pH 5.5 scanning the potential from −0.1 to +0.5 V.
10. Measure the E_p in each case and decide whether protons are involved in the AA electrochemical process.

2.5.2.4 Redox process of gallic acid
1. Prepare the WE by filling the hole of the Teflon holder (approximately 2 mm deep) with the carbon paste.
2. After compacting and polishing on a clean piece of paper, place it together with the RE and the AE in the cell stand.
3. Add 20 mL of sulfuric acid to the electrochemical cell and a magnetic bar.
4. Record the CV in sulfuric acid by scanning the potential between 0 and +0.65 V at a scan rate of 50 mV/s to check the cleanliness of the solution and/or electrode.
5. Add the appropriate volume of GA to the solution under stirring to obtain a final concentration in the cell of 1×10^{-4} M.
6. Stop the stirring and record the CV in the presence of GA and ascertain the nature of the process.

7. Vary the scan rate in the range of 10–500 mV/s (see Fig. 2.4B) and explain your observations, indicating the nature of the process (reversible vs. irreversible and diffusion vs. adsorption process).

2.5.3 Calibration curve of ascorbic acid

1. Prepare a new CPE and record the background CV in 0.01 M sulfuric acid.
2. Add increasing concentrations of AA in the range between 1 and 500 μM and record the CV after each addition (and after stirring a few seconds) at 50 mV/s.
3. Measure i_p for each concentration and construct the calibration curve. Indicate the linear range and the sensitivity of the methodology. Estimate the limit of detection and quantification of the method.

2.5.4 Determination of ascorbic acid in a dietary supplement

1. Dissolve a tablet in a beaker of appropriate volume with distilled water (AA will dissolve but the excipient may not be soluble in water).
2. Transfer carefully all the contents to a volumetric flask and fill it with distilled water until the mark.
3. Make an appropriate dilution of this solution.
4. Add 20 mL of the detection buffer to a clean electrochemical cell and record the background CV at 50 mV/s.
5. Add a small volume (e.g., 80 μL) of the most diluted sample solution to the cell, and after a few seconds of stirring, start recording the CV.
6. Add four concentrations of standard (pure) solution of AA prepared in sulfuric acid, so that the concentration added is within the linear range found before.

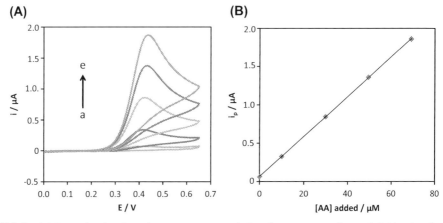

FIGURE 2.5   (A) Example of cyclic voltammograms recorded to determine ascorbic acid (AA) using the standard addition methodology. (B) Standard additions plot.

7. Repeat the measurements two more times with freshly prepared CPEs and clean solutions.
8. Calculate the concentration of AA in the tablet using both the standard additions (Fig. 2.5B, value of concentration that would intercept x-axis) and the external calibration methods and compare the results. Calculate the precision of the measurements.

2.6 Lab report

At the end of the experiment, write a lab report that includes an introduction with the relevance of this technique, the excitation and response signals, and the interest of the analyte. Comment and discuss the results and finish with main conclusions. Comment also important experimental points: e.g., solutions, instrumentation employed, or specific protocols, adding schematics when required. In the results and discussion section of the report, consider the following points:

1. Include graphical representation of the raw data obtained in the experiment. Be careful with the axis that should always include the magnitude and units.
2. Represent the graphs showing the effect of the scan rate and pH. Comment your observations, especially with respect to reversibility and diffusional/adsorptive behavior.
3. Plot the calibration curve and include the linear range, sensitivity, limit of detection, and limit of quantification.
4. Determine AA concentration of the samples, indicating the result with adequate number of significant figures (give the average and the standard deviation of the result).
5. Discuss the precision and trueness (bias) of the electrochemical methodology.

2.7 Additional notes

1. Working solutions of AA and GA should be prepared daily and protected from light. Stock solutions have to be kept at 4 °C protected from light.
2. The CPE preparation is of paramount importance. Carbon paste has to be thoroughly compacted and polished. This can be made by pressing hard the electrode on a clean satin office paper drawing an eight-figure several times.
3. The CPE surface can be renewed entirely or partially by rotating the piston a little bit and polishing the electrode again on the office paper.
4. The stirring has to be homogenous. Place the stir bar just below the WE and procure a homogenous and gentle stirring when adding solutions and in between measurements. In the first case a homogeneous solution should be obtained and in the second case the diffusion layer (where the concentration of analyte is depleted by electrolysis) should be renewed.
5. A cyclic voltammogram is always recorded in the background electrolyte when starting a study. If a peak is observed, discard the solution and the carbon paste. Prepare a new CPE and repeat the experiment.

6. The potential window is chosen according to the electroactive species. Students are encouraged to find the whole potential window available for this electrolyte/electrode, explaining what the limits of electroactivity are.
7. If the volume added to the cell is kept below 5%, its effect in the concentration is negligible, but students are encouraged to make calculations considering it to see the effect, especially for higher volumes. Additions have to be made carefully introducing the tip of the micropipette a little bit in the stirred solution and taking it out without removing solution.
8. Measurements could be also recorded in miniaturized thick or thin-film electrochemical cells. Then, specific connectors or hook clips should be employed to connect the electrochemical cell to the potentiostat. Place the cell in the connector and add 40 μL drop (or 5 μL in the case of a thin-film) of 0.01 M sulfuric acid, taking care of covering all the three electrodes and record the CV at 50 mV/s. Remove the drop by aspiration with a piece of absorbent paper and put another drop of a solution containing 1 mM of AA in 0.01 M sulfuric acid and record the CV at 50 mV/s.
9. Antioxidants AA and GA are used in this experiment, but other common redox probes of well-known electrochemical behavior such as potassium ferrocyanide and ferrocene (in the form of carboxylic acid or ferrocene methanol to solubilize it) could be employed. Alternatively, other electroactive organic molecules could be studied. Students are encouraged to search in the bibliography applications of interest in different fields (food, pharmaceutical, clinical, environmental, etc.).

2.8 Assessment and discussion questions

1. Draw excitation and response signals, explaining clearly why a peaked signal is obtained.
2. Explain the potential window chosen for the experiment.
3. Why three electrodes are employed? Explain each one's role.
4. Explain the effect of using a different electrode material, area, or scan rate on the analytical signal.
5. Enumerate the three criteria of reversibility.
6. Explain the differences between faradaic and capacitive currents.
7. Indicate the different mass transport mechanisms, in general and in this experiment.
8. Explain the differences between AA and GA when CVs are recorded at different scan rates.

References

[1] A.J. Bard, L.R. Faulkner, Electrochemical Methods: Fundamentals and Applications, 2nd ed., John Wiley and Sons, NY, 2001.
[2] G.-C. Yen, P.-D. Duh, H.-L. Tsai, Antioxidant and pro-oxidant properties of ascorbic acid and gallic acid, Food Chem. 79 (2002) 307–313.
[3] A.M. Pisoschi, A. Pop, A.I. Serban, C. Fafaneata, Electrochemical methods for ascorbic acid determination, Electrochim. Acta 121 (2014) 443–460.

[4] F. Li, J. Li, Y. Feng, L. Yang, Z. Du, Electrochemical behavior of graphene doped carbon paste electrode and its application for sensitive determination of ascorbic acid, Sens. Actuator. B Chem. 157 (2011) 110−114.

[5] O. Gilbert, B.E.K. Swamy, U. Chandra, B.S. Sherigara, Simultaneous detection of dopamine and ascorbic acid using polyglycine modified carbon paste electrode: a cyclic voltammetry study, J. Electroanal. Chem. 636 (2009) 80−85.

[6] V.X.G. Oliveira, A.A. Dias, L.L. Carvalho, T.M.G. Cardoso, F. Colmati, W.K.T. Coltro, Determination of ascorbic acid in commercial tablets using pencil-drawn electrochemical paper-based analytical devices, Anal. Sci. 34 (2018) 91−95.

Electrochemical behavior of the redox probe hexaammineruthenium(III) ($[Ru(NH_3)_6]^{3+}$) using voltammetric techniques

Estefanía Costa-Rama[1,2], *M. Teresa Fernández Abedul*[2]

[1]REQUIMTE/LAQV, Instituto Superior de Engenharia do Porto, Instituto Politécnico do Porto, Porto, Portugal; [2]Departamento de Química Física y Analítica, Universidad de Oviedo, Oviedo, Spain

3.1 Background

Voltammetry includes those electrochemical techniques in which the current (hence "amp") is measured against the potential (hence "volt") that is scanned. However, there are many possible excitation signals (E-t signals) that generate the different voltammetric techniques. The most common is cyclic voltammetry (CV) that employs a linear potential scan that is cycled (forward and reverse scans that can be repeated several times). This technique, commented in more detail in Chapter 2, is considered an excellent diagnostic technique. However, when sensitivity is a priority, alternative excitation waveforms could be employed. These are aimed to increase the ratio i_f/i_c (faradaic to nonfaradaic or capacitive current), i.e., to discriminate nonfaradaic currents improving the detection limits. The two most common alternatives are differential pulse voltammetry (DPV) and square wave voltammetry (SWV). In the first case, a voltammogram is obtained by scanning the potential with a sequence of pulses [1], with a double potential step being the pulse (Fig. 3.1A). Then, there is a forward step from an initial (E_1) to a final (E_2) potential that after the duration of the pulse (t_p), in a reverse step, comes back to a potential E_3 ($E_3 > E_1$). The amplitude of the pulses is constant (ΔE_p), but the values of the initial and final potentials are increased in an amount ΔE_s (step potential) each pulse

Laboratory Methods in Dynamic Electroanalysis
https://doi.org/10.1016/B978-0-12-815932-3.00003-6

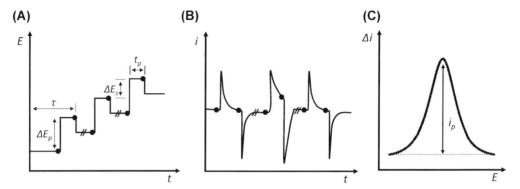

FIGURE 3.1   Schematics of the (A) excitation signal (dots indicate the points where current is sampled), (B) concentration profile and (C) response signal for differential pulse voltammetry. *Reprinted with modifications from H.A. Strobel, W.R. Heineman, Chemical Instrumentation: A Systematic Approach, third ed., John Wiley & Sons, 1989.*

(E$_3$ = E$_1$ + ΔE$_s$). The potential is then scanned with a scan rate (v) that is ΔE$_s$/τ, with τ being the pulse repetition period. The scan rate is usually low (1–10 mV/s).

The response signal for each pulse (i-t signal) is analogous to that obtained in chronoamperometry (see Chapter 25) (double potential step), but in this case, potential steps start in values where no electrolysis occur and move to a mass transport–controlled region. The current is sampled in two points of the step, and the difference between both values is recorded against the potential to which the pulse is stepped to obtain the voltammogram (Fig. 3.1C). This is the reason why this technique is named "differential." These two points are located just before the application of the pulse and near the end of the pulse. This sampling methodology is very advantageous because it distinguishes the two components of the current. After the potential step, the faradaic current decreases with the square root of the time, as indicated by Cottrell equation (for a planar electrode under a diffusion-controlled regime):

$$i_f = \frac{nFACD^{1/2}}{\pi^{1/2}t^{1/2}} = Kt^{-1/2}$$

where n is the number of electrons, F the Faraday's constant, A the area of the electrode, C the bulk concentration and D the diffusion coefficient. However, the nonfaradaic current (i$_c$) decays faster, exponentially with time, with a cell time constant R$_u$C$_d$ [2]:

$$i_c = \frac{E}{R_u}e^{-t/R_uC_d} = K'e^{-K''t}$$

where R$_u$ is the uncompensated resistance and C$_d$ is the capacitance of the double layer. Thus, discrimination between both currents is possible by sampling the current at a time where the capacitive current (nonfaradaic) has decayed considerably. Regarding the excitation signal, the pulse width is usually 30–50 ms to let the capacitive current to decay substantially but to maintain the faradaic current as high as possible. Pulses are applied at an interval of 0.5–5 s in such a way that the diffusion layer is renewed at the beginning of each measurement cycle.

The difference between these two sampled currents is plotted against the potential and constitutes the differential pulse voltammogram (Fig. 3.1C). In the first potential steps (no electrolysis), the current response has only nonfaradaic contribution because of the double-layer charging (Fig. 3.1B). When the potential is close to the formal potential ($E^{0'}$) in a reversible couple, substantial faradaic current appears because of the reduction (forward step)/oxidation (reverse step) of the species. The magnitude of these faradaic currents increases as the potential increases up to the limiting current region (mass transport–controlled region; i.e., diffusion-controlled region because voltammograms are recorded in unstirred solutions). Considering there is no convective contribution in this region, there is a decrease of the faradaic current and, at the end, only the double-layer charging contributes to the current. The plot is a peak-shaped voltammogram as shown in Fig. 3.1C. The main parameters of the voltammogram are the peak potential and peak current, very useful for qualitative and quantitative purposes, respectively.

The intensity of the peak current is proportional to analyte concentration and amplitude of the pulse, according to the Osteryoung–Parry equation [3] (developed for polarography, where i-E curves are recorded using dropping or static mercury electrodes):

$$i_p = \frac{n^2 F^2 A}{4RT} \left(\frac{D}{\pi t}\right)^{1/2} C \Delta E_p$$

where all terms have their usual meaning and *t* is the time between pulses. The rooted term in brackets is very similar to this in Cottrell equation, implying that diffusion of analyte to the electrode is an important determinant of the magnitude i_p [4]. As the technique is differential by nature, it is the area under the peak which is proportional to concentration, so this equation is merely an approximation, and in many cases using peak areas rather than peak heights enhances the accuracy. The current is also proportional to the amplitude, although peaks will become wider for higher amplitude values and this will decrease the ability to determine more than one analyte at a time.

An alternative to DPV is SWV that combines a pulse waveform with a current sampling methodology that decreases notoriously the nonfaradaic current. A fast scan rate that can achieve easily 1 V/s [4] is also a main characteristic of this technique. It was invented by Ramaley and Krause [5] and developed extensively by R. Osteryoung, J. Osteryoung et al. [6]. In Fig. 3.2, the excitation (E-t curve) and response (i-E curve, voltammogram) signals

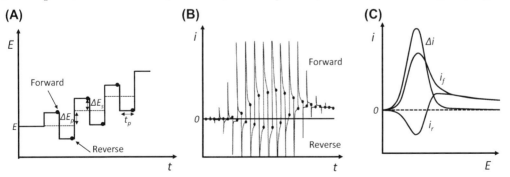

FIGURE 3.2 Schematics of the (A) excitation signal (dots indicate the points where current is sampled), (B) concentration profile and (C) response signal for square wave voltammetry. *Reprinted with modifications from A.J. Bard, L.R. Faulkner, Electrochemical Methods: Fundamentals and Applications, second ed., John Wiley & Sons: New York, 2001.*

are depicted, together with the response signal plotted against time (i-t curve). In SWV, a symmetrical pulse train of amplitude ΔE_p (measured with respect to the corresponding tread of the staircase), a pulse width t_p and period τ (being $\tau = 2t_p$ and $f = 1/\tau$ where f is the frequency) are superimposed on a staircase with a step height of ΔE_s [1]. The value of ΔE is typically about 50 mV, while ΔE_s is commonly about 5 mV and f varies over a wide range (from 1 to 500 Hz, for t_p varying from 500 to 1 ms, respectively). The result is a square wave excitation signal that scans the potential at a rate (v) that is $\Delta E_s/2t_p$ (i.e., $\Delta E_s/\tau$ or f ΔE_s). For $\Delta E_s = 10$ mV and $t_p = 1-500$ ms, the scan proceeds at 5 V/s to 10 mV/s, quick compared to most pulse methods and comparable to typical CV runs [2].

As in the rest of the pulsed techniques, the current is sampled at the end of the pulse to allow substantial decay of the nonfaradaic current, but here two types of faradaic currents are observed (for reversible processes): forward and reverse currents. These are due to the forward (in the direction of the staircase scan) and reverse (in the opposite direction) steps of the square wave, respectively. In this case, the diffusion layer is not renewed at the beginning of each measurement cycle [2]. In Fig. 3.2B, the current transients (i-t curve) and the current samples are represented. In the early cycles, as the staircase potential is too far positive (for reduction processes; far negative for oxidation processes) of $E^{0'}$, electrolysis is not occurring. In the middle, the staircase has moved into the region of $E^{0'}$, and the rate of electrolysis is a strong function of potential. Cathodic and anodic currents flow. At the end, as the potential is again far from $E^{0'}$, the electrolysis occurs at the diffusion-controlled rate regardless of potential [2]. Then, three different responses (voltammograms) are recorded, depending on the current that is plotted against the potential: forward, reverse, or net current, the last one as a difference between both (Fig. 3.2C). As these three currents can be recorded, this technique provides valuable information on the redox process and, similarly to what happened with CV, SWV is considered a useful diagnostic technique. Apart from this, the net peak current (Δi_p) is proportional to the concentration of analyte and can be used for analytical purposes. This can be described as:

$$\Delta i_p = \frac{nFAD^{1/2}C}{\pi^{1/2}t_p^{1/2}}\Delta\psi_p$$

where $\Delta\psi_p$ is a dimensionless current that depends on ΔE_p and ΔE_s. Therefore, the current depends on the amplitude, step potential, and frequency (through t_p). The higher ΔE_p, f or ΔE_s, the higher the net current, although the increase in peak width has to be taken into account.

SWV results very advantageous for a double enhancement of the sensitivity: this can increase not only for the already-commented discrimination of the nonfaradaic currents but also for obtaining a net current larger than either forward or reverse currents. It is slightly more sensitive than DPV because reverse pulses near $E^{0'}$ produce an opposite current (e.g., cathodic in forward and anodic in reverse or vice versa) that enlarges the net current.

Another advantage, when cathodic processes are recorded, is that it avoids oxygen interference and then purging with an inert gas is not required. [4].

In this chapter, a voltammetric (CV, DPV, and SWV) experiment is proposed with the aim of understanding the main parameters of the excitation and response signals. Hexaammineruthenium(III) is used as redox probe. It shows a simple redox reaction in which one electron is exchanged according to the reaction:

$$[Ru(NH_3)_6]^{3+} + 1e^- \rightleftharpoons [Ru(NH_3)_6]^{2+}$$

This experiment is adapted from [7] and is performed on screen-printed carbon electrodes (SPCEs), although conventional cells can be employed as well. The experiment can be completed in two laboratory sessions of 3–4 hours and is appropriate for undergraduate students of advanced Analytical Chemistry or Master students from different fields where analysis is required. With this experiment, the students will learn about different voltammetric techniques (the most common) in a comparative and critical way.

3.2 Electrochemical cell

For this experiment, screen-printed electrodes (SPEs) are used. The most common design of this kind of electrodes consists of a card (usually made of ceramic or plastic) that contains the three electrodes of the electrochemical cell (working electrode [WE], reference electrode [RE], and counter electrode [CE]) screen-printed on it. The SPEs used along this work (Fig. 3.3) have a WE (4-mm diameter) and a CE made of carbon and an RE made of silver ink. The dimensions of the ceramic card are $3.4 \times 1.0 \times 0.05$ cm. The small size of these SPCEs allows performing measurements using microvolumes (30–50 µL).

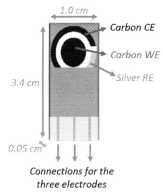

FIGURE 3.3 Schematics of the commercial screen-printed carbon electrode used along this work. *CE*, counter electrode; *RE*, reference electrode; *WE*, working electrode.

3.3 Chemicals and supplies

— *Components of background electrolyte:* Potassium chloride (KCl). A 0.1 M solution of KCl is employed as background electrolyte.
— *Analyte:* Hexaammineruthenium(III) chloride.
— *General materials, apparatus, and instruments:* Volumetric flasks, 100-µL and 1-mL micropipettes (with corresponding tips), microcentrifuge tubes, analytical weighing scale, edge connector (for the SPCEs) and potentiostat.
— Milli-Q water is used along this experiment for preparing solutions and washing.

3.4 Hazards

Hexaammineruthenium(III) chloride causes skin and eye irritation and may cause respiratory irritation. Students must handle it with care and avoid breathing the dust. They are required to wear lab coat, safety glasses and gloves.

3.5 Experimental procedure

3.5.1 Preparation of solutions

The following solutions are needed (estimate the required volume for each one):

— 0.1 M KCl solution. All the working solutions of [Ru(NH$_3$)$_6$]$^{3+}$ are prepared in this solution.
— 5 mM [Ru(NH$_3$)$_6$]$^{3+}$ stock solution. From this one, [Ru(NH$_3$)$_6$]$^{3+}$ solutions of the following concentrations should be prepared by dilution: 5, 10, 50, 100, and 500 µM.

3.5.2 Analytical signal and calibration curve obtained by cyclic voltammetry

First, a CV is recorded for the redox probe as follows:

1. Deposit 40 µL of 500-µM [Ru(NH$_3$)$_6$]$^{3+}$ solution on the SPCE covering the three electrodes of the electrochemical cell.
2. Record a cyclic voltammogram scanning the potential from +0.1 to −0.5 V at a scan rate of 50 mV/s (Fig. 3.4A).
3. Measure the intensity of the peak currents, cathodic (i_{pc}) and anodic (i_{pa}), and also the corresponding potentials (E_{pc} and E_{pa}) for the voltammogram obtained.

Then, using CV technique, perform a calibration plot for [Ru(NH$_3$)$_6$]$^{3+}$, between 5 and 500 µM, as follows:

1. Add 40 µL of a 5-µM [Ru(NH$_3$)$_6$]$^{3+}$ solution on the SPCE (in the same way as before).
2. Record a cyclic voltammogram from +0.1 to −0.5 V at a scan rate of 50 mV/s.
3. Wash the SPCE with Milli-Q water.

(A)

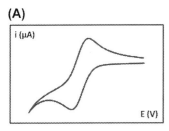

(B)

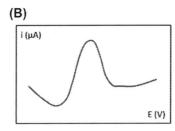

(C)

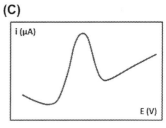

FIGURE 3.4   Schematics of: (A) cyclic, (B) differential pulse and (C) square wave voltammograms recorded in a $[Ru(NH_3)_6]^{3+}$ solution.

4. Repeat the steps 1, 2, and 3 using the $[Ru(NH_3)_6]^{3+}$ solutions of different concentrations (increasing values).
5. Repeat the steps 1, 2, 3, and 4 with two more SPCEs to evaluate the precision.
6. Obtain the calibration curve representing the intensity of the cathodic peak current vs. the $[Ru(NH_3)_6]^{3+}$ concentration (adjustment for the anodic peak current could be also evaluated). Obtain the linear range, sensitivity, limit of detection (LOD) and limit of quantification (LOQ). Calculate the reproducibility of the measurements for a 100-µM concentration of $[Ru(NH_3)_6]^{3+}$.

3.5.3 Calibration curve obtained by differential pulse voltammetry

A calibration plot for $[Ru(NH_3)_6]^{3+}$, in a range comprised between 5 and 500 µM, using DPV can be performed as follows:

1. Add 40 µL of a 5-µM $[Ru(NH_3)_6]^{3+}$ solution on the SPCE.
2. Record a DPV scanning the potential from 0.0 to −0.4 V with a 6-mV step potential, 25-mV pulse amplitude, pulse width of 0.01 s and pulse period of 0.5 s (Fig. 3.4B).
3. Wash the SPCE with water.
4. Repeat the steps 1, 2, and 3 using $[Ru(NH_3)_6]^{3+}$ solutions of different concentrations (increasing values).
5. Repeat the steps 1, 2, 3, and 4 with two more SPCEs to evaluate the precision.
6. Obtain the calibration curve representing the intensity of the peak current vs. the $[Ru(NH_3)_6]^{3+}$ concentration. Obtain the linear range, sensitivity, LOD and LOQ. Calculate the reproducibility of the measurements for a 100-µM $[Ru(NH_3)_6]^{3+}$ solution.

3.5.4 Calibration curve obtained by square wave voltammetry

A calibration plot for $[Ru(NH_3)_6]^{3+}$, in the range comprised between 5 and 500 µM, using SWV can be carried out as follows:

1. Add 40 µL of 5-µM $[Ru(NH_3)_6]^{3+}$ solution on the SPCE.
2. Perform an SWV scanning the potential from 0.0 to −0.4 V with a 6-mV step potential, 25-mV amplitude and 30-Hz frequency (Fig. 3.4C).
3. Wash the SPCEs with Milli-Q water.

4. Repeat the steps 1, 2, and 3 using the [Ru(NH$_3$)$_6$]$^{3+}$ solutions of different concentrations.
5. Repeat the steps 1, 2, 3, and 4 with two more SPCEs.
6. Obtain the calibration curve representing the intensity of the peak current vs. the concentration of [Ru(NH$_3$)$_6$]$^{3+}$. Obtain the linear range, sensitivity, LOD and LOQ. Calculate the reproducibility of the measurements for a 100 µM [Ru(NH$_3$)$_6$]$^{3+}$ solution.

3.6 Lab report

Write a lab report following the typical scheme of a scientific article. This report should include the following sections: introduction, experimental part, results, discussion and conclusions. It should also include images, diagrams and tables, and address the following points:

1. In the introduction, include the basis and the importance of the DPV and SWV techniques. Search and discuss recent publications that use these techniques with analytical purposes (e.g., as detection techniques for electrochemical (bio)sensors).
2. The advantages/disadvantages of SPEs vs. conventional electrochemical cells can be also addressed in the introduction section.
3. The excitation signals in each case (CV, DPV and SWV) could be depicted in a comparative way, with the main parameters that should be considered.
4. Include representative pictures/graphs representing the data obtained along the experiment: (i) CVs, DPVs and SWVs recorded for [Ru(NH$_3$)$_6$]$^{3+}$ solutions, (ii) calibration curves obtained using the three different electrochemical techniques. Discuss the differences obtained between the CVs, DPVs and SWVs.
5. Discuss critically the values obtained for the linear range, sensitivity, LOD and LOQ using the different electrochemical techniques.

3.7 Additional notes

1. Working solutions of [Ru(NH$_3$)$_6$]$^{3+}$ should be prepared daily.
2. The measurements can be performed in triplicate using the same SPCE to calculate the repeatability to allow the students comparing this value with the reproducibility.
3. CVs can be recorded in a 500-µM [Ru(NH$_3$)$_6$]$^{3+}$ solution (in 0.1 M KCl) at different scan rates (e.g., 10, 25, 50, 75, 100, 250 and 500 mV/s) to: (i) discuss the rate-limiting step of the electrochemical reaction by representing the peak currents (ip$_a$, ip$_c$) vs. the scan rate (a linear relationship indicates adsorption) and the square root of the scan rate (in this case, a linear relationship is indicative of diffusion); (ii) calculate both the difference between peak potentials and the ratio between anodic and cathodic peak currents at each scan rate and discuss the reversibility of the system; (iii) calculate the formal potential of the redox couple; (iv) calculate the electroactive electrode area using the Randles–Sevcik equation (diffusion coefficient in 0.1 M KCl for [Ru(NH$_3$)$_6$]$^{3+}$: 9.1 × 10^{-6} cm^2/s [8]), and (v) calculate the capacitance of the double layer at the formal potential using the cathodic curve.

4. In all the cases (CV, DPV and SWV), the potential should be scanned first in a solution of the background electrolyte (0.1 M KCl) to check the cleanliness and the lack of interferences.
5. A study and discussion of the influence that the main parameters of the excitation signal (e.g., amplitude, frequency, etc.) have over the response could result very interesting. Calculation of the scan rate can be made.
6. A comparison between normal pulse voltammetry (NPV) and DPV can be also made with didactic purposes.

3.8 Assessment and discussion questions

1. Draw the excitation signal, the response signal and the concentration profile for DPV and SWV.
2. Discuss the differences/advantages/disadvantages of CV, DPV and SWV.
3. Indicate the main parameters to be considered in the excitation signals of the different voltammetric techniques as well as those to be taken into account in the response signals. Also indicate the ones that have to be employed for qualitative or quantitative purposes.
4. DPV and SWV are preferred when sensitivity wants to be improved. Explain the reasons.
5. How do faradaic and nonfaradaic currents vary in DPV and SWV?
6. Where does the "differential" word come from in DPV?
7. Where is the current sampled in DPV and SWV?
8. Why CV and SWV are considered excellent diagnostic techniques?
9. In SWV, three different voltammograms are recorded. What are those?
10. Order the voltammetric techniques according to their usual scan rate.

References

[1] H.A. Strobel, W.R. Heineman, Chemical Instrumentation: A Systematic Approach, third ed., John Wiley & Sons, 1989.
[2] A.J. Bard, L.R. Faulkner, Electrochemical Methods: Fundamentals and Applications, second ed., John Wiley & Sons, 2001.
[3] P.E. Parry, R.A. Osteryoung, Evaluation of analytical pulse polarography, Anal. Chem. 37 (1965) 1634–1637.
[4] P.M.S. Monk, Fundamentals of Electroanalytical Chemistry, John Wiley & Sons, 2001.
[5] L. Ramaley, M.S. Krause, Theory of square wave voltammetry, Anal. Chem. 41 (1969) 1362–1365.
[6] J.G. Osteryoung, R.A. Osteryoung, Square wave voltammetry, Anal. Chem. 57 (1985) 101–110.
[7] D. Martín-Yerga, E. Costa Rama, A. Costa García, Electrochemical study and determination of electroactive species with screen-printed electrodes, J. Chem. Educ. 93 (2016) 1270–1276.
[8] F. Marken, J.C. Eklund, R.G. Compton, Voltammetry in the presence of ultrasound: can ultrasound modify heterogeneous electron transfer kinetics? J. Electroanal. Chem. 395 (1995) 335–339.

Anodic stripping voltammetric determination of lead and cadmium with stencil-printed transparency electrodes

Isabel Álvarez-Martos[1], Charles S. Henry[2],
M. Teresa Fernández Abedul[3]

[1]Interdisciplinary Nanoscience Center (iNANO). Aarhus University, Aarhus, Denmark;
[2]Department of Chemistry, Colorado State University, Fort Collins, CO, United States;
[3]Departamento de Química Física y Analítica, Universidad de Oviedo, Oviedo, Spain

4.1 Background

Electroanalytical stripping methods are extremely sensitive. Many species can be determined employing stripping methodologies but the main application, by far, is the determination of heavy metals. The high sensitivity is due to the combination of two main steps: (i) a preconcentration on the electrode surface and (ii) a measurement step that employs a sensitive electrochemical technique. The first step is the deposition of the metal on the surface of the electrode by electrolysis (e.g., a small portion of metal cation present in solution, M^{n+}, is reduced to M; in the case the electrode is mercury, it forms an amalgam, $M(Hg)$). Then, this is followed by the measurement (stripping step), which involves the dissolution or stripping of the deposited metal (in our example, the metal, M or $M(Hg)$, is reoxidized to M^{n+}, cation that diffuses back to the solution).

Regarding the preconcentration (or deposition) step, many strategies can be employed (see the summary in the scheme of Fig. 4.1), mainly electrochemical (reduction/oxidation) or adsorptive [1]. Once the metal (or the analyte) is on the electrode, stripping or redissolution can be electrochemical or chemical, depending if the redox reaction takes place on the

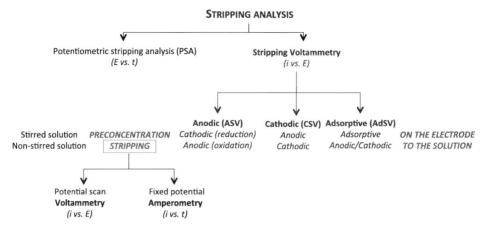

FIGURE 4.1   Schematics with the classification of the different stripping analysis techniques. (Potentiometric stripping analysis will not be considered in this book).

electrode surface or using a reducing/oxidizing agent. In stripping voltammetry, the metal is stripped anodically (sweeping the potential in the positive sense to produce its oxidation: anodic stripping voltammetry, ASV) or cathodically (sweeping the potential in the negative sense to produce its reduction: cathodic stripping voltammetry). In this case, a voltammogram (i-E curve) is recorded. On the other hand, the stripping analysis can be potentiometric (potentiometric stripping analysis, not considered in this book); it will not be considered in this book, devoted to dynamic electroanalysis. Hence, a potentiogram (E-t curve) records the change in potential because of the, e.g., oxidation of the deposit produced (using a chemical oxidant or an anodic current). Finally, in the case of the adsorptive stripping voltammetry or potentiometry, the preconcentration step consists of the adsorptive accumulation and further oxidation or reduction of the adsorbed species (e.g., organic molecule or metal in presence of a ligand forming a complex that adsorbs on the electrode) with a potential scan (in voltammetry) or employing a chemical agent or a current (in potentiometry).

In this experiment, adapted from Ref. [2], we are going to determine lead and cadmium using an anodic stripping voltammetric methodology. For this, the heavy metals will be deposited on the electrode surface by an electrochemical reduction process and then stripped anodically. Fig. 4.2 shows the excitation signal (E-t curve) in the accumulation (preconcentration or deposition) step and the stripping (or redissolution) step in ASV of the metals. In the accumulation step, a potential negative enough where the reduction of both metals happens is applied. After this, the potential is scanned in the positive direction, using an excitation (E-t curve) that can correspond to, e.g., linear sweep voltammetry (LSV), differential pulse voltammetry (DPV), or square wave voltammetry (SWV). DPV and SWV discriminate against interference from capacitive current, but SWV uses faster scan rates, and therefore it is being used most often [3]. Combining a preconcentration step with a sensitive measurement technique (such as DPV or SWV), limits of detection can be considerably decreased by one or two orders of magnitude.

If the conditions of the accumulation step are maintained constant, exhaustive electrolysis of the solution is not necessary, and by proper calibration and with fixed electrolysis times, the measured voltammetric response (e.g., peak current) can be employed to find the solution concentration [4]. Then, in this electrolytic step, the potential, the time, and the mass transport

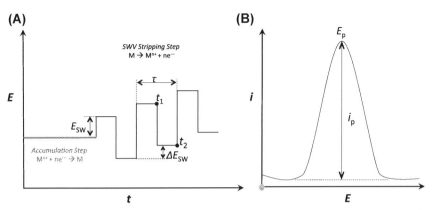

FIGURE 4.2　Anodic stripping voltammetry: (A) potential–time and (B) current–potential waveforms for the accumulation and stripping steps, where τ is the duration of each pulse, E_{SW} is the pulse amplitude, ΔE_{SW} is the step potential, i_p is the peak current, and E_p is the peak potential. *SWV*, square wave voltammetry.

have to be considered to get precise measurements. Therefore, the preconcentration (cathodic deposition) has to be made at a fixed potential for a fixed time. The potential is usually 0.3–0.5 V more negative than the formal potential ($E^{0'}$) for the least easily reduced metal ion to be determined (Cd^{2+} in this case) (see Fig. 4.2). In this way, species are reduced as rapidly as they are transported to the electrode surface. The duration of the deposition step depends on the concentration level. If concentrations are very low, higher times are required because the longer the time, the larger the amount of metal deposited on the electrode surface. Times used vary between 30 s and 10 min. Regarding the mass transport, apart from diffusion, forced convection is used to get the highest efficiency. Then, stirred solutions or rotating electrodes are commonly employed. In this experiment, no stirring is possible (drops of solutions are deposited on the electrodes and then very low volumes are employed), but on the other hand, these low volumes make diffusion distances much lower. Once the accumulation finishes, the stripping voltammogram is recorded. Peak potentials serve to identify metals (several metals can be stripped), and the intensity of the peak current (or peak height) is the analytical signal that can be correlated with the solution concentration of metal. In electrodes with high surface-to-volume ratio where a more efficient preconcentration occurs, total exhaustion of metals in solution results in sharper peaks and improved peak resolution for multianalyte determinations [1]. Integrated peak area can also be used to quantify metals and may, in complex sample matrices, improve quantification reproducibility.

Mercury electrodes (hanging mercury drop electrode or mercury film electrode), very convenient for stripping analysis of metals due to amalgam formation, are being substituted by others to avoid mercury toxicity. Certain carbon-based electrodes have also shown a good performance, especially with a metallic film able to preconcentrate effectively the metals. Bismuth and antimony are adequate alternatives to mercury film deposition for forming amalgams. Several methods of electrode modification can be found for bulk or surface modification, with drop-casting or electrodeposition as the more common for the last case (sputtering methodologies are also possible). Different sources of the metal (nanoparticles, oxides, etc.) can be employed, and in case electrodeposition is made, ex situ or in situ (depending on where the process is made, in a previous manner or simultaneously

FIGURE 4.3   Schematics with different strategies for bismuth film formation.

with the preconcentration of the analyte) can be followed (see the summary in Fig. 4.3). In this experiment, electrodes are modified with a bismuth film following an in situ electrochemical deposition. Bismuth can be added to the metal solution or drop-casted previously on the electrode.

Among heavy metals, lead and cadmium represent a concern for the public health. Human exposure may come from different contaminated sources (air in polluted atmospheres of cities supporting heavy traffic, water or food close to industrial areas, etc.). Drinking water is in the focus of many regulatory agencies: the accumulative behavior of these metals and continuous consumption are risk factors to be carefully considered. The World Health Organization establishes guideline values of 10 and 3 µg/L for lead and cadmium, respectively [5].

In the context of environmental and drinking waters, field analysis becomes very important. Here, low-cost electroanalysis is a valid approach (see Chapters 25–29, and the work with thick- and thin-film electrodes). Conductive inks can be deposited on flat surfaces where a drop of solution is added for measuring. Apart from glass or ceramic substrates, paper is becoming commonly used in electroanalytical devices. One can benefit not only from its porous nature but also from its hydrophobic properties when modified [6]. A different material, with hydrophobic properties, is transparency film, a low cost, more robust than paper, and compatible with aqueous solutions substrate. It was employed in this experiment modified with stencil-printed carbon ink electrodes.

Apart from the reduced size and low cost of devices, simplicity of procedures is also demanded. In this context, the integration of devices in food packages, sample containers, etc., is becoming a trending alternative. Integration of all the steps in a unique device is one of the aims of total analysis systems, which reduce time and cost of processes. Here, a design of a device to be integrated in the cap of a water sample container is also described.

This experiment can be completed in two laboratory sessions of 3e4 h (depending on the fabrication of the devices, they could be given to the students) and is appropriate for undergraduate students of Analytical Chemistry or Master students from different fields where analysis is required (e.g., environmental sciences). With this experiment, students will learn about sensitive electroanalytical determination of toxic heavy metals that can be performed in field with low-cost homemade platforms.

4.2 Electrochemical cell design

The electrochemical cell consists of a transparency film in which carbon ink electrodes are deposited following a stencil-printed procedure. CorelDRAW software is used to design the

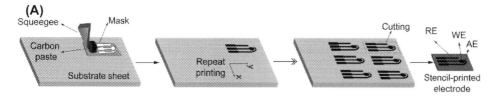

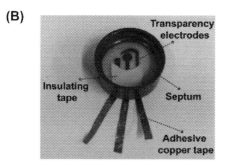

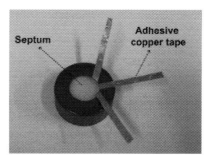

FIGURE 4.4 (A) Schematic drawing of the stencil-printing process to fabricate several low-cost electrodes on a transparency sheet. (B) Picture showing the integration of the transparency electrode onto the cap of a sample vial. (Inner and outer view) *AE*, auxiliary electrode; *RE*, reference electrode; *WE*, working electrode.

electrochemical cell, which consists of working electrode (WE), counter electrode (CE), and pseudo-reference electrode (RE). The design is then transferred to an A4 transparency sheet using a CO_2 laser cutter. This transparency sheet is then employed as stencil. After placing it over a new transparency sheet, carbon ink is printed and cured as commented in Section 4.5.2. A carbon ink circle-ended electrode located in the middle is the WE. The CE is a hook surrounding the WE, and a small rectangle acts as RE (see Fig. 4.4A). Adhesive cupper tape is used for connections to the potentiostat. An adhesive layer covers the connections and delimits the working area where solutions will be added.

The electrochemical cell can be inserted in the cap of a sample container for integrating sampling and measurement, as described in Section 4.5.2 and shown in Figure 4.4B.

4.3 Chemicals and supplies

— *Redox indicator*: Hexaammineruthenium(III) chloride.
— *Analytes*: Lead(II) nitrate and cadmium(II) nitrate.
— *Interfering species*: Iron(III) nitrate, copper(II) nitrate, cobalt(II) nitrate, nickel(II) nitrate, and zinc(II).
— *Components of background electrolytes*: Potassium chloride, citric acid, acetic acid, and sodium acetate.
— *Samples*: Tap and river water.
— *Other reagents*: Bismuth(III) nitrate.
— *Electrochemical cell preparation*: Polyester transparency sheets, carbon sensor paste (C10903P14), squeegee, adhesive glue, insulating tape, and adhesive copper tape.

— *General materials, apparatus, and instruments:* 1 mL microcentrifuge tubes, 100 μL and 1 mL micropipettes with corresponding tips, 5-mL sample containers with septum in the cap, 0.45 μm filter paper, weighing scale, pH meter, laser cutter, and potentiostat.
— Milli-Q water is employed to prepare all solutions.

4.4 Hazards

$Ru(NH_3)_6Cl_3$, $Bi(NO_3)_3$, $Pb(NO_3)_2$, and $Cd(NO_3)_2$ are harmful if swallowed and cause irritation to the eyes and skin. $Bi(NO_3)_3$, $Pb(NO_3)_2$, and $Cd(NO_3)_2$ are oxidizers and must be stored separately from other materials. Pb and Cd residues need to be disposed in a separate container. Students are required to wear lab coat, appropriate gloves, and safety glasses.

4.5 Experimental procedure

4.5.1 Solutions and sample preparation

The solutions needed are:

— 0.1 M KCl and 0.1 M acetate buffer pH 4.8 as background electrolytes.
— 2.5 mM $Ru(NH_3)_6Cl_3$ in 0.1 M KCl.
— 500 μg/L of $Bi(NO_3)_3$ in 0.1 M acetate buffer pH 4.8.
— 500 μg/L of $Pb(NO_3)_2$ in 0.1 M acetate buffer pH 4.8.
— 500 μg/L of $Cd(NO_3)_2$ in 0.1 M acetate buffer pH 4.8.

4.5.2 Fabrication of the electrochemical cell

To fabricate the electrochemical cell in transparency film, next steps have to be followed:

1. Fix the substrate (transparency sheet) on the table using an adhesive tape.
2. Place the stencil on top and print the carbon paste with the aid of a squeegee.
3. Remove the stencil carefully and repeat the process in the x-y directions of the sheet to prepare a batch of electrodes (see Fig. 4.4A).
4. Dry the electrodes at 37°C for 10 days to remove any remaining solvent and stabilize the surface.
5. Place an adhesive piece of paper with a drilled circle in the middle to delimit the electrochemical cell.

To integrate the electrochemical cell in the cap of a sample container (Fig. 4.4B), the next steps should be followed:

1. Cut the three-electrode system to fit in the vial caps (circular geometry) and fix it to the inner part of the septum with a small amount of adhesive glue.
2. Place adhesive copper tape to establish the electrical connections and extract them from the cap, through the septum.
3. Insulate the electrical connections using a piece of insulating tape.

4.5.3 Electrochemical characterization of the electrodes using hexaammineruthenium(III)

1. Place 70 μL of a 2.5 mM $[Ru(NH_3)_6]^{3+}$ solution on a piece of transparency film containing the electrochemical cell.
2. Using cyclic voltammetry, perform the potential scan between -0.2 and -0.8 V at scan rates of 10, 25, 50, 100, and 250 mV/s.
3. The anodic (i_{pa}) and cathodic (i_{pc}) peak currents will be plotted against the square root of the scan rate ($v^{1/2}$) and the scan rate (v) to determine if $[Ru(NH_3)_6]^{3+}$ displays a diffusion- or adsorption-controlled process.
4. The peak potential difference (ΔE_p) at 50 mV/s will be used to estimate the reversibility of the process, taking into account that for a fast reversible process $\Delta E_p < 59/n$ mV (see Chapter 2 on cyclic voltammetry).
5. If $[Ru(NH_3)_6]^{3+}$ exhibits a reversible and diffusion-controlled electron transfer reaction, calculate the electroactive surface area of the WE using the Randles−Sevcik equation:

$$i_p = (2.69 \times 10^5)n^{3/2}A\,C\,D^{1/2}\,v^{1/2}$$

where i_p is the peak current intensity (A), n is the number of electrons transferred in the electrochemical reaction, A is the electrode area (cm^2), C is the concentration of the analyte (mol/cm^3), and v is the scan rate (V/s). D is the diffusion coefficient of $[Ru(NH_3)_6]^{3+}$ in 0.1 M KCl, which is equal to 9.1×10^{-6} cm^2/s [7].

4.5.4 Quantitation of the amount of lead and cadmium deposited on the electrode surface

1. Add 70 μL of a mixture containing 100 μg/L of Pb(II) and 100 μg/L of Cd(II) on a piece of transparency film containing the electrochemical cell.
2. Apply a voltage of -1.4 V for 240 s to preconcentrate the metals on the electrode surface.
3. Use square wave anodic stripping voltammetry (SWASV) to scan the potential from -1.4 to -0.4 V under the following conditions: square wave amplitude of 25 mV, frequency of 20 Hz, step potential of 5 mV, and equilibration time of 10 s (see Chapter 3 on square wave voltammetry).
4. Estimate the amount of Pb(II) and Cd(II) deposited on the electrode surface (Γ_{ads}) by integrating the total charge under the voltammetric peaks (Q_{ads}) and using the following equation:

$$\Gamma_{ads} = Q_{ads}/nFA$$

where F is the Faraday constant (96,485 C/mol e$^-$).
5. Repeat this study with at least three independently prepared electrodes and estimate the interelectrode precision (i.e., variation of the anodic stripping currents between electrodes).

4.5.5 Bismuth effect on lead and cadmium stripping currents

1. Deposit 70 µL of a mixture containing 100 µg/L of Pb(II) and Cd(II) and increasing concentrations of Bi(III) in the range 0−10 mg/mL on a piece of transparency sheet containing the electrochemical cell for in situ electrodeposition (different concentrations of Bi(III) on different electrochemical cells).
2. Using the same protocol as in the previous section (i.e., preconcentration followed by SWASV), estimate if Bi(III) has any effect on the Pb(II) and Cd(II) stripping currents and evaluate what ratio of metals versus bismuth provides the highest stripping peak currents.

4.5.6 Analytical features of the sensor

1. Place 70 µL of mixtures containing increasing concentrations of Pb(II) and Cd(II) (to final concentrations of 1, 2, 5, 10, 20, 50, 100, and 200 µg/L) and the amount of Bi(III) found as optimal in the previous study, on different pieces of transparency films containing electrochemical cells.
2. Record SWASV curves to construct the calibration plots for Pb(II) and Cd(II).
3. Estimate the limit of detection (LOD) and limit of quantitation (LOQ) as well as the sensitivity of the methodology.

4.5.7 Selectivity assessment

1. Add 70 µL of a mixture containing 50 µg/L of Pb(II) and Cd(II), the amount of Bi(III) found as optimal, and increasing concentrations of interfering species, such as Cu(II), Co(II), Zn(II), Fe(III), and Ni(II) in the range 0.5−10 mg/L, on a piece of transparency film, containing the electrochemical cell (one for each concentration/interferent).
2. Use the SWASV curves to construct the calibration plots for each cation and establish if any of them have a significant effect on the Pb(II) and Cd(II) signals, indicating the tolerance ratio for each.

4.5.8 Determination of lead and cadmium in water samples

1. Place a 5 µL drop of 500 mg/L Bi(III) solution on top of the WE of the transparency-based electrochemical cell (that has been inserted in the cap) and let it dry at room temperature for surface electrode modification.
2. Collect and filter the water samples with a 0.45 µm filter paper to remove any particulate matter.
3. Mix 5 mL of the water sample with 68 mg of sodium acetate and 17 mg of citric acid salts to work under the same experimental conditions as before (they can be included in containers before sampling).
4. Turn upside down the container for the measurement (Fig. 4.5).
5. Add increasing concentrations of Pb(II) and Cd(II) to final concentrations of 1, 2, 5, 10, 20, 50, 100, and 200 µg/L.
6. Record SWASV curves to construct the calibration plots and estimate the amount of Pb(II) and Cd(II) in water samples by the standard addition methodology.

FIGURE 4.5 Schematics with the steps for lead and cadmium determination in real water samples using an adapted sampling container.

4.6 Lab report

At the end of the experiment, write a lab report that includes an introduction (discussing the importance of this sensitive technique and the main steps), experimental procedures, results obtained and discussion, finishing with main conclusions. In a more detailed manner, the following points should be considered:

1. Include pictures and schematics of the fabrication process. Indicate dimensions.
2. Plot schematically the excitation signal and the response signal.
3. Indicate if the process of the complex of ruthenium was diffusion- or adsorption-controlled, justifying with results the answer.
4. Explain the possibilities for modification with bismuth. Justify the optimal value for the ratio between bismuth and the analytes. Indicate the amount of metals adsorbed with and without modification.
5. Represent the calibration curves and include the linear range, sensitivity, LOD, and LOQ.
6. Discuss the selectivity of the devices in the view of the results obtained. A table with tolerance ratio values could be included.
7. Discuss the precision of the devices in terms of the RSD values.
8. Indicate the real sample analysis performed, where the samples were taken, and the device that was prepared for the determination. Give the results including average and standard deviation and the adequate number of significant figures. Check if the result meets the official regulations.

4.7 Additional notes

1. Transparency film is the flat substrate employed here, but different flat surfaces could be employed. In case a porous matrix is used (e.g., paper), the working area should be first delimited (e.g., with hydrophobic wax printing).
2. The cap of the vial that contains the electrodes (see Section 4.5.2 and 4.5.8) can also be employed as electrochemical cell instead the piece of transparency film. In this case, 1 mL of the background electrolyte or solution to be measured is added.

3. A laser cutter is employed for preparing the mask. In case it is not available, a manual cutter could be employed but precision could be affected.
4. Studies can be made on transparency sheets depositing a drop of solution covering the electrodes, or alternatively, a cap with integrated electrodes can be employed as electrochemical cell. In this case, a 1-mL volume of electrolyte could be added.
5. In the characterization of the electrochemical cell with the ruthenium complex, as the process is diffusive in nature, the same electrochemical cell can be used for the measurements performed with different scan rates. However, to renew the diffusion layer, it is good to wait ca. 1 minute between measurements.
6. SWV is employed for the stripping step because it is fast and sensitive. However, different scan formats could be evaluated, such as LSV or DPV.
7. Bismuth is electrochemically preconcentrated in situ at the same time than lead and cadmium. In the real sample determination, previous surface modification is employed for having the electrode prepared and not to require any additional solution. Different procedures of modification with bismuth (from different sources) could also be evaluated.
8. Lead and cadmium are determined in water samples. The standard additions methodology is employed for quantitation and possible values for additions are given. However, a first voltammogram should be recorded to ascertain the most adequate additions for the determination.
9. After taking the water sample in the container for heavy metal determination, it has to be turned upside down to wet the electrodes and perform the measurements. Standard additions have to be made by opening it. Ideally, a hole can be made in the cell to inject the standards through the septum with a syringe.
10. These cells are low cost and disposable. However, it can be reused if precision between several measurements indicates so. If the metal is totally stripped off, they can be washed with purified water and reused with a new drop. In the case of the cell in the cap, a higher volume of background electrolyte is employed and measurements can be taken successively without adding any further step.
11. This is a multianalyte methodology and more metals could be assayed simultaneously. Similarly, metals different from lead and cadmium could be determined.

4.8 Assessment and discussion questions

1. Explain the different approaches of stripping analysis.
2. Discuss why this methodology is so sensitive.
3. Potential and time are very important parameters in the preconcentration step. What happens if they are decreased/increased?
4. Draw the excitation and response signals employed in the determination.
5. A peaked response is obtained. Explain clearly why and the reason of its sharpness.
6. Indicate how the determination is made if external calibration or standard addition methodologies are employed.

7. Establish if the sensor fulfills the legal requirements fixed by the World Health Organization (WHO).
8. Discuss some other designs you consider of interest as well as improvements of this one in the cap.

References

[1] J. Wang, Analytical Electrochemistry, second ed., Wiley-VCH, 2000, pp. 75−84.
[2] D. Martín-Yerga, I. Álvarez-Martos, M.C. Blanco-López, C.S. Henry, M.T. Fernández-Abedul, Point-of-need simultaneous electrochemical detection of lead and cadmium using low-cost stencil-printed transparency electrodes, Anal. Chim. Acta 981 (2017) 24−33.
[3] Chemical Instrumentation: A Systematic Approach, third ed., H.A. Strobel, W.R. Heineman, 1989, pp. 1137−1141.
[4] A.J. Bard, L.R. Faulkner, Electrochemical Methods, Fundamentals and Applications, second ed., J.Wiley & Sons, 2001, pp. 458−464.
[5] http://www.who.int/water_sanitation_health/publications/drinking-water-quality-guidelines-4-including-1st-addendum/en/.
[6] A.C. Glavan, D. Christodouleas, B. Mosadegh, H.D. Yu, B. Smith, J. Lessing, M.T. Fernández-Abedul, G.M. Whitesides, Folding analytical devices for electrochemical ELISA in hydrophobic R^H paper, Anal. Chem 86 (2014) 11999−12007.
[7] P.M. Hallam, C.E. Banks, A facile approach for quantifying the density of defects (edge plane sites) of carbon nanomaterials and related structures, Phys. Chem. Chem. Phys 13 (2011) 1210−1213.

Adsorptive stripping voltammetry of indigo blue in a flow system

Estefanía Costa-Rama[1,2], M. Teresa Fernández Abedul[2]

[1]REQUIMTE/LAQV, Instituto Superior de Engenharia do Porto, Instituto Politécnico do Porto, Porto, Portugal; [2]Departamento de Química Física y Analítica, Universidad de Oviedo, Oviedo, Spain

5.1 Background

Adsorptive stripping voltammetry (AdSV) is a variation of stripping voltammetry which refers to the stripping of species spontaneously adsorbed on the surface of the working electrode without needing a previous electrolysis step [1]. In Chapter 4, anodic stripping voltammetry is presented for determination of lead and cadmium. A cathodic electrodeposition of metals allows further quantification through their anodic stripping. In this case, adsorption is the mechanism for the preconcentration on the electrode surface. For the stripping step, different voltammetric techniques, such as linear sweep, differential pulse, or square wave voltammetry, can be used. In this experiment, alternating current voltammetry (ACV) is used as detection technique for the stripping step. Hence, the potential program imposed on the working electrode is a direct current (dc) mean value, E_{dc}, which is scanned slowly with time, plus a sinusoidal component, E_{ac}. The alternating current (ac) signal thus causes a perturbation in the surface concentration, around the concentration maintained by the dc potential ramp [2]. The measured responses are the magnitude of the ac component of the current at the frequency of E_{ac} and its phase angle with respect to E_{ac}. The ac voltammogram shows a peak, which height is proportional to the concentration of the analyte and, for a reversible reaction, to the square root of the frequency [2]:

$$i_p = \frac{n^2 F^2 A \omega^{1/2} D^{1/2} C \, \Delta E}{4RT}$$

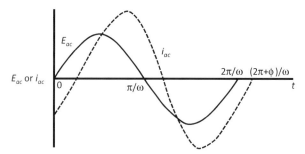

FIGURE 5.1 Relationship between alternating current and voltage signals at frequency ω. *Reprinted with modifications from A.J. Bard, L.R. Faulkner, Electrochemical Methods: Fundamentals and Applications, second ed., John Wiley & Sons: New York, 2001.*

where the term ΔE is the amplitude and ω is the frequency. In Fig. 5.1 the relationship between the alternating current and voltage signals at frequency ω is shown.

The detection of the ac component allows separating between the faradaic and capacitive components of the current because the phase shift of the faradaic component, relative to the applied sinusoidal potential, is different from this of the capacitive. Then, it could be decreased if a phase-sensitive detection is employed.

In this experiment, the detection is integrated in a flow injection analysis (FIA) system to inject the product of an enzyme-linked immunosorbent assay (ELISA) for interleukin-10 (IL-10) quantification. FIA, also commented in Chapters 9 and 28, consists of the injection of a fixed volume of sample into a nonsegmented continuous laminar flow of a carrier solution that transports the sample to the detector [3,4]. Thus, the FIA system allows simple and flexible configuration that helps to the automatization of analysis, decreasing analysis time and human errors. The configuration in this case is basic, as no reaction occurs inside the system. Then, peristaltic pump, injector (rotary valve), tubing, and detector are the components of the FIA system. The two main configurations for the flow cell are wall jet and thin layer. In the first one, the sample enters the cell perpendicularly to the working electrode, meanwhile in the thin-layer model, the flow passes over the electrode, entering laterally from one side and leaving by the opposite.

ELISAs are among the most common immunoassays that are performed every day. The use of microplates allows parallelization of assays, and the employ of enzymes as labels (there are also many other possible [5–7], e.g., metallic nanoparticles or organic redox compounds, see Chapter 20) has resulted in very sensitive methodologies. In this case, among the several immunoassay formats that exist, a sandwich type has been chosen. It is based on the interaction of the antigen with a specific antibody immobilized onto the surface of the wells. Then, an enzyme-labeled antibody is added and it binds with the antigen forming a "sandwich." A substrate is later enzymatically converted into an electroactive product. Alkaline phosphatase is an enzyme of particular interest because of its broad substrate specificity and high turnover number. Several substrates have been employed: phenyl, p-nitrophenyl, p-aminophenyl, naphthyl phosphate, etc. The product of the ideal substrate has to give rise to a sensitive oxidation at low potentials, which enhances selectivity, together with the easy elimination from the electrode of the products generated, which increases precision. 3-Indoxyl phosphate (3-IP) fulfills all these requirements and has been proposed as an appropriate alkaline phosphatase substrate for enzyme immunoassays with voltammetric detection [8]. The adsorption behavior of indigo allows the use of an adsorptive stripping voltammetric methodology, which enhances sensitivity and selectivity.

The enzymatic reaction (Fig. 5.2) comprises the hydrolysis of the phosphate moiety by alkaline phosphatase (AP), the formation of the unstable enol product and its subsequent oxidation in air to give indigo blue. Indigo blue is nonsoluble in aqueous solutions and it can be solubilized adding sulfuric acid that converts it into indigo carmine.

In this experiment, adapted from Ref. [10], an ELISA is performed for detection of IL-10. This is an antiinflammatory cytokine that plays a relevant role in the infection process by regulation of the immune response [11]. Elevated levels of IL-10 indicate an acute stage of inflammation and are also found in patients of malignant tumors such as melanoma, colorectal, ovarian, breast, or gastric carcinomas [12]. The enzyme AP is used as label in the ELISA assay and it catalyzes the hydrolysis of 3-IP to indigo blue. This enzymatic product, once converted into indigo carmine after stopping the reaction with sulfuric acid, is introduced into the FIA system that possesses an electrochemical cell with a carbon paste working electrode. Indigo is adsorbed on the surface of the working electrode and, then, it is quantified by stripping voltammetry.

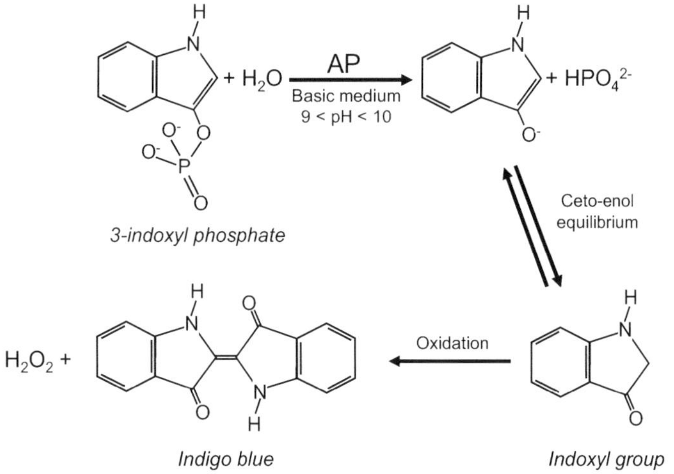

FIGURE 5.2   Alkaline phosphatase (AP) hydrolysis of 3-indoxyl phosphate to produce indigo blue.

5.2 Chemicals and supplies

Enzyme-linked immunosorbent assay

— 0.05 M carbonate buffer pH 9.6 containing 0.1% NaN_3 (coating solution).
— 0.1 M Tris-HCl buffer with 150 mM NaCl (TBS) and 0.05% Tween 20 (TW), 0.05% NaN_3, and 2% casein pH 7.4 (blocking solution).
— 0.1 M TBS 7.4 containing 0.02% NaN_3 and 0.05% TW (washing buffer).
— Monoclonal anti-IL-10 antibody (in 0.1 M phosphate buffer, 150 mM NaCl pH 7.4).
— Il-10 (dilutions prepared in 0.1 M TBS, 0.05% NaN_3, 0.05% TW, 1% casein, pH 7.4).
— Biotinylated anti-IL-10 antibody (in 20 mM TBS, 0.1% BSA pH 7.4).
— AP labeled with streptavidin (prepared in 10 mM Tris-HCl buffer, 50 mM NaCl, 1.5 mM $MgCl_2$, 0.2% TW pH 7.4).
— 3-IP (dilutions prepared in 0.1 M Tris-HCl buffer pH 9.8 with 10 mM $MgCl_2$, stored at 4°C protected from light).
— H_2SO_4 (concentrated and 1 M solutions).
— Microtiter plates.
— Stirrer with microtiter adapter.
— Incubator.

Detection

— Peristaltic pump with two channels
— Six-port rotary valve with 20- and 100-μL loops.
— PVC tubes.
— Thin-layer flow cell.
— Graphite powder.
— Paraffin oil.
— Ag/AgCl/saturated KCl as reference electrode.
— Stainless steel tube as counter electrode.
— Potentiostat and computer system.

General chemicals and materials

— Indigo carmine (diluted in 0.1 M H_2SO_4)
— 100-μL eight-channel micropipette and corresponding tips
— 1.5-mL microcentrifuge tubes, pH-meter, and analytical balance

Ultrapure water is employed throughout the work to prepare solutions and to clean electrodes and glassware.

5.3 Hazards

Sodium azide is very toxic if swallowed or in contact with skin. Read its safety data sheet and handle it in a fume hood. Students are required to wear a lab coat, appropriate gloves and eye protection, as well as to handle all the reagents with care, especially concentrated acids and bases as well as sodium azide.

5.4 Flow injection analysis electrochemical system

First, the FIA system (Fig. 5.3) is prepared connecting the pump, the valve, and the flow cell by the PVC tubing. The sample in introduced in the FIA system by aspiration from the well of the microtiter plate. More information about FIA systems can be found in Chapters 9 and 28.

The electrochemical thin-layer flow cell was equipped with a carbon paste electrode (geometric area of 9.6 mm^2) and consists of two methacrylate blocks and a PVC spacer put together with four screws. The carbon paste was prepared by intimately mixing graphite powder (1 g) and paraffin oil (0.36 mL). This carbon paste is packed in the corresponding place of the flow cell and the surface can be entirely or partially renewed with the help of a piston inserted in the electrode body. The reference and counter electrodes are located downstream in a syringe coupled to the cell outlet. The counter electrode is a hollow stainless steel tube that leads to the waste.

5.5 Experimental procedures

5.5.1 Enzyme-linked immunosorbent assay procedure

To perform the ELISA assay, represented in Fig. 5.3A, several steps have to be followed:

1. Coat each well with 100 μL of anti-IL-10 antibody (6 μg/mL) and incubate overnight at room temperature. Wash the plate with washing solution four times using 200 μL per well.
2. Add 200 μL of the blocking buffer and incubate at 37°C for 3 h and rinse again carefully.

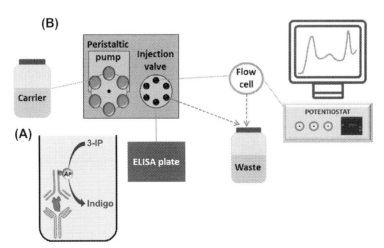

FIGURE 5.3 Schematic representation of (A) the enzyme-linked immunosorbent assay (ELISA) and (B) the flow injection analysis (FIA) system. *3-IP*, 3-indoxyl phosphate; *AP*, alkaline phosphatase.

3. Deposit 100 µL of IL-10 (analyte) to each well and incubate overnight at room temperature and then, wash.
4. Add 100 µL of biotinylated anti-IL-10 antibody (1 µg/mL), keep at room temperature for 2 h, and perform the washing step.
5. Incubate with streptavidin-AP (1:1000 diluted) for 1 h at room temperature and rinse again.
6. Add 100 µL of 0.2 mM 3-IP solution to each well. The enzymatic reaction occurs for 2 h at 35°C and then it is stopped by adding 150 µL of H_2SO_4 concentrated.
7. Wait for 10 min and then aspire the content of the well, which contains the generated indigo, using a capillary connected to the tubing of the peristaltic pump until the loop of the injection valve is filled.
8. Inject the sample into the FIA system ("load" position) to carry out the voltammetric detection.

5.5.2 Flow procedure and voltammetric detection

1. Pump a stream of 1 M H_2SO_4 (carrier) at 2 mL/min through the FIA system and confirm the absence of bubbles.
2. Pretreat the surface of the working electrode by applying +1.3 V for 30 s.
3. After having filled the loop, change the injection valve from "load" to "inject" position to introduce indigo into the flow and change the potential to −0.3 V. In this step, indigo is adsorbed on the surface of the working electrode.
4. Pass the carrier through the cell for a fixed time (e.g, 15 s).
5. Stop the flow and scan the potential in the positive direction to +0.9 V in the ACV format (e.g., 75 Hz of frequency, 35 mV of amplitude, and 0 degree phase angle) to obtain the corresponding voltammogram.

5.5.3 Electrochemical behavior of indigo

As indigo is the product of the enzymatic reaction that is measured in the FIA system, it is interesting to know its electrochemical behavior (see additional note 2 in Section 5.7). To know which is the analytical signal, inject 100 µL of a 5-µM indigo carmine solution in the FIA system (using the injection valve) and record the corresponding ACV (record first an ACV after injecting the carrier to obtain the background).

Two well-defined peaks must be obtained (Fig. 5.4). Choose one of the peaks for the quantification and justify it.

5.5.4 Calibration curve of IL-10

Obtain a calibration curve for IL-10 in the concentration range from 0.2 to 1500 pg/mL. Represent the intensity of the peak current vs. the concentration (Fig. 5.5) and also the log—log graph. Calculate the linear range, the sensitivity, the limit of detection, and the limit of quantification.

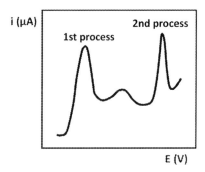

FIGURE 5.4 Example of the alternating current voltammogram obtained for indigo carmine. *Reprinted with modifications from M.J. Bengoechea Álvarez, C. Fernández Bobes, M.T. Fernández Abedul, A. Costa-García, Sensitive detection for enzyme-linked immunosorbent assays based on the adsorptive stripping voltammetry of indigo in a flow system, Anal. Chim. Acta 442 (2001) 55–62.*

5.6 Lab report

At the end of the experiment, write a lab report including introduction, experimental part (reagents, materials, equipment, and protocols), results and discussion, and conclusions. The following points should be considered to write the report:

1. In the introduction, explain the purpose and the basis of the experiment. Include an overview on the different types of immunoassays. Explain the advantages and disadvantages of the use of labels in immunosensors (vs. label-free approaches) and how the label here used allows obtaining the analytical signal.
2. Detail the protocol followed including schemes.
3. Discuss the results obtained and include graphs and tables with representative raw data.
4. For the calibration plot, indicate the values obtained for the figures of merit.
5. Discuss the incidences that happened during the experiment and the main conclusions.

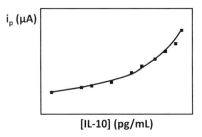

FIGURE 5.5 Example of a calibration plot obtained for interleukin-10 (IL-10). *Reprinted with modifications from M.J. Bengoechea Álvarez, C. Fernández Bobes, M.T. Fernández Abedul, A. Costa-García, Sensitive detection for enzyme-linked immunosorbent assays based on the adsorptive stripping voltammetry of indigo in a flow system, Anal. Chim. Acta 442 (2001) 55–62.*

5.7 Additional notes

1. 3-IP solutions must be prepared the day of use and kept away form light.
2. The electrochemical behavior of indigo carmine (product of the enzymatic reaction after stopping with sulfuric acid) could be studied employing a conventional cell of three electrodes and recording the corresponding cyclic voltammograms (CVs) [13]. Alternatively, screen-printed electrodes could be also employed. The comparison between electrodes is interesting (e.g., diffusional behavior on screen-printed electrodes and adsorptive behavior on carbon paste electrodes). In Fig. 5.6, an example of the CV recorded in 0.1 M $HClO_4$ on screen-printed electrodes is presented.
3. Screen-printed electrodes can be used also in a flow cell [14]. As indigo presents a diffusional behavior on these electrodes and the flow cleans the electrode surface, even being disposable, they can be reused.
4. ACVs of indigo carmine show two peaks at around −0.1 and +0.7 V. The first one is due to the oxidation of leucoindigo to indigo and the second one to the conversion of indigo to dehydroindigo. The first one is chosen for quantification because it shows a higher peak current and appears at lower potential. A voltammogram has to be recorded always in the background for comparison.
5. The different parameters of ACV should be optimized (especially the phase angle).
6. The calibration curve (i_p vs. [IL-10]) is fitted to a log plot to obtain a linear equation in the whole range of concentrations (from 0.2 to 1500 pg/mL). Different functions can be evaluated.
7. A commercial protein matrix could be used to validate the methodology, spiking it with different concentrations of IL-10.
8. A spectrophotometric detection could be used to compare the results obtained by the electrochemical detection using p-nitrophenylphosphate (pNPP), instead of 3-IP, as enzyme substrate. When pNPP reacts with the enzyme AP, a yellow color is developed allowing the calibration by measuring the absorbance at 405 nm. To carry out the spectrophotometric detection, 100 μL of 1-mM pNPP solution (prepared in 0.1 M Tris-HCl buffer pH 9.8 containing 10 mM of $MgCl_2$) is added to each well. The

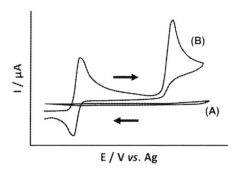

FIGURE 5.6 Cyclic voltammetric responses recorded on (A) 0.1 M $HClO_4$ and (B) indigo carmine solution in 0.1 M $HClO_4$. *Reprinted with modifications from M.J. Bengoechea Álvarez, M.T. Fernández Abedul, A. Costa García, Flow amperometric detection of indigo for enzyme-linked immunosorbent assays with use of screen-printed electrodes, Anal. Chim. Acta 462 (2002) 31−37.*

enzymatic reaction occurs for 30 min at 35°C and then it is stopped by adding 150 μL of 1 M NaOH into each well. Finally, absorbances are read at 405 nm.

9. Immunoassays (ELISA or other format) can be performed for other analytes (e.g., pneumolysin [8,9]) using AP-labeled antibodies.

10. In this experiment, an immunoassay with introduction of the product in a flow system is employed. The flow system acts as a linkage between the batch immunoassay and the detection system. The coupling of the flow system implies an important advantage when compared with batch protocols of immunoassays that use AP as label and detect the enzymatic product of 3-IP: (i) The electrode pretreatment commonly necessary for obtaining a reproducible methodology is simpler because the flow acts cleaning its surface; (ii) on the other hand, the accumulation step is reduced to the passage of the substrate loop by the electrode so that analysis time is decreased, and (iii) as the measurement is performed after the adsorption in fresh electrolyte, the influence of interfering species can be eliminated. However, AdSV and ACV could be demonstrated in a simpler way without requiring an immunoassay and a flow system as exemplified for other analytes such as melatonin [15]. This analyte (considered in Chapter 9) presents a reversible pair, whose current is increased after accumulation. It seems that the system can be a very interesting alternative and very appropriate also to study adsorption processes.

11. The experiment combines immunoassay/flow/adsorptive stripping/ACV. Discussion of possible combinations with only two or three elements can be very interesting.

12. 3-IP could be employed also as a substrate for other enzyme, also employed commonly in enzymatic immunoassays, horseradish peroxidase [16].

5.8 Assessment and discussion questions

1. Explain the basis of the electrochemical technique employed for the detection and suggest other possibilities.
2. What is the format of the immunoassay employed and what are the main steps?
3. Draw a scheme for the FIA system.
4. What is the difference between indigo blue and indigo carmine?
5. Explain the electrochemical behavior of indigo carmine.
6. Compare the analytical characteristics of this method with others found in the bibliography for the same or similar analytes, using other labels and methodologies.
7. Look for other electrochemical immunosensors that use alkaline phosphatase as label and discuss their differences.

References

[1] A.J. Bard, L.R. Faulkner, Electrochemical Methods: Fundamentals and Applications, second ed., John Wiley & Sons, New York, 2001.
[2] J. Wang, Analytical Electrochemistry, second ed., Wiley-VCH, New York, 2000.
[3] A.A. Kulkarni, I.S. Vaidya, Flow injection analysis: an overview, J. Crit. Rev. 2 (2015) 19−24.
[4] M. Trojanowicz, K. Kołacińska, Recent advances in flow injection analysis, Analyst 141 (2016) 2085−2139.

[5] C. Kokkinos, A. Economou, M.I. Prodromidis, Electrochemical immunosensors: critical survey of different architectures and transduction strategies, TrAC Trends Anal. Chem. 79 (2016) 88–105.

[6] F.S. Felix, L. Angnes, Electrochemical immunosensors — a powerful tool for analytical applications, Biosens. Bioelectron. 102 (2018) 470–478.

[7] E.C. Rama, A. Costa-García, Screen-printed electrochemical immunosensors for the detection of cancer and cardiovascular biomarkers, Electroanalysis 28 (2016).

[8] C. Fernández-Sánchez, M.B. González-García, A. Costa-García, 3-Indoxyl phosphate: an alkaline phosphatase substrate for enzyme immunoassays with voltammetric detection, Electroanalysis 10 (1998) 249–255.

[9] C. Fernández Bobes, M.T. Fernández Abedul, A. Costa-García, Pneumolysin ELISA with adsorptive voltammetric detection of indigo in a flow system, Electroanalysis 13 (2001) 559–566.

[10] M.J. Bengoechea Álvarez, C. Fernández Bobes, M.T. Fernández Abedul, A. Costa-García, Sensitive detection for enzyme-linked immunosorbent assays based on the adsorptive stripping voltammetry of indigo in a flow system, Anal. Chim. Acta 442 (2001) 55–62.

[11] M. Saraiva, A.O. Garra, The regulation of IL-10 production by immune cells, Nat. Rev. Immunol. 10 (2010) 170–181.

[12] A.O. Garra, F.J. Barrat, A.G. Castro, A. Vicari, C. Hawrylowicz, Strategies for use of IL-10 or its antagonists in human disease, Immunol. Rev. 223 (2008) 114–131.

[13] C. Fernández-Sánchez, A. Costa-García, Voltammetric studies of indigo adsorbed on pre-treated carbon paste electrodes, Electrochem. Comm. 2 (2000) 776–781.

[14] M.J. Bengoechea Álvarez, M.T. Fernández Abedul, A. Costa García, Flow amperometric detection of indigo for enzyme-linked immunosorbent assays with use of screen-printed electrodes, Anal. Chim. Acta 462 (2002) 31–37.

[15] J.L. Corujo-Antuña, S. Martínez-Montequín, M.T. Fernández-Abedul, A. Costa-García, Sensitive adsorptive stripping voltammetric methodologies for the determination of melatonin in biological fluids, Electroanalysis 15 (2003) 773–778.

[16] P. Fanjul-Bolado, M.B. González-García, A. Costa-García, 3-Indoxyl phosphate as an electrochemical substrate for horseradish peroxidase, Electroanalysis 16 (2004) 988–993.

CHAPTER

6

Enhancing electrochemical performance by using redox cycling with interdigitated electrodes

*Diego F. Pozo-Ayuso, Mario Castaño-Álvarez,
Ana Fernández-la-Villa*

MicruX Technologies, Gijón, Asturias, Spain

6.1 Background

Electrochemistry is the branch of Chemistry concerned with the interrelation of electrical and chemical effects. A large part of this field deals with the study of chemical changes caused by the passage of an electric current and the production of electrical energy by chemical reactions. In fact, the field of electrochemistry covers a huge array of different phenomena (e.g., electrophoresis and corrosion), devices (electrochromic displays, electroanalytical sensors, batteries, and fuel cells), and technologies (the electroplating of metals and the large-scale production of aluminum and chlorine) [1]. Then, electrochemistry now plays an important role in a vast number of fundamental research and applied areas. These include, but are not limited to, the exploration of new inorganic and organic compounds, biochemical and biological systems, corrosion, energy applications involving fuel cells and solar cells, and nanoscale investigations [2].

Thus, Electroanalysis is a very important field in Analytical Chemistry that has demonstrated its potential for solving real-life analytical problems by understanding the fundamentals of electrode reactions and the principles of electrochemical methods [3]. The high-performance, small-size, low-power requirements, and low-cost electrochemical devices have led to many important detection systems.

The main tool of Electroanalytical Chemistry is the electrode, where a reduction−oxidation (redox) reaction takes place. Depending on the experiment (voltammetry, amperometry, conductimetry, impedance spectroscopy, etc.) to be accomplished, different electrode configurations, designs, materials, etc., could be employed. The most typical configuration consists of a three-electrode approach: working electrode (WE), reference electrode (RE), and auxiliary electrode (AE). The electrodes have evolved from the classical dropping mercury electrode, going through traditional solid electrodes and reaching the most novel thick- and thin-film—based electrodes. This natural evolution brings advantages in terms of miniaturization, integration, and cost for high-mass production.

Likewise, miniaturization is a growing trend in the field of Analytical Chemistry [4]. Thus, microelectrodes are very small-size electrodes with at least one dimension not greater than 25 μm. Microelectrodes provide unique electrochemical properties such as small capacitive-charging currents, reduced ohmic (iR) drop, and steady-state diffusion currents. Moreover, the small size of the microelectrodes permits measurements on very limited solution volumes. Microelectrodes also enable the use of different materials, geometric shapes, and configurations. In this sense, interdigitated array (IDA) microelectrodes have demonstrated to be a very useful tool in electroanalytical applications.

IDA electrodes consist of a pair of array microelectrodes that mesh with each other. Each set of microelectrodes can be connected to the potentiostat individually to carry out redox cycling as it is shown in Fig. 6.1. Interactions between the individual electrodes in the microelectrode array enable the regeneration of the electroactive substance through redox cycling.

In this way, a redox species, which is transformed (oxidized or reduced) at one electrode (generator), can diffuse across a small gap and be regenerated (reduced or oxidized) at an adjacent electrode (collector). The regenerated species diffuses back to the original electrode. Thus, the particular analytical performance of IDA microelectrodes leads to the improvement of the signal-to-noise ratio, enhancing the detection sensitivity for reversible or quasi-reversible charge transfer reactions [5]. Besides, IDA electrodes are available in two main configurations: linear band and ring arrays (Fig. 6.6).

Microelectrodes can be useful in applications such as the study of electrochemical reaction mechanisms and kinetics, trace electrochemical analysis, electrochemical reactions in solutions of very high resistance, in vivo measurements on biological objects, multichannel (bio)sensors, and detection in flowing liquids (flow injection analysis, high-performance liquid chromatography, capillary electrophoresis). Therefore, IDA-based approaches can become a powerful, simple, rapid, and cost-effective analytical tool for environmental, food, and clinical analysis compared with available conventional ones [6].

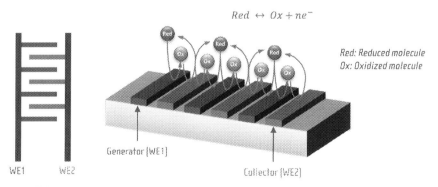

$$Red \leftrightarrow Ox + ne^-$$

Red: Reduced molecule
Ox: Oxidized molecule

Generator (WE1)

Collector (WE2)

WE1   WE2

FIGURE 6.1   Redox cycling performed at interdigitated array microelectrodes.

This lab experiment is focused on undergraduate and postgraduate students of different areas of science (Chemistry, Biotechnology). The students will be able to learn about the principles and performance of IDA microelectrodes with different redox compounds.

6.2 Chemicals and supplies

Reagents

- *Redox probes*: Potassium hexacyanoferrate (II) ($K_4[Fe(CN)_6]$), ferrocenylmethyl alcohol (FcMeOH), acetaminophen (APAP), epinephrine (EP), dopamine (DA), and p-aminophenol (pAP).
- *Preparation of electrolytes*: Potassium chloride (KCl), sulfuric acid (H_2SO_4), phosphoric acid (H_3PO_4), sodium hydroxide (NaOH), hydrochloric acid (HCl).
- *Preparation of solutions*: Methanol.

Instrumentation and materials

- Bipotentiostat.
- Drop-cell interface.
- Thin-film interdigitated microelectrodes (gold and platinum).
- *Lab apparatus for preparation of solutions*: pH meter, magnetic stirrer, analytical balance.
- *Volumetric material*: Flasks, pipettes, vessels, micropipettes, etc. All the material necessary for the preparation of the solutions should be of analytical reagent grade.
- Syringes and syringe filters (0.1−0.45 μm) used for removing particles from working solutions and samples.

Solutions

- *Background electrolytes* (BGEs): Different BGEs are used in the evaluation of the interdigitated electrodes:
  - 0.1 M KCl
  - 0.1 M HCl
  - 0.1 M H_2SO_4
  - 0.1 M Phosphate buffer (PB) of different pH values: 2, 7.4, and 12.
- *Stock and standard solutions* (prepare 1 mL of each solution):
  - 10 mM stock solutions of acetaminophen, epinephrine, dopamine, and p-aminophenol should be prepared in 10 mM HCl.
  - 10 mM stock solution of potassium hexacyanoferrate (II) should be prepared in deionized water, and 10 mM stock solution of ferrocenylmethyl alcohol should be prepared in methanol.
  - 1 mM standard solutions of potassium hexacyanoferrate (II), ferrocenylmethyl alcohol, acetaminophen, epinephrine, dopamine, and p-aminophenol are prepared from the stock solutions by dilution in each BGE.

6.3 Hazards

Avoid the use of potassium hexacyanoferrate (II) in acid solutions because it can release a toxic gas, cyanhydric acid.

Acid and alkaline solution preparations should be carried out under a fume hood. Protective garment and gloves should be worn at all times.

6.4 Electrochemical system setup

The system setup for using the electrodes consists in a bipotentiostat and an electronic interface (drop-cell) where the electrode is placed. In Fig. 6.2, an example of the basic electrochemical setup is shown.

Thin-film IDA band electrodes (Fig. 6.3) consist of two individually addressable microelectrode arrays strips with an interdigitated approach. Each WE contains 15 microelectrodes with 10 μm width. The gap between each interdigitated microelectrode is 10 μm. The RE and AE are also integrated on chip. The electrodes are fabricated on gold (150 nm) or platinum (150 nm) on a glass substrate (10 × 6 mm). A SU-8 insulating layer (3−4 μm thickness) is used for delimiting the electrochemical cell and electrode pads for electrical connections.

6.5 Experimental procedure

6.5.1 Electrode precleaning

Metal surface of thin-film electrodes should be cleaned to get the best electrochemical signals. This is made by a simple electrochemical pretreatment.

Using the drop-cell setup, a cyclic voltammetry (CV) experiment is performed between −1.5 and +1.5 V with a scan rate of 0.1 V/s (at least 10 cycles) for platinum-based electrodes and between −1.0 and +1.0 V with a scan rate of 0.1 V/s (at least 12 cycles) for

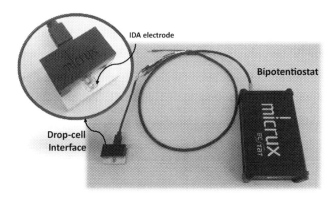

FIGURE 6.2   Electrochemical system setup. *IDA*, interdigitated array.

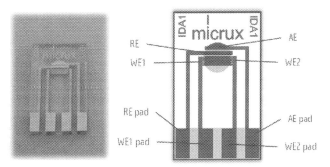

FIGURE 6.3   Thin-film interdigitated array (IDA) electrode layout $10 \times 6 \times 0.7$ mm. *RE*, reference electrode; *AE*, auxiliary electrode; *WE*, working electrode.

gold-based electrodes. Place a 5-µL drop of the BGE (chloride free for gold electrodes) on the electrode surface (be sure the sample drop covers all the electrodes of the cell) for the CV precleaning experiment. Be careful the solution drop is not driven to the PCB (Printed Circuit Board) of the connector. In IDA electrodes, the precleaning procedure is just performed in one of the WEs (WE1).

6.5.2 Electrochemical measurements

After cleaning the electrode surface, remove carefully the initial drop (BGE) from the electrode. A small piece of lab paper can be used to remove the drop avoiding the contact with the electrode surface.

For single-mode voltammetry experiments, place a drop (5 µL or less) of the BGE or standard solution on the electrode to perform the measurement. In the potentiostat software, the different parameters for each experiment have to be configured. Finally, start the voltammetric experiment and save the resulting curves.

For dual-mode experiments, enable the bipotentiostat option and configure the WE2 at a fixed potential. After that, start the electrochemical experiment and record and save the resulting curves for WE1 and WE2.

6.5.3 Electrochemical behavior of redox systems by cyclic voltammetry

CV is generally used to study the electrochemical properties of an analyte in solution as well as its behavior on a specific electrode surface.

The resulting cyclic voltammograms depend on the nature of the species (electroactivity) and BGE (composition, pH, ionic strength) as well as the electrode surface.

In these experiments, the CVs with the IDA electrodes are performed in single mode (just WE1 is used) to:

1. Study the electrochemical behavior of different benchmark redox compounds: potassium hexacyanoferrate (II), ferrocenylmethyl alcohol, acetaminophen, epinephrine, dopamine, and p-aminophenol.

2. Study the effect of the BGE in the electrochemical behavior of the compounds. Different analytes are evaluated in KCl, HCl, H_2SO_4, and PB (with different pH values).
3. Study the effect of the electrode material (Au and Pt) on the electrochemical behavior of the compounds. Each redox compound/BGE is studied in gold- and platinum-based IDA electrodes. For gold electrodes, chloride-based BGEs (KCl and HCl) must be avoided.
4. Study the effect of the scan rate on the electrochemical behavior of the compounds. Evaluate the scan rate between 10 and 500 mV/s.

For voltammetric experiments, configure in the potentiostat software, the starting potential, vertex potentials, step potential, and scan rate. For the first experiments, use a wide potential window (e.g., between −0.5 and +1.25 V) to allocate the electrochemical processes for each compound in different BGEs. A potential step of 5 mV and a scan rate of 50 mV/s can be used for initial experiments.

After the first experiments, when the potential window for each compound/BGE/electrode is well-defined, the scan rate study can be performed for each compound/BGE/electrode.

In Fig. 6.4, an example of cyclic voltammograms recorded at single mode for different redox compounds is shown.

6.5.4 Interdigitated array microelectrodes performance by cyclic voltammetry

IDA electrodes are an interesting tool to enhance the sensitivity of the electrochemical detection of redox active compounds. They can be used in single- or dual mode. In single mode, just one WE is used, in the same way as a single electrode (see Section 6.5.3). In dual mode, electrochemical measurements are carried out using a bipotentiostat to control the two WEs. In a CV experiment, the potential of WE1 (generator) is scanned, meanwhile the potential of WE2 (collector) is held at a fixed potential.

In this experiment, the behavior of the benchmark redox compounds is studied in dual mode, using different BGEs. Then, with this aim

1. Study the electrochemical behavior in dual mode of the different benchmark redox compounds: potassium hexacyanoferrate (II), ferrocenylmethyl alcohol, acetaminophen,

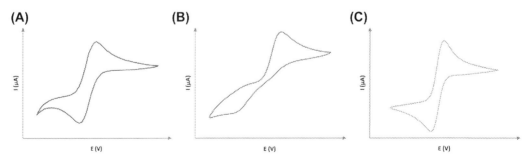

FIGURE 6.4 Cyclic voltammograms recorded with interdigitated array electrodes at single mode for different redox compounds: (A) potassium hexacyanoferrate (II), (B) dopamine, and (C) ferrocenylmethyl alcohol.

epinephrine, dopamine, and p-aminophenol. The behavior should be studied in the BGEs providing better results in previous single-mode experiments for each redox compound.

2. Study the effect of the electrode material (ED-IDA1-Au and ED-IDA1-Pt) in dual-mode experiments.

In dual mode, the fixed potential at WE2 is selected taking into account the cyclic voltammograms accomplished in single-mode configuration.

Experiments performed in this section can be combined with the experiments from Section 6.5.3. Thus, single- and dual-mode studies can be carried out simultaneously for each redox compound/BGE/electrode.

In Fig. 6.5, an example of cyclic voltammograms recorded at single- and dual mode for different redox compounds is shown.

6.6 Lab report

At the end of the experiment, write a lab report including an introduction, experimental (materials, equipment, and protocols), results and discussion, and conclusions sections. The following points should be considered in the report:

1. Sketch the cyclic voltammograms at single mode for the benchmark compounds in the different BGE solutions using platinum and gold thin-film electrodes.
2. Sketch the cyclic voltammograms at dual mode for the benchmark compounds in the different buffer solutions using platinum and gold thin-film microelectrodes.
3. Include the comparison between the cyclic voltammograms at single- and dual mode for each benchmark redox compounds.
4. Determine the redox cycling ($R_c = i_{a\ (dual\text{-}mode)}/i_{a\ (single\text{-}mode)}$) and efficiency ($i_{collector}/i_{generator}$) of the IDA electrodes for each compound/BGE.

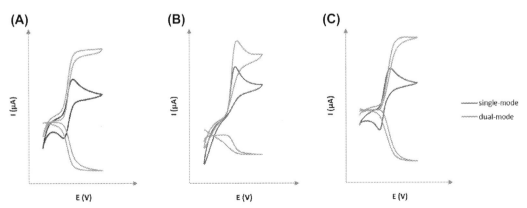

FIGURE 6.5 Cyclic voltammograms with interdigitated array electrodes at single- (red (black in print version)) and dual mode (blue (gray in print version)) for different redox compounds: (A) potassium hexacyanoferrate (II), (B) acetaminophen, and (C) p-aminophenol.

I. Dynamic electroanalytical techniques

6.7 Additional notes

1. Stock solutions should be kept on dark and stored at 4°C.
2. Be careful the solution drop is not driven to the PCB in the drop-cell connector. Solutions could damage the electronics parts.
3. The electrode surface should be cleaned before starting the first experiments. In this pretreatment, the different particles adhered to the electrode surface are removed with the hydrogen and oxygen gas generated. After the pretreatment, the electrodes could be used for several measurements depending on the sample. Thin-film electrodes could be reused after a new precleaning process.
4. The CV should be recorded firstly in each BGE (without the analyte) and then in the standard solutions of the different redox compounds. One electrode can be used for multiple experiments, but it is recommended to perform the pretreatment (see Section 6.5.1) between experiments with different BGEs and/or analytes.
5. In all the experiments, at least, three consecutive scans should be performed for each compound in different BGEs and electrode materials to evaluate the precision.
6. In the dual-mode approach, the collector potential should be set at a value lower than the redox potential if the redox species is in the reduced form in solution.
7. The number of pair of electrodes and the gap between fingers and the electrode area are of paramount relevance. Students are encouraged to evaluate the behavior of the redox probes on different configurations.
8. The configuration of interdigitated electrodes can vary. Here, a band IDA design is presented, but others are also possible. In this context, interdigitated ring arrays (Fig. 6.6) could also be evaluated [4].

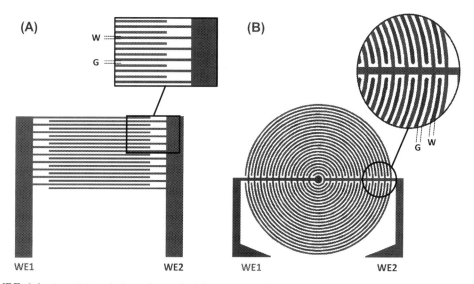

FIGURE 6.6 Interdigitated electrodes with different approaches: (A) band and (B) ring. Microelectrodes with different widths (W) and gaps (G) are available.

6.8 Assessment and discussion questions

1. What are the electrochemical processes for each compound?
2. What information could be obtained from a cyclic voltammogram?
3. Indicate the main criteria for process reversibility and comment which electrochemical processes of the benchmark compounds are reversible.
4. List the key advantages of microelectrodes and of IDA microelectrodes.
5. What are the main factors that affect the performance of IDA electrodes?

References

[1] A.J. Bard, L.R. Faulkner, Electrochemical Methods. Fundamentals and Applications, second ed., John Wiley & Sons, Inc., New York, NY, USA, 2001.
[2] C.G. Zoski, Handbook of Electrochemistry, first ed., Elsevier B.V., Amsterdam, The Netherlands, 2007.
[3] J. Wang, Analytical Electrochemistry, third ed., John Wiley & Sons, Inc., Hoboken, New Jersey, USA, 2006.
[4] A. Ríos, A. Escarpa, B. Simonet, Miniaturization of Analytical Systems: Principles, Designs and Applications, John Wiley & Sons, Inc., West Sussex, UK, 2009.
[5] E.O. Barnes, A. Fernández-la-Villa, D.F. Pozo-Ayuso, M. Castaño-Alvarez, G.E.M. Lewis, S.E.C. Dale, F. Marken, R.G. Compton, Interdigitated ring electrodes: theory and practice, J. Electroanal. Chem. 709 (2013) 57–64.
[6] O. Niwa, Y. Xu, H.B. Halsall, W.R. Heineman, Small-volume voltammetric detection of 4-aminophenol with interdigitated array electrodes and its application to electrochemical enzyme immunoassay, Anal. Chem. 65 (1993) 1559–1563.

Amperometric detection of NADH using carbon-based electrodes

Rebeca Miranda-Castro, Noemí de los Santos Álvarez,
M. Jesús Lobo-Castañón

Departamento de Química Física y Analítica, Universidad de Oviedo, Oviedo, Spain

7.1 Background

Amperometry is a dynamic electroanalytical technique in which the excitation signal applied to the working electrode (WE) is a constant potential. It is selected such that all the electroactive substance (the analyte) at the electrode surface is instantly electrolyzed (oxidized or reduced) and the electron transfer is not the rate-limiting step of the electrode process. Measurements are performed under hydrodynamic conditions achieved in this case by stirring the solution where the WE, the reference (RE), and counter (CE) electrodes are immersed (Fig. 7.1). The current flowing in the cell as a consequence of the electron transfer is monitored as a function of time and constitutes the analytical signal.

Consider a planar electrode immersed in a cell containing the electroactive species R that is oxidized at the WE in a reversible process: $R \rightleftarrows O + ne^-$. The electrode process is characterized by the following stages:

1. Transport of the electroactive (analyte) to the electrode surface
2. Electron transfer as dictated by the electrochemical reaction

During oxidation of R, the electrochemical reaction gives rise to the release of electrons to the electrode. The flow of electrons (current) measured will depend on the potential applied to the WE that is selected high enough (oxidation) to achieve the maximum charge transfer process, so the limiting step will be the mass transfer to the electrode surface. In addition, because of mixing, the analyte is present at a constant concentration throughout the bulk solution except for a narrow layer, of thickness δ, at the electrode surface, the Nernst diffusion layer, through which transport takes place by diffusion. When the molecules reach the electrode surface, they will be oxidized, and the consumption of the analyte creates a

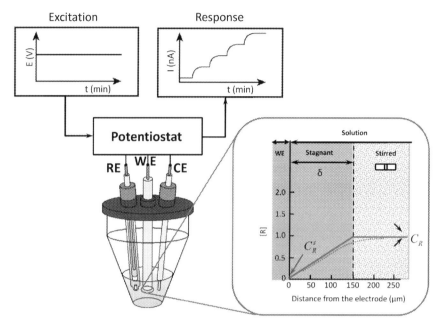

FIGURE 7.1   Instrumental setup used for the amperometric measurements. Inset shows the concentration profile at the electrode surface as a consequence of the electrochemical reaction that takes place at the applied potential. *CE*, counter electrode; *RE*, reference electrode; *WE*, working electrode.

concentration gradient, the driving force for diffusion. The instantaneous current we obtained is given by differentiating the Faraday's law:

$$ i = \frac{dQ}{dt} = nF\frac{dN}{dt} $$

where Q is the charge passing at each time, N is the number of moles of R transformed on the electrode surface, n the number of electrons involved, and F the Faraday constant. The instantaneous current thus depends on the transport rate of R, which is proportional to the concentration gradient at the electrode surface as given by the Fick's first law (written for a one-dimensional mass transfer along the x-axis through an imaginary plane with an area of 1 cm^2):

$$ J_x = -D_R\left(\frac{\partial C_x}{\partial x}\right) $$

where J_x is the flux of R (mol/cm^{2}s) at a distance x from the electrode surface, D_R is the diffusion coefficient (cm^2/s), and $(\partial C_x/\partial x)$ is the concentration gradient at a distance x from the electrode surface. The negative sign reflects the decrease in the R concentration that is being converted into O at the electrode surface. From the above equations, the current we measured is written as:

$$i = nFAD_R\left(\frac{\partial C_x}{\partial x}\right)$$

with A being the surface area of the electrode (cm^2). If we assume a linear concentration gradient through the diffusion layer (Fig. 7.1, inset), this current can be written as a function of the bulk concentration of R (C_R) and the concentration of R at the electrode surface (C_R^S):

$$i = nFAD_R\frac{C_R - C_R^S}{\delta}$$

At the fixed applied potential selected for the amperometric measurements, essentially all R reaching the electrode surface is transformed to O, and C_R^S is zero. Under these conditions, we obtain the largest mass transfer rate of R, and the current we measured is called the limiting current, i_L, which is directly proportional to the surface area of the electrode and the analyte concentration (C_R). This current is used for quantification purposes:

$$i_L = nFA\left(\frac{D_R}{\delta}\right)C_R$$

In this experiment, the amperometric detection of the coenzyme nicotinamide adenine dinucleotide (NADH) is performed. NADH and its oxidized form (NAD$^+$) or phosphorylated parent compounds (NADPH and NADP$^+$) are essential in the living cell system because they play a major role as the main redox carriers in the biological electron transfer chain. Known since the beginning of the 20th century, they participate in more than 400 reactions involving dehydrogenase and other enzymes [1].

NADH is a combination of two mononucleotides, adenosine monophosphate and nicotin-amide mononucleotide, and it is a soluble cosubstrate, that is, it is not permanently bound to the active site of the enzyme. The structure of coenzymes and general enzymatic reaction are depicted in Fig. 7.2.

Monitoring of the enzymatically generated or consumed NADH allows the determination of not only a large variety of important substrates (e.g., ethanol, glucose, lactate, etc.) but also the enzymatic activity of many dehydrogenases. Their wide applicability in clinical and food analysis has contributed to the development of optical and electrochemical methods for its determination.

On metal and carbon surfaces, the electrochemical oxidation of NADH requires high overvoltages (up to +1.6 V) due to a slow kinetics and strong adsorption of NAD$^+$ on the electrode surface. Under these circumstances, electron transfer mediators help to reduce the overpotential to moderate values where interferences are minimized [1].

N9-substituted adenine derivatives (adenosine, AMP, ADP, ATP, coenzyme A, and others including the cofactor NAD$^+$ itself) are a group of compounds whose oxidation at high potentials, typically +1.2 V, yields a common electroactive species (Fig. 7.3A), which once adsorbed on the electrode surface presents electrocatalytic activity toward the oxidation of

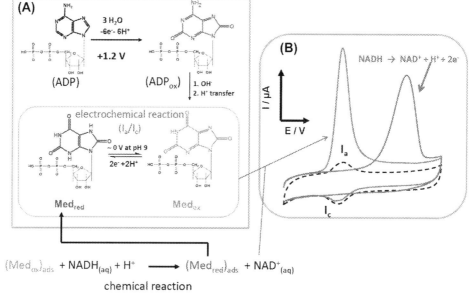

FIGURE 7.2   General scheme of the enzyme reactions in which NADH is involved.

FIGURE 7.3   (A) Mechanism of the electrochemical generation of the electrocatalyst species (Med$_{ox}$) from an N9-substituted adenine derivative (ADP). In the redox process, the reduced form of the mediator is highlighted in blue (black in print version) for the reduced form and green (gray in print version) and the oxidized form in green. Below, the reaction between the NADH and the electrocatalyst on the electrode surface is depicted. (B) Cyclic voltammogram of the redox process of oxidized ADP (dashed black line), unmediated (magenta, or dark gray in print version), and electrocatalytic (green, or light gray in print version) oxidation of NADH.

NADH on carbon-based electrodes such as carbon paste (CP) and pyrolytic graphite (PG) electrodes [2,3]. The oxidized adenine moiety (ADP$_{ox}$ in Fig. 7.3A) has a diimine structure able to catalyze the oxidation of the coenzyme. More recently, another species, also containing the catalytic diimine structure, has been proposed to the light of the analogous catalytic effect of guanine derivatives [4]. A common catalytic species for both purine bases has been established containing no primary amine groups (Med_{ox} in Fig. 7.3A). The formation of this compound requires a nucleophilic attack of a hydroxide ion to the amine-bearing carbon and further proton transfer with NH_2 elimination. The effect of the electrocatalysis is a reduction of the potential at which the oxidation of NADH takes places and an increase in the peak current at this potential (Fig. 7.3B). Using this group of electrocatalysts, concentrations in the low range of nM are routinely measured.

Preparation of the electrocatalyst species is very simple from a solution of the precursor (the adenine-containing molecule) when PG electrodes are used or mixing the precursor in the CP at an adequate percentage. Cyclic voltammetry (CV) is the best and fast way of generating the catalyst that remains adsorbed on graphite or is entrapped within the CP.

In this experiment, the electrocatalytic effect is studied on graphite electrodes using adenine diphosphate (ADP) as a precursor. After generation of the electrocatalyst by CV, NADH will be measured amperometrically (Fig. 7.4) at the potential selected according to cyclic voltammograms to ensure the limiting current is achieved.

The experiment can be completed in a 3-hour laboratory session being appropriate for junior/senior-level undergraduate courses in Chemistry. Students will learn about the chemical modification of renewable carbon-based electrodes and the use of mediators to improve the electron transfer kinetics, providing an excellent system for understanding electrocatalytic reactions, which play a central role in Electrochemistry and sensing applications. Besides, they will take an insight on detection at trace levels using amperometry.

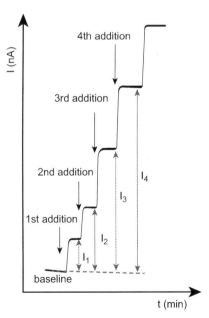

FIGURE 7.4   Typical current versus time (amperometric signal) obtained at a fixed potential.

7.2 Chemicals and supplies

— Adenosine-5′-diphosphate (ADP), β-NADH, phosphoric acid, and sodium hydroxide
— Graphite electrode, Ag|AgCl|KCl(sat) RE, and Pt CE
— Electrochemical cell and stand
— Small magnetic bars and magnetic stirrer
— Potentiostat
— Fine sandpaper and micropipettes
— N_2 gas to remove the oxygen from the solution

7.3 Hazards

Students are required to wear protective gloves, lab coat and glasses, especially when handling concentrated acids or bases.

7.4 Experimental procedure

7.4.1 Preparation of solutions

1. Prepare a 0.1 M phosphate buffer solution (PB) pH 9.
2. Prepare a 0.05 M ADP stock solution in PB.
3. Prepare a 0.1 M NADH stock solution in PB.
4. Prepare serial dilutions from the NADH stock solution in the same buffer to be able to reach nM range in the electrochemical cell.

7.4.2 Direct unmediated oxidation of NADH on graphite electrodes

1. Polish the graphite electrode on wet fine sandpaper.
2. Sonicate it in purified water for 2 min.
3. Add 20 mL of the PB solution to the electrochemical cell.
4. Put the electrochemical cell in its stand on a magnetic stirrer. Place the three electrodes in the cell and connect them to the potentiostat.
5. Deoxygenate the solution with N_2 for 10 min.
6. Sweep the potential between -0.2 and $+1.4$ V at 50 mV/s until a stable featureless cyclic voltammogram is obtained (background voltammogram).
7. Add the appropriate volume of NADH stock solution to obtain 0.1 mM NADH in the electrochemical cell.
8. Stir the solution with the help of a magnet for 20 s and then sweep the potential between -0.2 V and $+0.5$ V.
9. Record CVs at five different scan rates from 10 to 250 mV/s between -0.2 and $+0.5$ V.
10. Measure the current and the peak potential and plot the values against the scan rate to study the reversibility and adsorption/diffusion nature of the redox process (see Chapter 2).

7.4.3 Mediated oxidation of NADH

1. Repeat steps 1–6 to record a new background voltammogram with a freshly polished graphite electrode.
2. Add the appropriate volume of the ADP stock solution to obtain 0.5 mM ADP in the cell and stir the solution.
3. Stop the stirring and sweep the potential between −0.2 and +1.4 V at 50 mV/s 10 times to generate the electrocatalyst compound.
4. Remove the electrode from the cell and wash the electrode with water.
5. Add a fresh 20-mL aliquot of the PB to an empty cell and put the WE there.
6. Record CVs at five different scan rates from 10 to 250 mV/s between −0.2 and +0.3 V.
7. Measure the current and the peak potential and plot the values against the scan rate to study the reversibility and adsorption/diffusion nature of the redox process (see Chapter 2).
8. In the same solution, add the appropriate volume of the NADH stock solution to obtain a 0.1 mM concentration in the cell.
9. Stir the solution for 20 s and then sweep the potential between −0.2 V and +0.5 V.
10. Note the differences in the CV with respect to the unmediated oxidation of NADH.
11. Select the potential for the amperometric measurement to ensure that NADH is immediately oxidized on the surface of the ADP-modified electrode.

7.4.4 Amperometric measurement of NADH

1. Repeat steps 1–5 of the previous section to prepare a freshly ADP-modified electrode.
2. Set the potential at the value chosen in step 11, start the stirring and then the amperometric measurement.
3. When the background current is stable, add without stopping the stirring, the appropriate volume of NADH solution to obtain a 10 nM concentration in the cell.
4. After the current is stabilized at a higher value, repeat the addition of NADH.
5. Keep adding increasing amounts of NADH to cover a concentration range up to 0.1 mM.
6. Plot the current measured at each concentration assayed against the true NADH concentration in the cell, that is, taking into account the correction due to the volume added.
7. Estimate the figures of merit, the linear range, and the limit of detection.

7.5 Lab report

At the end of the experiment, write a lab report that include an introduction with the importance of the detection of NADH, the relevance of the use of mediators and the mechanism of the electrochemical generation of the electrocatalyst species from a precursor. Comment the technique, with the corresponding excitation and response signals. Include information on important experimental points: e.g., solutions, instrumentation employed or specific protocols, adding schematics when required. In the results and discussion section of the report, consider the following points:

1. Include graphical representation of the raw data obtained along the experiment. Be careful with the axis that should always include magnitude and units.
2. Show and comment the graphs for the direct unmediated oxidation of NADH on graphite electrodes as well as for the mediated oxidation of NADH.
3. Plot the calibration curve and include the linear range, sensitivity, limit of detection and limit of quantification, showing also the amperogram recorded.

7.6 Additional notes

1. NADH solutions are light sensitive so they must be protected from light and prepared daily.
2. Identical experiment can be carried out with CP electrodes composed of a mixture of graphite powder and silicone (1 g of spectroscopy grade graphite powder and 0.323 g of silicone high vacuum grease). Unmodified CP will be needed for the direct unmediated oxidation of NADH. To generate the catalyst species, a 5% of ADP should be added to the CP and then apply the CV procedure except there is no ADP in solution but in the electrode material. Electrode transfer to a background electrolyte solution after electrocatalyst formation is not needed.

7.7 Assessment and discussion questions

1. What are the advantages of using an electrocatalyst to detect NADH electrochemically?
2. Estimate the reduction in overpotential of the oxidation of NADH achieved with the mediator.
3. Explain how the amperometric current is measured and the selection of the potential applied.

References

[1] N. de-los-Santos-Álvarez, P. de-los-Santos-Álvarez, M.J. Lobo-Castañón, A.J. Miranda-Ordieres, P. Tuñón-Blanco, NADH-based electrochemical sensors, in: C.A. Grimes, E.C. Dickey, M.V. Pishko. (Eds.), Encyclopedia of Sensors. American Scientific Publisher, vol. 6, 2006, pp. 349–378.
[2] M.I. Alvarez-González, S.A. Saidman, M.J. Lobo-Castañón, A.J. Miranda-Ordieres, P. Tuñón-Blanco, Electrocatalytic detection of NADH and glycerol by NAD(+)-modified carbon electrodes, Anal. Chem. 72 (2000) 520–527.
[3] N. de-los-Santos-Álvarez, P. Muñiz-Ortea, A. Montes-Pañeda, M.J. Lobo-Castañón, A.J. Miranda-Ordieres, P. Tuñón-Blanco, A comparative study of different adenine derivatives for the electrocatalytic oxidation of beta-nicotinamide adenine dinucleotide, J. Electroanal. Chem. 502 (2001) 109–117.
[4] N. de-los-Santos-Álvarez, P. de-los-Santos-Álvarez, M.J. Lobo-Castañón, R. López, A.J. Miranda-Ordieres, P. Tuñón-Blanco, Electrochemical oxidation of guanosine and adenosine: two convergent pathways, Electrochem. Commun. 9 (2007) 1862–1866.

Chronoamperometric determination of ascorbic acid on paper-based devices

Estefanía Núñez-Bajo[1], M. Teresa Fernández Abedul[2]

[1]Department of Bioengineering, Royal School of Mines, Imperial College London, London, United Kingdom; [2]Departamento de Química Física y Analítica, Universidad de Oviedo, Oviedo, Spain

8.1 Background

Electroanalytical Chemistry is becoming of tremendous relevance in an era where information is obtained, with increased frequency, in places different from traditional labs. There are many examples of clinical, food, or environmental samples that require in situ analysis. Then, not only devices have to be miniaturized, low-cost, easy-to-handle, autonomous, reusable/disposable, etc., but also the methodologies employed have to be simple, intuitive, and friendly, but yet sensitive enough, producing accurate and precise results. In this context, chronoamperometry is a simple electroanalytical technique that results very useful for decentralized determinations. Concerning electrochemical techniques, cyclic voltammetry (Chapter 2) is of paramount relevance to know the electrochemical behavior of a specific analyte; square wave voltammetry (Chapter 3) to increase the sensitivity, especially when fast electron transfer (reversible) processes are involved, and electrochemical impedance spectroscopy (Chapters 11 and 12) to characterize surfaces and perform label-free bioassays. In this context, chronoamperometry, where an i-t curve is recorded, is very appropriate when simplicity is a priority. Because a potential scan is not required, the instrumentation (this needed to apply potential steps and record the current) is notoriously simplified.

This technique can be employed for the measurement of the number of electrons transferred, electrode surface areas, diffusion coefficients, rate constants of coupled chemical reactions, or the concentration of material adsorbed on an electrode surface. In this case, we are going to use this technique for electroanalytical purposes, i.e., determining the concentration

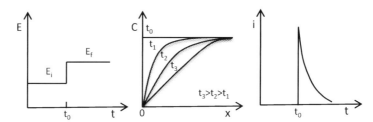

FIGURE 8.1 Schematics of the (A) excitation signal, (B) concentration profile, and (C) response signal for single-step chronoamperometry.

of the analyte, ascorbic acid. As commented in Chapter 2, ascorbic acid or vitamin C is one of the most important vitamins in human diet as it acts as a powerful antioxidant. The main samples are food (fruits and vegetables), vitamin formulations, and biological fluids (urine and serum). It is an important additive, as acting as electron donor prevents the oxidation of other food compounds that can cause unwanted changes.

In more detail, chronoamperometry is a potential step electroanalytical technique (opposed to voltammetry where a potential scan is made). Then, the potential is stepped from one value to another and the current generated is measured. The solution can be stirred (Chapter 7), or as in this case where a paper-based electrochemical cell is employed, it can remain quiescent. As usual, in electroanalytical techniques, it is very useful to depict the excitation and the response signals (Fig. 8.1). Then, a potential step is applied from an initial potential E_i to a final value E_f. If the experiment finishes at this potential, this is a single potential step, but if the potential moves to a different value (*e.g.*, the initial one), this is a double potential step experiment (with forward and reverse steps) [1].

During this potential step, a faradaic current flows (due to the electron transfer) if the step is high enough to produce the transfer, as well as a capacitive current, due to the charge of the interface. The potential applied to the electrochemical cell controls the surface concentrations of the redox species in solution. In many potential step techniques, the values of the initial and final potentials are carefully chosen so that essentially 100% conversion occurs at the electrode surface [1]. If an oxidation is recorded, the initial potential should be sufficiently negative (referred to the formal potential, $E^{0'}$) to warranty that no measurable amount of electroactive species is being oxidized when E_i is applied to the cell. When the potential moves to E_f, and its value is positive enough to produce complete oxidation, the concentration of the oxidizable species drops to ca. 0 at the electrode surface. In other words, there is a change in potential from a value where no electrolysis occurs to a value in the mass transfer—controlled region [2]. Then, a concentration gradient is established near the electrode in such a way that the electroactive species diffuses to the electrode to be oxidized. After oxidation, it diffuses back to the solution.

The instantaneous current obtained is given by differentiating the Faraday's law:

$$i = \frac{dQ}{dt} = nF\frac{dN}{dt}$$

where Q is the charge passing at each time, N the number of moles transformed on the electrode surface, n the number of electrons involved, and F the Faraday's constant. This instantaneous current thus depends on the mass transport rate, which is proportional to the concentration gradient at the electrode surface as given by the Fick's first law. Then, the current measured can be written as

$$i = nFAD\left(\frac{\partial C_x}{\partial x}\right)$$

where A is the surface area of the electrode (cm^2), D the diffusion coefficient (cm^2/s), and ($\partial C_x/\partial x$) the concentration gradient at a distance x from the electrode surface. The area is here included because the Fick's law considers the one-dimensional flux of the electroactive species through an imaginary plane with an area of 1 cm^2. Assuming a linear concentration gradient through the diffusion layer, this current can be written as a function of the difference between the bulk concentration (C) and the concentration at the electrode surface (C^S):

$$i = nFAD\frac{C - C^S}{\delta}$$

When the fixed applied potential selected for the amperometric measurements is chosen so that all the species reaching the electrode are transformed, the current depends on the bulk concentration (C) and the thickness of the diffusion layer, as the concentration at the electrode surface is zero. This current can be employed for quantitative purposes.

The mass transport regime is very important. If there is no convective transport, and as the electrolysis continues, the profile extends further into solution increasing the thickness of the diffusion layer current (δ). Then, the current decays, as happens in the experiment commented in this chapter (also in Chapter 25). However, if the chronoamperogram is recorded in a convective regime (e.g., stirred solution), then the thickness of the diffusion layer remains constant and the current attains a steady state (as in Chapter 7).

The name of the technique, chronoamperometry, stands by the measurement of the current (hence "amp") that is recorded with time (hence "chrono"). The current–time response signal for the forward potential step is described for a planar electrode and linear diffusion by Cottrell equation:

$$i = \frac{nFACD^{1/2}}{\pi^{1/2}t^{1/2}} = Kt^{-1/2}$$

with the current inversely proportional to the square root of time. Two issues have to be considered here: (i) the response contains also a charging (also named nonfaradaic or capacitive) current. As the potential is kept constant during the step, it decays very quickly (exponentially with time), usually in milliseconds or less [1]. The behavior of this current with time is:

$$i = \frac{E}{R_S}e^{-t/R_uC_d}$$

where R_u is the uncompensated resistance and C_d is the capacitance of the double layer. Then, the current decays exponentially with a cell time constant ($\tau = R_uC_d$). Then, the current for charging the double-layer capacitance drops to 37% of its initial value at $t = \tau$ and to a 5% of its initial value at $t = 3\tau$ [2]. (ii) Cottrell equation stands for a planar electrode with linear diffusion. In the case of porous electrodes, a more complicate regime is present because linear diffusion in planar zones is combined with thin-layer diffusion in narrow channels or even radial diffusion in pointed areas. On the other hand, at longer times, convective disruption of the diffusion layer may occur and this results in currents larger than those predicted by the Cottrell equation. In water and other fluid solvents, diffusion-based measurements for times longer than 300 s are difficult, and even measurements longer than 20 s may show some convective effects [2]. However, empirically, a linear relationship can be found between instantaneous current (current measured at a fixed time) and bulk concentration. Thus, this can be used for quantitative determination of the analyte.

In this chapter, a chronoamperometric experiment is performed for the determination of ascorbic acid using low-cost paper-based electrodes. The electrochemical cell here employed is explained in more detail in Chapter 25. The use of paper in electrochemical cells has many advantages, especially because of its porosity. Then, conductive inks can be deposited, as did Henry's group [3] in a seminal work of the integration of electrochemical detection on paper microsystems. Moreover, paper allows the use of very low volumes of sample and reagents and also the storage of bioreagents. Whatman Grade 1 chromatographic paper, the most employed for microfluidic systems, is here used for a simple and low-cost determination of ascorbic acid.

This experiment can be completed in two laboratory sessions of 3–4 hours and is appropriate for undergraduate students of advanced Analytical Chemistry or Master students from different fields where analysis is required. With this experiment, the students will learn not only about chronoamperometry, one of the simplest electrochemical techniques, but also about the work with low-cost paper-based electrochemical cells.

8.2 Electrochemical cell design

The electrochemical cell consists of a paper-based working electrode (WE) combined with two-wire counter (CE) and reference (RE) electrodes. The paper-based strip can contain single [4] (Fig. 8.2) or multiple cells [4]. The WE consists of a circular paper area, defined by wax printing, modified with carbon ink. Wire electrodes (RE and CE) are part of a gold-plated commercial connector header. This connector header also provides the necessary connection for the paper-based WE. Samples or standards are deposited on the opposite side of the ink and therefore the wire acting as connection of the WE does not contact RE and CE. Ionic

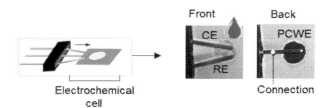

FIGURE 8.2   Schematics of the paper-based electrochemical cell containing the three electrodes and the gold-plated connector header, including pictures with top and bottom views. *CE*, counter electrode; *WE*, working electrode; *RE*, reference electrode.

contact is established when the solution wets all the three electrodes. To connect electronically the header to the potentiostat, a commercial connector is employed as interface.

8.3 Chemicals and supplies

— *Components of background electrolyte:* Acetic acid and sodium hydroxide. A 0.1 M acetate buffer solution pH 3.5 is employed as background electrolyte.
— *Analyte:* L-ascorbic acid. Solutions are prepared in the background electrolyte.
— *Samples:* Orange, apple, or multifruit juice. Regarding the samples, the only pretreatment required is the dilution with buffer solution to enter in the linear range of the calibration curve.
— *Electrochemical cell preparation:* Carbon paste, dimethylformamide (DMF), Whatman chromatographic paper grade 1, gold-plated connector headers.
— *General materials, apparatus, and instruments:* 1-mL microcentrifuge tubes, 100-μL and 1-mL micropipettes and corresponding tips, wax printer, hot plate, oven, analytical balance, pH meter, and potentiostat.
— Milli-Q water is employed for preparing solutions and washing.

8.4 Hazards

Acetic acid, used for the preparation of the buffer, is corrosive and causes serious burns. Students are required to wear a lab coat, appropriate gloves, and safety glasses.

8.5 Experimental procedure

8.5.1 Fabrication of the electrochemical cell

1. Print the chromatographic paper with the desired pattern (a circular area of a 6-mm diameter) employing a wax printer. Then, diffuse the wax by heating in a hot plate during 1 min at 100°C.
2. Deposit 2 μL of carbon ink (25% (w/w) of carbon paste diluted in DMF) on the circular area and let dry for 12 h at room temperature or for 1 h at 70°C.
3. Separate three wires from the connector header and form a clip to support the paper. The one in the middle is the connection of the WE. Bend the wires that will act as RE and CE (those at the ends) one against the other and separate them from the plane where WE is. Thus, insert the paper WE in the clip that wires acting as RE and CE form with the wire for WE connection.

The paper-based WE is single use, while only one connector header is required all along this experiment if it is washed with Milli-Q water after each use.

8.5.2 Choice of the potential step by cyclic voltammetry

Chronoamperometry is performed measuring with time the intensity of the current produced after the application of a potential step, from a potential where no oxidation occurs to a potential positive enough to produce ascorbic acid oxidation (ideally maximum concentration gradient). To select these potentials, a cyclic voltammogram is first recorded. With this aim:

1. Prepare the electrochemical cell as commented in Section 8.5.1.
2. Prepare a 0.1 M acetate buffer solution pH 3.5 and a 1 mM ascorbic acid solution.
3. Add 10 μL of the buffer solution on the paper, covering also RE and CE wires.
4. Record the CV by scanning the potential between -0.1 and $+0.7$ at a scan rate of 50 mV/s in the background electrolyte to check the cleanliness of the solution and/or electrochemical cell.
5. Record (in a different paper-based electrochemical cell) another CV but adding now 10 μL of 1 mM ascorbic acid solution.
6. Decide the best potential for the chronoamperometric step.

8.5.3 Calibration curve with chronoamperometric readout

1. Prepare the electrochemical cell as commented in Section 8.5.1.
2. Prepare a 0.1 M acetate buffer solution pH 3.5 and ascorbic acid solutions (in the acetate buffer solution) with concentrations ranging from 0.05 to 1 mM.
3. Add 10 μL of the buffer solution on the paper, covering also RE and CE wires.
4. Apply the initial potential (e.g., 0.0 V) for 10 s and then the final potential (e.g., +0.6 V) for 50 s and record the intensity of the current with time (as in the example in Fig. 8.3).
5. Measure the current at a fixed time (e.g., 40 s).
6. Repeat steps 3–5 but adding the ascorbic acid solutions of increasing concentration in different paper-based electrochemical cells. The connector can be reused after washing with Milli-Q water.
7. Represent the intensity of the current at the fixed time chosen versus the concentration of ascorbic acid to obtain the calibration curve.
8. Calculate the equation of the line that best fits the data.

To know the precision of the calibration curve, it could be repeated in different days and performed by different students.

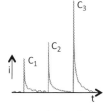

FIGURE 8.3 Example of chronoamperograms recorded with paper-based electrochemical cells in solutions with different concentrations of ascorbic acid.

8.5.4 Determination of ascorbic acid in fruit juices

The fruit juice sample has to be diluted with buffer solution to be included in the linear range of the calibration curve. The measurement procedure is the same as the one followed for ascorbic acid standards in the calibration curve. Once obtained the chronoamperogram, the intensity of the current (measured at the same time than the standards) is converted into concentration of ascorbic acid.

If matrix effects are suspected, the standard addition methodology can be followed. With this aim, solutions of the sample diluted with buffer and with increasing additions of ascorbic acid standards are prepared. The procedure for measuring is the same as this indicated for performing the calibration curve.

In all the cases, several replicates are measured to obtain the average and the standard deviation of the result.

8.6 Lab report

Write a lab report following the typical scheme of a scientific article. Thus, the report should include a short introduction, experimental part, results and discussion, and conclusions. To include diagrams, tables, and images is highly recommended. The following points should be considered:

1. In the introduction, include the current importance of this electrochemical technique in different fields of Analytical Chemistry. Alternative simple platforms that are employed with chronoamperometric readout can also be discussed.
2. Include representative pictures of the device and graphs representing the data obtained along the experiment: (i) CV recorded in the background electrolyte and in the ascorbic acid solution, (ii) chronoamperograms for different ascorbic acid concentrations and calibration curve obtained representing the intensity of the current at a fixed time for the different concentrations, and (iii) several chronoamperograms for the same concentration of ascorbic acid to show the precision of the methodology.
3. Discuss the values obtained for the linear range, sensitivity, and limit of detection (they can be compared to others found in the literature).
4. Determine the ascorbic acid concentration in the samples and report the results including the average and standard deviation. Compare the results obtained by different (groups of) students with the aim of calculating the precision of the results.

8.7 Additional notes

1. Working solutions of ascorbic acid should be prepared daily and protected from light. Stock solutions have to be kept at 4°C protected from light.
2. A buffer solution of pH 3.5 is employed throughout the experiment but is also interesting to evaluate the effect of the pH on the analytical signal. Then, buffer solutions with pH comprised between 3.5 and 5.5 could be prepared. The variations in the values of the peak potential and peak current can be discussed.

3. Ascorbic acid can be determined in different samples (e.g., other food samples or pharmaceuticals). In case a value is indicated in the sample, a statistical t-student test could be made to compare the values obtained.
4. In real sample analysis, possible interferences have to be considered (also here) and discussed. A cyclic voltammogram should be made in the diluted sample to know if processes of possible interferences are present. Alternatives can be discussed.
5. The analytical signal is the intensity of the current measured at a fixed time. This will be chosen to get the maximum precision. To increase it, sometimes the current is averaged for the last 5 or 10 s.
6. All the measurements can be performed in triplicate to allow the students to observe the precision. An interesting study could be to prepare several devices to study their repeatability/reproducibility.
7. By measuring the area under the i-t curve, the charge can be obtained. Therefore, coulometric readout is also possible. If total electrolysis occurs, calibration is not required because using Faraday's law ($Q = nFN$, where n is the number of electrons, F the Faraday's constant and N the number of moles), the amount of ascorbic acid electrolyzed can be directly determined (if it occurs with close to 100% current efficiency). To know if the electrolysis is total, a calibration curve can be made in such a way that the representation of Q versus NF should give the number of electrons involved.
8. The degree of agreement with Cottrell equation can be evaluated by representing the current, i, versus $t^{-1/2}$. The higher the linearity, the higher the agreement. Discussion on the nonfaradaic component can also be made. For example, if $R_u = 1\ \Omega$ and $C_d = 20\ \mu F$, then the cell time constant, τ, is 20 μs, and double-layer charging is 95% complete in 60 μs [2].
9. A very simple design is employed here for the electrochemical cell. In case paper or connectors are not available, alternative designs can be discussed (e.g., transparency film with conductive ink as WE and two stainless steel pins introduced in a drop of electrolyte as RE and auxiliary electrode).
10. Multiplexed paper devices can be also employed [5] for performing simultaneous measurements, but in this case a multipotentiostat have to be employed [4].
11. A sampler can be included to automate sampling and advance toward real lab-on-paper devices [5]. Discussion on the integration of the different steps of the analytical process is encouraged.
12. Because a very low-cost device is employed, calculation of its price could be an interesting issue. The use of these platforms for decentralized analysis can be discussed.

8.8 Assessment and discussion questions

1. Draw the concentration profile response (C-x, where x is the distance to the electrode surface) to a single- and double-potential step for an oxidizable and a reducible species, indicating the evolution with time for an unstirred solution.
2. Draw the excitation signal for a single-potential step and the corresponding chronoamperometric readout.

3. How are the initial and final potential values chosen? Discuss what would happen if higher or lower values were employed.
4. Explain the analytical signal (which can be employed for quantitative purposes) in chronoamperometry.
5. Discuss what happens with the nonfaradaic current during a chronoamperometric experiment.
6. Indicate the meaning of all the parameters involved in Cottrell equation and the limitations to this i-t behavior.
7. Explain how coulometric readout is also made possible using an i-t curve from a chronoamperometric experiment.
8. Discuss the effect of the pH on the redox process of ascorbic acid and the effect this would have on the chronoamperometric measurements.

References

[1] H.A. Strobel, W.R. Heineman, Chemical Instrumentation: A Systematic Approach, third ed., Wiley, New York, 1989.
[2] A.J. Bard, L.R. Faulkner, Electrochemical Methods: Fundamentals and Applications, second ed., Wiley, New York, 2000.
[3] W. Dungchai, O. Chailapakul, C.S. Henry, Electrochemical detection for paper-based microfluidics, Anal. Chem. 81 (2009) 5821–5826.
[4] O. Amor-Gutiérrez, E. Costa Rama, A. Costa-García, M.T. Fernández-Abedul, Paper-based maskless enzymatic sensor for glucose determination combining ink and wire electrodes, Biosens. Bioelectron. 93 (2017) 40–45.
[5] O. Amor-Gutiérrez, E. Costa Rama, M.T. Fernández-Abedul, Sampling and multiplexing in lab-on-paper bioelectroanalytical devices for glucose determination, Biosens. Bioelectron. 135 (2019) 64–70.

CHAPTER

9

Electrochemical detection of melatonin in a flow injection analysis system

Andrea González-López, M. Teresa Fernández Abedul

Departamento de Química Física y Analítica, Universidad de Oviedo, Oviedo, Spain

9.1 Background

One of the most common applications of amperometry (apart from its combination with biosensors or methodologies with chronoamperometric readout, as commented in Chapters 7 and 8), is the detection for liquid chromatography and flow injection analysis (FIA) [1]. In these applications, a small volume of sample is injected into a flowing stream. In liquid chromatography, the sample passes through a chromatographic column where analytes are separated. Combination with other separation techniques such as capillary electrophoresis (see Chapters 13–15) is also possible. In FIA, introduced by Ruzicka and Hansen in 1975 [2], the sample is also injected but there is no column to separate the analytes (see also Chapters 5 and 28). As they are based on the injection of a reproducible volume of sample into a nonsegmented continuous laminar flow of a carrier solution and delivery to the detector without intervention of the operator, FIA methodologies are considered automated. A valve is employed to inject the sample in such a way that the flow of the stream is not disturbed. The electrochemical detector can present different cell geometries and flow arrangements.

The general requirements for an electrochemical detector (three-electrode cell) in a flow system are: (1) well-defined hydrodynamics, (2) low dead volume, (3) high mass transfer rate, (4) high signal/noise ratio, (5) robust design, and (6) reproducible electrode responses [3]. The main cell arrangements are thin-layer and wall-jet cells. On the other hand, various electrode geometries, such as: (1) tubular, (2) planar electrode with parallel flow, (3) planar electrode with perpendicular flow, and (4) wall jet, are possible [3]. Advances in microfluidics have introduced new geometries and approaches, including microelectrodes [4] and

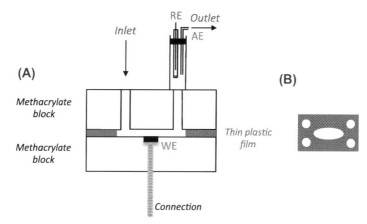

FIGURE 9.1 Schematics of the: (A) thin-layer cell employed in this work and (B) thin plastic separator. *AE*, auxiliary electrode; *RE*, reference electrode; *WE*, working electrode.

even porous and conductive materials that allow the flow of electrons and solutions simultaneously (*electrofluidics* [5]). In this chapter, a thin-layer cell (see Fig. 9.1) where the working electrode (WE) is positioned in a thin channel through which the liquid flows is employed [1].

Within the basic thin-layer design format, there are different choices as to the placement of the reference electrode (RE) and auxiliary electrode (AE) [6]. The design here employed (see Fig. 9.1A) is very simple, with both electrodes, RE and AE, located downstream of the WE. Although this produces a nonuniform current distribution across the electrode surface (because of the relative location of WE and AE) and uncompensated resistive drop to the RE (because of the relative placement of WE and RE), good responses are obtained and potentially interfering products that could be formed at the AE cannot reach the WE.

The simplest electrochemical technique used with these cells is amperometry, where the WE potential is fixed at a value where the analyte is oxidized or reduced. Then, the analyte is detected through its redox process, as it sweeps over the electrode in the thin channel. A peak current response is obtained in the fiagram (i-t curve) and the concentration of the analyte can be quantified measuring either the peak height (i_p) or integrating the peak area to obtain the charge (Q) [1]. Although voltammetry in flow systems is very interesting regarding selectivity, the detection limits are much higher in the voltammetric mode because of the large background current, which arises partly from double-layer charging and slow faradaic surface processes occurring with changing potentials [7]. Although this could be improved with the use of fast techniques (such as square wave voltammetry, see Chapter 3), the best sensitivity is always associated with an electrode operating at a fixed potential (amperometry) in an electrolyte of unchanging composition [7].

In electroanalysis, the mode of mass transport of the species to the electrode surface is of paramount importance. Mainly, diffusion, convection, and migration can be considered. The addition of electrolyte (background or supporting electrolyte) nearly eliminates the contribution of migration to the mass transfer of the electroactive species. On the other hand, in flow systems, the electrode remains stationary but the flow generates a convective system. Then, the concentration-distance profile (concentration of the electroactive species vs. the distance to the electrode surface) is similar to this of a stirred solution, where diffusion and convection occur. Initially, when a potential (high enough to produce a redox process) is applied to a reversible system (electrochemical reaction proceeds rapidly in both directions [1]), there is a change in concentration according to Nernst equation. Then, there is a depletion of the substance being oxidized (or reduced) in the immediacies of the electrode, and a stagnant layer of the electroactive species having a thickness δ is generated. This layer is also named diffusion layer because this is the predominant mass transport mechanism operating in it. When the potential is high enough to attain the maximum rate of electron transfer, the substance that arrives to the electrode surface is electrolyzed and then the gradient of concentration (C−C^s, C being the concentration in solution and C^s the concentration at the electrode surface in an approximated linear gradient) is maximum (C^S is zero as the analyte is rapidly electrolyzed). Here, we can have two different scenarios: (1) if there is no any convection, the thickness of this layer increases to renew, by diffusion, the substance depleted at the electrode surface or (2) if there is forced convection (e.g., rotating electrodes, stirring or flowing solutions as in this case), a homogeneous concentration of the analyte out of this stagnant layer is maintained. Thus, its thickness remains constant. At this moment, in convective systems, a limiting current is obtained:

$$i_L = nFCDA/\delta$$

where n is the number of electrons, F the Faraday's constant, A the electrode area, D the diffusion coefficient, C the concentration, and δ the thickness of the diffusion layer. For a flow system, with a channel height b and an average volume flow rate v (cm^3/s), the limiting current for a planar electrode with a parallel flow in channel [8] is:

$$i_L = 1.47nFC\left(\frac{DA}{b}\right)^{2/3} v^{1/3}$$

Therefore, this current is proportional to the concentration as is, in turn, the peak current of the corresponding fiagram. To know the potential at which the limiting current is achieved, a hydrodynamic voltammogram (HDV) should be made. This is a voltammogram (i-E curve, current response to a range of potentials applied to an electrochemical cell) under forced convection. To obtain it, a sequence of discrete potentials that covers the desired potential range is applied. Then, the steady-state current is measured at each potential, and the current is plotted versus the potential. For maximum sensitivity, the potential of the electrode is held on the limiting current region (plateau) of the HDV.

A single potential step, from an initial potential value (usually the open circuit potential) to a value where the limiting current is achieved, is commonly applied. However, in situations where electrode fouling occurs, the potential can be cycled between anodic and cathodic limits or other types of potential programs can be used to obtain reproducible behavior by oxidation (or reduction) and desorption of impurities from the electrode surface (pulsed amperometric detection). In this experiment, where the flow helps to clean the electrode and produces reproducible signals, a single potential step is employed.

Melatonin, N-acetyl-5-methoxytryptamine, is a hormone (Fig. 9.2) mainly produced in the pineal gland, which concentration depends on the circadian rhythms and it is produced at night. This hormone has influence on a variety of physiological and behavioral processes

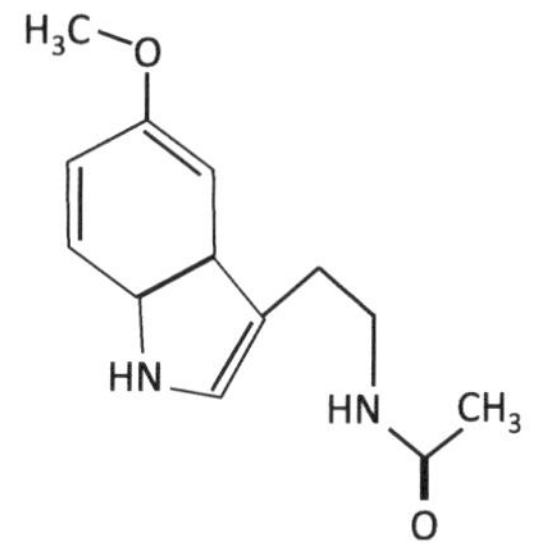

FIGURE 9.2   Molecular structure of melatonin.

including neurological, psychiatric, or reproductive [9]. Some disorders, such as anxiety and seasonal depression, are also related to it [10]. It may even delay the aging process because of its ability to scavenge free radicals and antioxidant activity [11]. Admittedly, it is quite important in mammalian life. Then, its detection is of paramount relevance and that is why several analytical methods have been developed for its determination in different matrices mainly pineal secretion, plasma, and pharmaceuticals. Because of the complexity of the matrix in these samples, most of them include a separation method, with high-performance liquid chromatography being the most employed.

The most common detection principles are fluorometric or electrochemical, with low detection limits in both cases. However, considering that separation is not always needed, the cost and maintenance of chromatographic systems could be avoided if a simple and cheap flow electrochemical methodology is employed. Moreover, the electrochemical behavior is very appropriate [12] and analysis time can be greatly reduced because of the high sample throughput of flow systems. On the other hand, the possibility of automation is very important when dealing with pharmaceutical analysis and quality control.

In this experiment, adapted from Ref. [13], a flow system with amperometric detection is employed for determination of melatonin. The electrochemical behavior is firstly studied using a simple flow system with conventional components (Fig. 9.3): (1) propulsion system, (2) injector, (3) tubing, (4) detector consisting of a thin-layer flow cell coupled to a

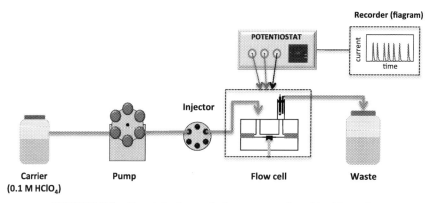

FIGURE 9.3   Flow injection analysis system employed in this work.

potentiostat, and (5) recorder/computer system (a computer can act as an active/passive interface that can control flow, injection, and detection parameters and record, store, and process data. Alternatively, a recorder can be employed and then manual control over the flow parameters [also over those of the potentiostat] has to be made). In this case, and because no reaction is required, the flow system is employed as a means to transport the sample to the detector, in a low-dispersion system.

This experiment can be completed in three laboratory sessions of 3 h and is appropriate for undergraduate students of advanced Analytical Chemistry or Master students from different fields where analysis is required. With this experiment, students will learn about the components of an FIA system, the main variables that influence the analytical signal, and the principles of quantitative analysis employing FIA. The students will also learn how to use cyclic voltammetry (CV) to know the electrochemical behavior of a species and develop the most appropriate FIA methodology.

9.2 Electrochemical thin-layer cell

The thin-layer cell (Fig. 9.1A) is constituted by two methacrylate blocks fixed together with four strews. A thin plastic film defines the thickness and flow geometry of the channel through which solution flows (Fig. 9.1B).

A carbon paste electrode (CPE) of approximately 7 mm^2 of geometric area is included. It is made of a hollow cylindrical channel tube in which an inserted stainless steel rod acts as the electric contact between the carbon paste and the potentiostat. At the same time, it works as a piston that allows renewing the surface. This is made manually by smoothing it on a clean paper before use. A coupled downstream compartment contains the RE (Ag/AgCl/saturated KCl) and a stainless steel waste tube acting as AE.

9.3 Flow injection analysis system

The basic FIA system (Fig. 9.3) consisted of the following elements:

— A peristaltic pump with corresponding tubing to provide a 1 mL/min constant flow of carrier solution (for a defined flow system, it depends on the rotation speed of the pump drums and the tubing diameter).
— A six-port rotary valve to inject a constant and reproducible volume of sample in a short time without disturbing the flow carrier (equipped with a 100-μL loop).
— Tubing to connect all the components. Diameter and length are important parameters to control sample dispersion. The material is also important depending on the type of solutions.
— The electrochemical detector consisting of a thin-layer flow cell (Section 9.2) connected to the potentiostat.
— A recorder (or computer) is employed to plot the i-t curve or fiagram, including the analytical signal that correlates to analyte concentration.

9.4 Chemicals and supplies

Reagents

— *Analyte*: Melatonin standards.
— *Background electrolyte*: 0.1 M perchloric acid and Britton—Robinson buffers covering the pH range 2—11 (prepared from an acid mixture of acetic, phosphoric, and boric acid. pH values are adjusted with sodium hydroxide).
— *Electrode preparation*: Paraffin oil and spectroscopic grade graphite powder. Carbon paste is prepared by intimately mixing 0.36 mL of paraffin oil with 1 g of graphite.
— Milli-Q water is employed for washing and preparing solutions.

Materials, apparatus, and instrumentation

— *CV*: 20-mL glass electrochemical cell, CPE (Teflon cylindrical holder, carbon paste, and stainless steel rod), Ag/AgCl/saturated KCl RE, platinum wire, potentiostat, and x-y recorder (or computer).
— *FIA system*: Peristaltic pump, tubing, six-port rotatory valve, electrochemical flow cell (see Section 9.2), potentiostat, and recorder (or computer).
— *Sample preparation*: Mortar, filter paper (7—11 μm), and volumetric flask.
— *General materials and apparatus*: Volumetric flasks, 100- and 1000-μL micropipettes and corresponding tips, magnetic bars and stirrer, analytical balance, and pH meter.

9.5 Hazards

Concentrated acids are corrosive and cause serious burns. Students are required to wear a lab coat, appropriate gloves, and safety glasses.

9.6 Experimental procedure

9.6.1 Preparation of solutions

— A volume of 1 L of 0.1 M perchloric acid solution.
— Stock solutions of melatonin were prepared in 0.1 M perchloric acid. Before being transferred to a volumetric flask, the initial solution is heated at 50°C for 45 min to favor the solubility. Once prepared, it is stored at 4°C protected from light. Dilutions were daily made in 0.1 M perchloric acid.

9.6.2 Cyclic voltammetric measurements

9.6.2.1 Redox processes of melatonin

1. Prepare the WE by filling the hole of the Teflon holder (approximately 2 mm deep) with the carbon paste.
2. After compacting and polishing on a clean piece of paper, place it together with the RE and the AE in the cell stand.
3. Add 20 mL of 0.1 M $HClO_4$ solution to the electrochemical cell and a magnetic bar.
4. Attach the cell to the stand and connect the three electrodes to the potentiostat.

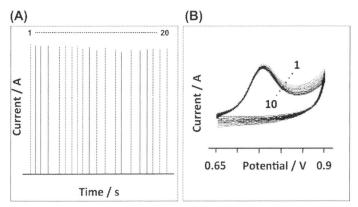

FIGURE 9.4 Successive signals recorded for a 10^{-6} M melatonin solution with (A) amperometric detection in a flow system and (B) cyclic voltammetry (after following electrochemical pretreatment).

5. Stir the solution (e.g., 500 rpm) for 30 s with electrolysis at +1.6 V.
6. Change the potential to +0.2 V and maintain it for 30 s under stirring.
7. Switch off the stirring and allow the solution to quiesce for 15 s.
8. Record two CVs by scanning the potential between +0.2 and +1.1 V at a scan rate of 50 mV/s in the background electrolyte to check the cleanliness of the solution/electrode.
9. Add the appropriate volume of melatonin to the solution under stirring to obtain a final concentration in the cell of 1×10^{-5} M.
10. Repeat steps 5–7 to clean/activate the electrode always between measurements.
11. Record the CV by scanning the potential between +0.2 and +1.1 V at a scan rate of 50 mV/s.
12. Obtain the peak potentials (E_p) and peak currents (i_p) and ascertain the nature of the processes (oxidation/reduction and reversible/irreversible) and characterize the electrochemical behavior of melatonin.
13. Choose the process that is going to be the basis of the analytical signal in the flow system.

9.6.2.2 Scan rate study

1. Follow the steps 1–8 in the previous section to prepare the electrochemical setup and confirm that there is no any impurity in the system.
2. Record the CVs (without stirring) between +0.6 and +0.9 V, varying the scan rate between 10 and 500 mV/s (stir the solution between measurements) without changing the solution or the WE.
3. Plot i_p versus the square root of scan rate ($v^{1/2}$) and versus v. Decide if the process recorded is an adsorption or diffusion-controlled process.

9.6.3 Flow injection analysis with amperometric detection

9.6.3.1 Hydrodynamic curve

CV is an electrochemical technique recorded on stationary solutions. In this case, amperometric detection is performed under forced convection in a flow system. However, CV is a diagnostic technique that can be indicative of the most adequate potential to perform the amperometric detection. Despite this, it is convenient to obtain an HDV to confirm this.

1. Prepare the WE by filling the hole of the methacrylate holder (approximately 2 mm deep) with the carbon paste.
2. After compacting with a spatula and polishing on a clean piece of paper, put the thin plastic film in between the two pieces of methacrylate and close the blocks by fixing the four screws.
3. Place the syringe containing RE and AE at the outlet of the flow cell.
4. Connect all the components of the FIA system and, using 0.1 M $HClO_4$ as carrier stream, confirm there is no any leaking.
5. Fix the flow rate at 3 mL/min.
6. Apply a detection potential of $+0.60$ V and wait until a stable baseline is obtained.
7. Make three successive injections of a 10^{-6} M melatonin solution.
8. Change the potential detection to $+0.65$ V and wait until the baseline is stable.
9. Make three successive injections of a 10^{-6} M melatonin solution.
10. Repeat steps 8–9 for $+0.70$, $+0.75$, $+0.80$, $+0.85$, $+0.90$, $+0.95$ V, and $+1.0$ V of detection potential.
11. Measure the peak height (i_p) and plot the values against the detection potential to obtain the HDV.
12. Decide which is the best detection potential, at which the maximum velocity of mass transfer is attained and the intensity is not affected by background electrolysis.

9.6.3.2 Effect of the flow rate

In a flow system, the flow rate is a parameter that must be optimized because dispersion of the sample, sensitivity, precision, and sample throughput depend on it. To study its effect:

1. Follow steps 1–4 of Section 9.6.3.1.
2. Fix the flow rate at 0.5 mL/min.

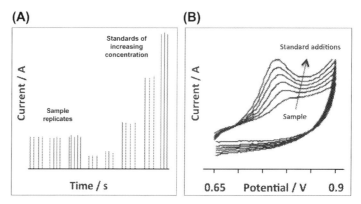

FIGURE 9.5 (A) Fiagrams corresponding to melatonin determination in three aliquots (the first three groups of signals) of a melatonin capsule employing a calibration curve; (B) Cyclic voltammograms corresponding to melatonin determination in an aliquot of a capsule following the standard additions method.

3. Apply the detection potential chosen in Section 9.6.3.1 and wait until a stable baseline is obtained.
4. Make three successive injections of a 10^{-6} M melatonin solution.
5. Change the flow rate to 1 mL/min and make three successive injections of a 10^{-6} M melatonin solution.
6. Repeat step 5 for 1.5, 2, 2.5, 3, 3.5, and 4 mL/min.
7. Measure the peak height (i_p) and plot the values against the flow rate.
8. Observe what happens with the peak width.

Higher flow rates produce shorter analysis times and improve the cleaning of the electrode surface. Decide which is the best flow rate, at which the maximum intensity is obtained precisely with shorter analysis time.

9.6.3.3 Precision studies

In the case of flow systems, the continuous passage of the carrier stream could produce an effective cleaning and then an activation of the electrode is not necessary. This, in turn, saves time because electrode treatment stages can be omitted. For melatonin, the passage of the 0.1 M $HClO_4$ stream at 3 mL/min is enough for ensuring an adequate precision of the signals and no treatment of the electrode is necessary. Then:

1. Confirm the precision of the signals by performing 20 successive injections of a 10^{-6} M melatonin solution (see Fig. 9.4A as orientation).
2. Calculate the relative standard deviation of the peak height.

9.6.3.4 Calibration curve

The relationship between the peak current and the concentration has to be studied to know the analytical features of the methodology. With this aim:

1. Perform injections of solutions with varying concentration of melatonin in the range 10^{-8} and 5×10^{-5} M (each concentration in triplicate).
2. Plot the peak height versus the concentration to obtain the calibration curve.
3. Calculate the sensitivity as the slope of the equation representing the linear relationship between the peak height and the concentration.
4. Calculate the limit of detection and quantification as the concentration corresponding to a signal that is 3 and 10 times, respectively, the standard deviation of the intercept (or estimate).

9.6.3.5 Sample preparation and measurement

In this chapter, melatonin is determined in pharmaceutical capsules. Once it has been demonstrated that there are no matrix effects, determination can be done using the external calibration curve. For sample preparation and measurement:

1. Pulverize a tablet with the help of a mortar.
2. Solve it in 0.1 M perchloric acid by heating (50°C) for 45 min.
3. Filter it through a filter paper (7–11 μm).

4. Transfer the filtrate quantitatively to a 500-mL volumetric flask and make it up with 0.1 M perchloric acid.
5. According to the calibration curve, prepare an appropriate dilution to inject in the flow system.
6. Inject the diluted solution (in triplicate) in the flow system (see Fig. 9.5A as orientation).
7. Repeat all the steps with two more melatonin tablets.
8. Use/perform a calibration curve to calculate the concentration of melatonin in the tablets.

9.7 Lab report

At the end of the experiment, students should write a lab report including an introduction, experimental, results and discussion sections, and the main conclusions. The following points should be considered:

1. Include schematics and pictures of the electrochemical cell employed for voltammetric and amperometric measurements.
2. Include raw data (voltammograms and fiagrams) that you consider interesting to show a behavior or an effect (e.g., a CV of melatonin showing the different processes, fiagrams showing the precision, etc.).
3. Indicate which is the process of melatonin that is the basis of the analytical signal employed for its determination and comment if it is diffusive or adsorptive in nature.
4. Represent the graphs with the HDV as well as those showing the effect of the flow rate and the calibration curve.
5. Indicate the linear range, sensitivity, limit of detection, and limit of quantitation.
6. Discuss the precision of the system in terms of the relative standard deviation of the peak height. Compare it with that of the signals obtained by CV.
7. Determine the concentration of melatonin in the samples and give the results appropriately, with an adequate number of significant figures. Perform statistical tests (e.g., t-student) to evaluate the accuracy.
8. If the methodology has been validated with a fluorometric methodology (see Section 9.8), comment the results and perform statistical tests to evaluate if differences between both are significant.

9.8 Additional notes

1. Stock solutions of melatonin must be stored at 4°C protected from light. Dilutions must be daily made in 0.1 M perchloric acid.
2. Carbon paste can be monthly prepared. The paste is stored protected from light. At least, the carbon paste of the WE has to be renewed daily.
3. The CPE preparation is of paramount importance. Carbon paste has to be thoroughly compacted and polished. This can be made by pressing hard the electrode on a clean satin office paper drawing an eight figure several times.
4. The CPE surface can be renewed entirely or partially by rotating the piston a little bit (either in the pen electrode or in this of the flow cell) and polishing the electrode again on the office paper.

5. In CV, the stirring has to be homogenous. Place the stir bar just below the WE and procure a homogenous and gentle stirring to obtain precise measurements.

6. A cyclic voltammogram is always recorded in the background electrolyte when starting a study. If a peak is observed, discard the solution and the carbon paste. Prepare a new CPE and repeat the experiment.

7. The potential window is chosen according to the processes of the electroactive species. Students are encouraged to find the whole potential window available for this electrolyte/electrode, explaining what the limits of electroactivity are and also the most adequate for the process that will be employed as the basis of the flow methodology.

8. Solid electrodes use to need pretreatments to clean or activate the surface and obtain sensitive and precise measurements. In this case an electrochemical procedure based on the application of an anodic potential (+1.6 V), is proposed, but mechanical (polishing), chemical (acid or basic solutions or organic solvents), and other electrochemical (anodic or cathodic) or mixed procedures could be employed. In case the precision between measurements is not adequate, students are encouraged to revise the potential and the time of pretreatment.

9. The scan rate study is useful to ascertain the diffusive or adsorptive nature of a process (observing the behavior of i_p), but it is also indicative of the reversibility of a process, observing the relationship between E_p and v (in reversible systems, E_p does not change with v).

10. The RE for the flow system can be easily made anodizing a silver wire on a saturated KCl solution. Silver chloride forms then on the surface of silver. The RE is completed by inserting the anodized wire on a syringe body with filter paper wetted with saturated KCl positioned at the outlet and filled with saturated KCl.

11. The RE syringe can be included in a bigger one through the rubber of the piston, as well as the hollow stainless steel tube acting as AE. This RE–AE syringe is connected to the outlet of the flow cell. Then, the solution that leaves the detector is forced to go to waste by passing through the hollow stainless steel wire. Tubing is connected at the end of the hollow tube to lead the solution to waste.

12. Attention has to be paid to position the hollow AE above the RE. Then, RE is wetted with the solution and produces an adequate potential measurement.

13. The flow rate could be checked collecting the solution at the outlet in a graduate test tube. The value is calculated by dividing the volume by the time required to collect it.

14. In flow systems, several measurements are made for each parameter (potential, flow rate, concentration, etc.). Usually injections are made in triplicate. Then, all the points could be included in the graph (e.g., three for one value of parameter). Alternatively, the mean value is plotted with error bars indicating its standard deviation. In this last case, it is more appropriate to perform a high number of injections.

15. When performing the HDV, after changing the potential, a stable baseline has to be obtained before injecting.

16. Regarding the precision, it is interesting to know this of the CVs recorded without pretreatment and after anodic activation (see Fig. 9.4B). Similarly, the precision can be studied at different flow rates.

17. Sample throughput can be determined (h^{-1}) measuring the time between the injection and the return to the baseline (e.g., in precision studies).

18. The determination of melatonin in tablets could also be performed by CV using the standard addition methodology (Fig. 9.5B). Students are encouraged to compare, critically, both methodologies.
19. Validation of the methodology can be made with fluorometric measurements. Then, melatonin standards must be freshly prepared in ethyl acetate and kept in darkness below 0°C before use. They must be rapidly processed to prevent melatonin decomposition. Tablets have to be pulverized with the help of a mortar, dissolved in ethyl acetate and stirred at 700 rpm for 2 min. After filtering through filter paper (7—11 μm), the filtrate is quantitatively transferred to a flask and volume was made up to 100 mL with ethyl acetate. The quantification can be achieved by the standard addition method in such a way that final concentrations are included between 10 and 100 ng/mL. The excitation and emission wavelengths are set at 285 and 336 nm, respectively.

9.9 Assessment and discussion questions

1. Why FIA methodology is considered as automated methodology?
2. Indicate the components (and their role) of this FIA system.
3. Indicate the flow cell employed (configuration and electrodes) and draw a scheme.
4. What are the advantages and disadvantages of the location of RE and AE?
5. Explain why peaked signals are obtained in the fiagrams and indicate the main parameters.
6. Why is it necessary to pretreat an electrode in CV and why is this not done in amperometric FIA?
7. Explain why amperometric FIA methodologies are usually more sensitive than voltammetric techniques.
8. Indicate how the detection potential is chosen.
9. What happens if the detection potential is not at the plateau of the HDV?
10. What is the relationship between flow rate and sample throughput? What are the advantages and disadvantages of using high flow rates?
11. Why methodologies have to be validated? How is this done in this case?

References

[1] H.A. Strobel, W.R. Heineman, Chemical Instrumentation: A Systematic Approach, third ed., J. Wiley and Sons, 1989.
[2] J. Ruzicka, E.H. Hansen, Flow injection analyses. Part I. A new concept of fast continuous flow analysis, Anal. Chim. Acta 78 (1975) 145—157.
[3] H. Gunasingham, B. Fleet, Hydrodynamic voltammetry in continuous-flow analysis, Electroanal. Chem. 16 (1990) 89—180.
[4] R.G. Compton, A.C. Fisher, R.G. Wellington, P.J. Dobson, A.P. Leigh, Hydrodynamic voltammetry with microelectrodes: channel microband electrodes; theory and experiment, J. Phys. Chem. 97 (1993) 10410—10415.
[5] M.M. Hamedi, A. Ainla, F. Guder, D.C. Christodouleas, M.T. Fernández-Abedul, G.M. Whitesides, Integrating electronics and microfluidics on paper, Adv. Mat. 28 (25) (2016) 5054—5063.

[6] S.M. Lunte, C.E. Lunte, P.T. Kissinger, in: P.T. Kissinger, W.R. Heineman (Eds.), Laboratory Techniques in Electroanalyticals Chemistry, second ed., Marcel Dekker, New York, 1996.

[7] A.J. Bard, L.R. Faulkner, Electrochemical Methods, Fundamentals and Applications, second ed., J. Wiley and sons, New York, 2001.

[8] J.M. Elbicki, D.M. Morgan, S.G. Weber, Theoretical and practical limitations on the optimization of amperometric detectors, Anal. Chem. 56 (1984) 978−985.

[9] R.J. Reiter, Pineal melatonin: cell biology of its synthesis and of its physiological interactions, Endocr. Rev. 12 (1991) 151−180.

[10] F. Waldhouser, B. Ehrart, E. Förster, Clinical aspects of the melatonin action: impact of development, aging, and puberty, involvement of melatonin in psychiatric disease and importance of neuroimmunoendocrine interactions, Experientia 49 (1993) 671−681.

[11] L.C. Manchester, A. Coto-Montes, J.A. Boga, L.P.H. Andersen, Z. Zhou, A. Galano, J. Vriend, D.-X. Tan, R.J. Reiter, Melatonin: an ancient molecule that makes oxygen metabolically tolerable, J. Pineal. Res. 59 (2015) 403−419.

[12] J.L. Corujo-Antuña, S. Martínez-Montequín, M.T. Fernández-Abedul, A. Costa-García, Sensitive adsorptive stripping voltammetric methodologies for the determination of melatonin in biological fluids, Electroanalysis 15 (2002) 773−778.

[13] J.L. Corujo-Antuña, E.M. Abad-Villar, M.T. Fernández-Abedul, A. Costa-García, Voltammetric and flow amperometric methods for the determination of melatonin in pharmaceuticals, J. Pharm. Biomed. Anal. 31 (2003) 421−429.

CHAPTER

10

Batch injection analysis for amperometric determination of ascorbic acid at ruthenium dioxide screen-printed electrodes

David Hernández-Santos, Pablo Fanjul-Bolado,
María Begoña González-García

Metrohm Dropsens, Parque Tecnológico de Asturias, Edificio CEEI, Asturias, Spain

10.1 Background

The development of automated systems with high speed and good precision has become essential in Analytical Chemistry owing to the increasing demand for fast analyses of a large number of samples. For this purpose, methods based on flow injection analysis (FIA) or continuous flow systems have received particular attention. FIA methods are being extensively used for several environmental, industrial, and chemical applications, mainly because it provides improvements in versatility, reliability, and sample throughput, coupled with high reproducibility in comparison with wet (batch) chemical assays [1]. However, in 1991, Wang and Taha [2] introduced a technique named batch injection analysis (BIA). In BIA systems with electrochemical detection, a sample plug is injected from a micropipette tip directly on the electrode surface that is immersed in a large-volume blank electrolyte solution. The detector records a transient, peak-shaped response that reflects the passage of the sample zone over its surface. The magnitude of the peak thus reflects the concentration of the injected analyte. Such dynamic measurements performed under batch operation yield an analytical performance similar to that obtained under well-established FIA conditions, but without using additional components such as pumping systems, injection valves, and tubes. In this way, problems with air bubbles and leaks are overcome [3]. So, the BIA system is considered by some researchers as a tubeless FIA system.

The first BIA cells reported in the bibliography used three independent conventional electrodes. The combination of BIA cells with screen-printed electrodes (SPEs) has been carried out recently (BIASPE) [4,5]. Similar to BIA systems, SPEs also presented some desirable features useful in portable analytical systems, such as low cost (large-scale production), disposability, rapid responses, simplicity, and robustness (working, counter, and pseudo-reference electrodes are printed on a chemically inert substrate). The proposed system shows several requirements of a portable system that include user-friendly, high-speed quantitative and qualitative analysis, excellent cost-effectiveness, and low-power requirements. The coupling of the BIASPE cell with battery-powered accessories (electronic micropipette, potentiostat, and tablet or laptop computer) can be considered as a robust portable electroanalytical system.

In Fig. 10.1 a BIASPE system setup is shown. The SPE is placed on the bottom of the cell (Fig. 10.1A). The cell has a tank (maximum volume of 100 mL) with a hole in the base that exactly matches the electrochemical cell of the SPE (see Fig. 10.1B). An O-ring placed around the hole and between the base of the BIA cell and the electrode prevents leaks of the electrolyte. After the tank is filled with the electrolyte, the BIA cell is closed with a cover that had two holes, one of them to put the electronic micropipette and the other one to put a rod stirrer controlled by a stirrer interface connected to the computer. The adapter for micropipette assures that the position of the tip of the pipette is centered on the electrode surface and the distance between the tip of the micropipette and the electrode surface remains constant. These two parameters contribute significantly to the reproducibility of the procedure (Fig. 10.1C).

When compared with a standard FIA system, the electronic micropipette (20−200 μL) works as pump and injection valve. Therefore reproducible flow rate and injection volume (electronically controlled) are ensured and problems with air bubbles in the system are

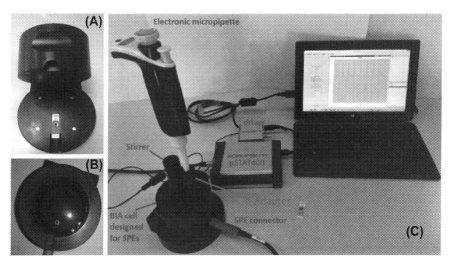

FIGURE 10.1   (A) Overview of the opened BIASPE cell with detail of the positioning of the screen-printed electrode (SPE) (B) top view of the batch injection analysis (BIA) cell, and (C) full setup of BIASPE portable system including potentiostat, BIA cell, electronic micropipette, stirrer, and computer.

avoided. The micropipette can be programmed with user-friendly and intuitive computer software allowing automatic aliquot dispensing after being programmed.

In this experiment a BIASPE system as that shown in Fig. 10.1C is used with SPEs of ruthenium oxide. These electrodes have been already used as transducers for detection of ascorbic acid [6]. This molecule presents an irreversible oxidation process on these electrodes (at around +0.09 V) at lower potential than this obtained with graphite electrodes.

Although other electrochemical techniques can be used, in this experiment amperometry is the technique employed to carry out the measurement, one of the most widely used for the development of (bio)sensors and bioassays.

The experiment is directed to undergraduate or Master students of courses related to the development of analytical methodologies (Chemistry, Biotechnology). It will give them, during a 3-h laboratory session, an appreciation of the usefulness of a BIA cell combined with SPEs as well as the main advantages and disadvantages when compared with FIA systems or the main possibilities of using this simple: hydrodynamic system in analytical chemistry.

10.2 Chemicals and supplies

Reagents

— Phosphate buffer saline (PBS) (0.04 M phosphate, 4.5% NaCl, pH 7.4).
— Ascorbic acid: A 0.1 M solution should be prepared in 0.1 M sulfuric acid. Dilutions should be prepared in 0.04 M PBS pH 7.4.
— Glucose: Solutions of this compound should be prepared in 0.04 M PBS pH 7.4.
— Fructose: Solutions of this compound should be prepared in 0.04 M PBS pH 7.4.
— Milli-Q purified water is employed to prepare all solutions.

Instrumentation and materials

— A potentiostat equipped to perform cyclic voltammetry and chronoamperometry.
— SPEs of ruthenium oxide from Metrohm DropSens (DRP-810), with auxiliary electrode of carbon and pseudo-reference electrode of silver.
— A specific connector acting as interface between the SPE and the potentiostat.
— A BIA cell including an electronic micropipette of 20—200 μL (Metrohm DropSens (DRP-BIASPE02) as that shown in Fig. 10.1.
— A Teflon rod stirrer moved by DC motor connected to a stirrer controller and the computer (DRP-STIRBIA from Metrohm DropSens). The controller box has a switch with three positions, 1, 2, and 3 corresponding to 500, 1500, and 3000 rpm, respectively. It is possible to use a conventional magnetic bar and stirrer but the magnetic bar should be placed in the center of the BIA cell, avoiding the contact with the electrode surface.
— Microcentrifuge tubes of 1.5 mL, 500 mL (for PBS), and 50 mL (for ascorbic acid standard) volumetric flasks, 100 mL graduate cylinder, and 100 and 1000 μL micropipettes and tips for these micropipettes and for the electronic one (DRP-DTIPD200).

10.3 Hazards

Students are required to wear a lab coat, appropriate gloves, and safety glasses. Special care should be taken when handling concentrated acids or bases.

10.4 Experimental procedure

10.4.1 Preparation of the BIA cell

1. Open the BIA cell and put a ruthenium oxide electrode as it is shown in Fig. 10.1A.
2. Close the tank of the BIA cell and make sure that the electrode is centered on the hole of the bottom of the tank (see Fig. 10.1B).
3. Fill the tank with 80 mL of 0.04 M PBS pH 7.4 buffer (see Fig. 10.1B) using the graduate cylinder. It is not necessary to add an exact volume because this parameter does not affect the reproducibility of the peak currents. If during the filling of the tank, an air bubble is found around the electrode surface, it can be removed aspirating with a micropipette. It is possible to work with a smaller volume of PBS into the tank, for example, 50 mL. This value depends on the concentration of analyte and the volume and number of injections, taking into account that PBS contained in the tank will not be replaced during the experiment.
4. Put the cover of the tank. Then place the micropipette adapter in the corresponding hole (see Fig. 10.1C) and the rod stirrer connected to the stirrer controller box (and this one to the computer).
5. Switch on the stirrer and place the rpm selector in position 2. If it is necessary to work in an unstirred solution, switch off the stirrer.
6. Connect the electrode card to the potentiostat using the adequate cable (this depends on the potentiostat used, for Metrohm DropSens potentiostats, a DRP-CAST cable)
7. Select in the potentiostat the amperometric technique and the desired parameters (applied potential, interval time, run time, etc.) to record the BIA amperometric response.
8. Put the micropipette into the adapter (see next section)
9. Start recording the amperogram. *Caution:* Put a run time large enough to finish each assay before the end of the amperometric record; it is better to have more time and stop record when the assay finishes.
10. Wait until a stable background is achieved. Then, perform the first injection using the electronic micropipette.

10.4.2 Preparation of the electronic micropipette

This is an important part of the procedure because the electronic micropipette works as pump and injection valve, so the management of this device should be carefully done.
Caution: Before its use, the battery of the micropipette should be charged.
The micropipette has six dispensing modes: reverse, mixed, forward, pipet, repetitive, and programmed, the latter three being the most useful for BIA assays. In this experiment, the

mode repetitive is used. It consists on the injection of a fixed volume several times. The injection will only be done when the yellow bottom is pushed. Then, students should control the injection time. Taking into account the maximum volume of the micropipette (200 μL) and the volume to be dispensed (e.g., 40 μL), then the number of possible injections is obtained (200/40 injections in this case). The micropipette calculates automatically the maximum number of injections for a fixed volume of injection, but it is possible to select a smaller number of injections (in the last example it is possible to select also 1, 2, 3, or 4 injections).

The micropipette has 6 speeds of injection, numbered from 1 to 6. This is a very important parameter because the height of the peaks can be affected by this value. The higher the number in the micropipette, the higher is the speed obtained.

Follow the next steps to do injections of analyte (ascorbic acid standards or orange juice samples):

1. Select the repetitive mode in the micropipette and later the desired volume and number of injections. Finally, select the speed of injection (for more detail, check the manual of the micropipette). To start the experiment, the next values are recommended:
   − Injection volume: 40 μL
   − Number of injections: 5 (maximum value for this injection volume)
   − Injection speed: 4
2. Put an adequate tip in the micropipette.
3. Push the yellow bottom. In the display of the micropipette appears the word "Aspirate." Introduce the tip into the ascorbic acid solution, of desired concentration, and press the yellow bottom. If the repetitive mode has been selected, the micropipette aspirates a little more than 200 μL. Now, in the display it appears the word "Discard." Push again the yellow bottom without retiring the tip from the solution. Now the total volume into the tip is exactly 200 μL and in the display of the micropipette the volume and the number of injections that the micropipette will do will appear.
4. Place carefully the micropipette with the filled tip into the adapter of the BIA cell until the top.
5. When the baseline is stable, press the yellow bottom to dispense the analyte. For each injection, the yellow bottom should be pressed. It is recommended that the next injection is done when the baseline is recovered.

Fig. 10.2 shows an example of 15 consecutive additions for a concentration of 0.2 mM of ascorbic acid.

10.4.3 Optimization of parameters that affect the analytical signal

10.4.3.1 Optimization of the potential of detection (hydrodynamic curve)

Ascorbic acid presents an irreversible process on ruthenium electrodes around +0.09 V (vs. silver pseudo-reference). When working in a BIA system, as well as in an FIA system, it is necessary to optimize the potential applied to detect a redox species.

With this aim, using a concentration of 0.02 mM of ascorbic acid in 0.04 M PBS pH 7.4, vary the detection potential between 0.0 and +0.4 V (for example, 0 V, +0.1 V, +0.2 V, +0.3 V, and +0.4 V). For this study fix an injection volume and a speed of 40 μL and 4, respectively. Put the stirrer speed selector in position 2 (1500 rpm). Make 3 injections for each potential.

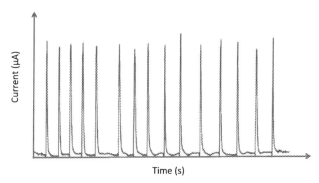

FIGURE 10.2   Fifteen consecutive injections of a 0.2 mM ascorbic acid solution in 80 mL of 0.04 M phosphate buffer saline pH 7.4. Injection volume: 40 µL; number of injections: 5 (3 aspirations have been done); injection speed: 4; detection potential: +0.2 V; stirrer speed: 2 (1500 rpm).

Caution: For each potential applied, it is necessary to obtain a stable background before the first injection. After the last injection, wait until the background achieves a stable level and then stop recording. Save the amperogram. Select the next potential and start recording again, applying a new potential.

Plot the mean of peak height versus the potential applied. Include also the deviation for each measurement using error bars. From this graph (hydrodynamic curve), select the optimum detection potential.

10.4.3.2 Optimization of speed and volume of injection

Both parameters affect the analytical signal and, depending on the concentration assayed, the effects can be more remarkable. For example, if a very low concentration is expected, a higher speed of injection has to be employed if working in a stirred solution. Lower speed can be used for high concentrations.

On the other hand, the higher the volume dispensed, the higher the peak height obtained, but the number of injections is limited to a lower number (the maximum volume of the micropipette is 200 µL). Baseline recovering will take more time for higher dispensing volumes, so the number of injections per hour will be lower. However, this effect is minimized if working in a stirred solution.

For evaluating the influence of these two parameters, test two concentrations of ascorbic acid (0.02 and 0.5 mM in 0.04 M PBS pH 7.4):

1. Stir the solution fixing the position 2 in the stirrer speed selector (1500 rpm).
2. Fix the speed of dispensing of the micropipette in position 4, the volume of dispensing in 20 (maximum number of injections: 10), 40 (maximum number of injections: 5), or 100 µL (maximum number of injections: 2) for the two concentrations of ascorbic acid. Do three injections for each volume, except volume 100 µL (only two injections are possible).
3. Repeat the last step fixing a speed of dispensing in position 2.
4. Plot the mean of peak height versus the dispensing volume and versus the dispensing speed for each concentration. Include also the deviation for each measurement using error bars. From these graphs, select the best dispensing volume and speed.

10.4.3.3 *Effect of stirrer speed*

The stirrer used in this experiment has a selector that allows selecting three different speeds: 500 rpm (position 1), 1500 rpm (position 2), and 3000 rpm (position 3). It is very interesting to see the effect on the shape, width, and height of the peaks when working in an unstirred solution or when the electrolyte (PBS) is stirred during the dispensing. To study this effect, follow the next steps:

1. Turn off the stirrer.
2. Put a tip in the micropipette. Select repetitive mode and the next parameters:
   - Dispensing volume: 40 µL
   - Dispensing number: 3
   - Dispensing speed: 3
3. Aspirate a 2 mM ascorbic acid solution. Discard the excess of volume.
4. Place carefully the micropipette into the cell through the adapter.
5. Start recording the amperogram applying the detection potential selected in Section 10.4.3.1 (fix a long time of measurement and an interval time of 0.1 s).
6. When the baseline achieves a stable current level, start with the dispensing protocol.
7. Make three injections waiting a period of time large enough to recover the background level.
8. After the third addition, remove carefully the micropipette without stopping the recording of the amperogram.
9. Turn on the stirrer.
10. Select the position 1 (you can see that the baseline is recovered faster).
11. Repeat the steps from 3 to 8 (except step 5).
12. Finally, repeat again the procedure selecting position 3 in the stirrer.

The study could be repeated for a smaller concentration of ascorbic acid (0.02 mM). An example of the effect of stirrer speed is shown in Fig. 10.3:

From the results obtained, select a stir speed that allows recovering the background line faster with a lower noise.

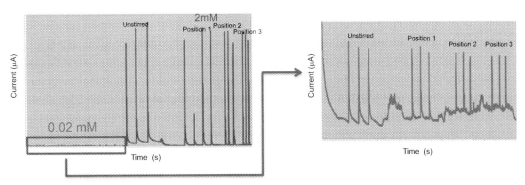

FIGURE 10.3 Batch injection analysis amperogram obtained for injections of 0.02 and 2 mM ascorbic acid solutions under stirring conditions.

10.4.4 Calibration plot of ascorbic acid

Using the experimental conditions optimized in Section 10.4.3, make consecutive injections of increasing concentrations of ascorbic acid comprised between 0.01 and 1 mM. Make five injections for each concentration of ascorbic acid.

Plot the mean of peak height versus the concentration of ascorbic acid. Include also the deviation for each measurement using error bars. From this graph, estimate the linear range, the sensitivity (as the slope of the calibration curve), the limit of detection (as the concentration corresponding to a signal that is three times the standard deviation of the intercept [or of the estimate] in the lower range of concentrations), and the limit of quantification (as the concentration corresponding to a signal that is 10 times the standard deviation of the intercept or estimate).

An example of the analytical signals obtained and a calibration plot are shown in Fig. 10.4.

10.4.5 Interference evaluation

To study the selectivity of the method, two potential interferences such as fructose and glucose are evaluated. They are chosen as interferences because of its usual presence in orange juice samples. To carry out this study, make 3 injections of a 0.5 mM solution of glucose, followed by 3 injections of a 0.5 mM solution of fructose and other 3 injections of binary mixtures of glucose and ascorbic acid (both in 0.5 mM concentration) and fructose and ascorbic acid (0.5 mM in each compound). Finally make 3 injections of a mixture of the 3 compounds in a concentration of 0.5 mM.

Discuss the effect of the presence of these interferences based on the results obtained.

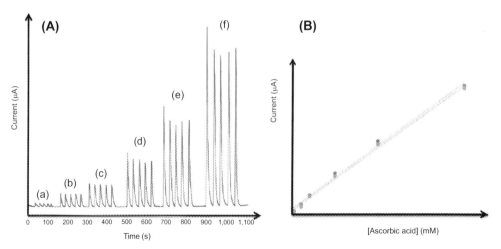

FIGURE 10.4 (A) Batch injection analysis amperogram of five injections of different concentrations of ascorbic acid increasing from (a) 0.01 to (f) 1 mM; (B) the corresponding calibration plot.

10.4.6 Determination of ascorbic acid in orange juices

Taking into account the ascorbic acid content labeled in the orange juices, dilute them with 0.04 M PBS pH 7.4 to a final concentration that is included in the linear range of the calibration plot. Calculate the ascorbic acid concentration in orange juice by using the external calibration curve. Several replicates should be measured to obtain the average and standard deviation of the results.

10.5 Lab report

Write a lab report following the typical style of a scientific article. It should include an abstract, a brief introduction explaining the purpose of the experiment, a detailed experimental section, results and discussion, and conclusions. Include tables, graphics, or figures wherever necessary. The following points should be taken into account:

1. In the introduction, include a revision of the different applications of BIA systems with electrochemical techniques. Discuss the main advantages and disadvantages of the BIA systems compared with FIA systems.
2. In the experimental section, explain the protocols to prepare solutions in detail, the equipment used, and the real samples that are going to be analyzed.
3. In the results and discussion section, include figures with representative raw data (chronoamperograms) and results presented in tables and graphs (e.g., hydrodynamic curve, calibration curve) paying special attention to the significant figures in each case.
4. Discuss the values obtained considering the expected results and the incidences during the course of the experiment.

10.6 Additional notes

1. PBS solution can be stored at 4°C for at least 1 week. However, the ascorbic acid stock solution should be prepared daily.
2. The tips used for the electronic micropipette should be those recommended by the supplier. Do not use other tips from other suppliers. The reason is that the distance from the tip to the electrode should be kept constant.
3. If the rod stirrer is not available, it is possible to use a conventional magnetic stirrer placing a magnetic bar stir in the BIA cell. In this case, especial care should be taken because the magnetic bar has to be in the center of the BIA cell without touching the electrode surface. Then, the effect of the stirrer speed on the analytical signal can be done, but taking into account the rpm values employed.
4. Another kind of SPEs could be used.
5. The electronic micropipette has software that allows programming the protocol of dispensing. It should be interesting that the students program the micropipette with the final optimized protocol. For this, check the manual.

6. It is not necessary to change the electrolyte into the BIA cell except if the liquid overflows. It is possible to use less volume of electrolyte (60 instead of 80 μL as it is recommended).
7. The sample throughput (number of samples per hour) can be also estimated.

10.7 Assessment and discussion questions

1. Explain why the typical signal of the BIA cell is a transient response and the reason of the shape of the peaks.
2. Is it possible to use a BIA cell with other electrochemical techniques? Explain with an example.
3. Indicate the influence of the different operational parameters.
4. What are the parameters that affect sample throughput?

References

[1] J. Ruzicka, E.H. Hansen, Flow-injection Analysis, Wiley, New York, 1981.
[2] J. Wang, Z. Taha, Batch injection analysis, Anal. Chem. 63 (1991) 1053—1056.
[3] M.S.M. Quintino, L. Angnes, Batch injection analysis: an almost unexplored powerful tool, Electroanalysis 16 (2004) 513—523.
[4] E.M. Richter, T.F. Tormin, R.R. Cunha, W.P. Silva, A. Pérez-Junquera, P. Fanjul-Bolado, D. Hernández-Santos, R.A.A. Muñoz, A compact batch injection analysis cell for screen-printed electrodes: a portable electrochemical system for on-site analysis, Electroanalysis 28 (2016) 1856—1859.
[5] T.F. Tormin, R.R. Cunha, W.P. Silva, R.A. Bezerra da Silva, R.A.A. Muñoz, E.M. Richter, Combination of screen-printed electrodes and batch injection analysis: a simple, robust, high-throughput, and portable electrochemical system, Sens. Actuators, B 202 (2014) 93—98.
[6] J. Wu, J. Suls, W. Sansen, Amperometric determination of ascorbic acid on screen-printing ruthenium dioxide electrode, Electrochem. Commun. 2 (2000) 90—93.

Impedimetric aptasensor for determination of the antibiotic neomycin B

Noemí de los Santos Álvarez, Rebeca Miranda-Castro,
M. Jesús Lobo-Castañón

Departamento de Química Física y Analítica, Universidad de Oviedo, Oviedo, Spain

11.1 Background

Electrochemical impedance spectroscopy (EIS) is an electrochemical technique with applications in virtually all areas of Electrochemistry. This experiment illustrates how this technique can be used for the transduction of the recognition event in biosensors, in a label-free approach. Measurements are performed in a typical three-electrode electrochemical cell containing a solution with both forms of a redox couple (for example, ferro/ferricyanide) as a probe, so that the potential of the working electrode is fixed at a value determined by the ratio of oxidized and reduced forms of the couple according to the Nernst equation. An excitation signal, consisting of a small-amplitude alternating electrical (AC) potential, is applied to the cell varying the frequency of the AC source over a wide range (usually 10^6 to 10^{-3} Hz), and the impedance of the cell is registered as a function of the imposed frequency (Fig. 11.1A).

Impedance (Z) is a measure of the ability of an electric circuit to impede the flow of an AC current. In a linear (or pseudolinear) system, which is achieved by applying small-amplitude changes of potential from the equilibrium value, the current response to a sinusoidal potential will be a sinusoid of the same frequency but shifted in a certain phase angle (θ). If the excitation signal is represented by $E = E_0 \sin(\omega t)$ (where E_0 is the amplitude of the excitation signal and ω is the angular frequency, measured in radians per second and related to frequency, f, in Hz, by $\omega = 2\pi f$), the response signal will be $I = I_0 \sin(\omega t + \theta)$.

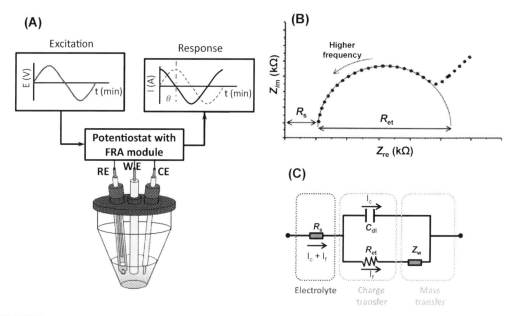

FIGURE 11.1 (A) Experimental setup (FRA: frequency response analysis). (B) Nyquist plot of the impedance spectrum. (C) Schematic representation of the equivalent circuit of an electrochemical cell in an electrochemical impedance spectroscopy measurement (Randles circuit). I_f, faradaic current; I_c, capacitive current; CE, counter electrode; RE, reference electrode; WE, working electrode, and other symbols as explained in the text.

An expression analogous to the Ohm's law allows to calculate the impedance of the system as

$$Z = \frac{E}{I} = \frac{E_0 \sin(\omega t)}{I_0 \sin(\omega t + \theta)} = Z_0 \frac{\sin(\omega t)}{\sin(\omega t + \theta)}$$

Therefore, to characterize an impedance both its magnitude or module (Z_0) and its phase (θ), as well as the frequency ($f = 2\pi\omega$) at which it is measured have to be specified. It is usual to express these magnitudes in terms of complex notation. Taking into account the Euler's relationship $\exp(j\theta) = \cos\theta + j\sin\theta$, where j is the imaginary unit ($\sqrt{-1}$), the excitation signal is represented as $E = E_0 \exp(j\omega t)$ and the current response as $I = I_0 \exp j(\omega t - \theta)$. With this notation, the impedance is represented as a complex number, composed of a real (Z_{re}) and an imaginary part (Z_{im}):

$$Z = Z_0 \exp(j\theta) = Z_0(\cos\theta + j\sin\theta) = Z_{re} + jZ_{im}$$

A plot of Z_{im} ($E_0 \sin\theta$) in y-axis and Z_{re} ($E_0 \cos\theta$) in x-axis is known as the Nyquist plot and it is the usual way to depict the impedance spectrum (Fig. 11.1B). Each point in this plot corresponds to the impedance at a different frequency, and higher frequency values are represented toward the left [1]. This spectrum gives information about the capacitance and the interfacial properties of conductive or semiconductive electrodes.

The electrochemical reactions that take place in the electrode–solution interface can be modeled according to different equivalent circuits, with the Randles circuit being the most used. This equivalent circuit consists in four elements (Fig. 11.1C): R_s: the ohmic solution resistance, Z_w: the Warburg impedance or impedance related to diffusion of ions toward the electrode surface, C_{dl}: the double-layer capacitance, and R_{et}: the electron transfer resistance. The two former ones represent general properties of the solution and diffusion characteristics of the electroactive species in that solution; while the latter two depend on the properties of the electrode–solution interface [1]. The Nyquist plot for a cell modeled by the Randle's cell circuit includes a semicircle region at higher frequencies, followed by a straight line. The solution resistance R_s can be calculated by the intercept of the semicircle with the real axis at high frequency, whereas the electron transfer resistance R_{et} is obtained by measuring the semicircle diameter (Fig. 11.1B).

In faradaic impedimetric sensors, the analytical signal typically monitored is the change in the electron transfer resistance, R_{et}. Its magnitude is related to the electron transfer kinetics of the electroactive molecule present in solution at a fixed concentration. A change in the electrode surface because of the presence of the target (e.g., recognition and concomitant capture of the target present in solution) will give rise to a decrease or an increase in the electron transfer rate through the electrode interface, thereby increasing or decreasing the R_{et}, respectively, without the need of any labeled reagent. The R_{et} value matches the diameter of the semicircle in the Nyquist plot (Fig. 11.1B).

This experiment shows the construction and evaluation of a label-free impedimetric aptasensor for the detection of the aminoglycoside antibiotic neomycin B. This molecule is produced by *Streptomyces fradiae*. Aminoglycoside antibiotics contain two or more amino sugars connected by glycosidic bonds (Fig 11.2). They can inhibit the growth of some particular bacteria and, consequently, they were used in veterinary medicine to deal with bacterial

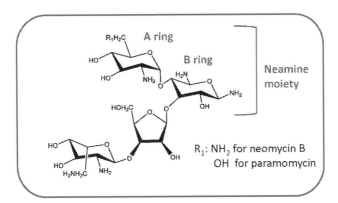

FIGURE 11.2   Structure of aminoglycoside antibiotics.

infections in animals. However, like other aminoglycoside antibiotics, neomycin B is potentially toxic and it may cause ototoxicity and nephrotoxicity. The European Union have established maximum residue limits in foodstuffs of animal origin [2]; therefore, it is very important to control neomycin B in products such as meat, milk, and eggs.

In the absence of analytically profitable electrochemical and spectroscopic properties (check Fig. 11.2), direct detection of neomycin B is hampered. However, the binding between neomycin B and its specific receptor could be successfully monitored by EIS.

In this experiment, the specific detection of neomycin B will be carried out by using an oligonucleotide-based synthetic receptor, particularly, a short RNA sequence selected in vitro (so-called aptamer) able to distinguish neomycin B from other aminoglycoside antibiotics with similar structure [3]. However, RNA stability is compromised by the ubiquitous RNases (a family of enzymes present in virtually all living cells that can degrade RNA molecules). For this reason, all nucleotides are modified with a methoxy group at the sensitive 2′-OH position of the ribose (fully 2′-O-methylated RNA).

On the other hand, neomycin B is a low molecular size molecule (<700 Da); therefore, the change in R_{et} when it is entrapped on the electrode surface is hardly measurable (the higher the molecular size of the target, the larger the change in R_{et}). This issue can be circumvented by implementation of a displacement assay format, by making full use of the higher size of the RNA receptor [4]. Displacement assay format involves the immobilization of the target molecule onto the electrode surface and subsequent attachment of the specific receptor through affinity recognition. Then, if the target is present in the sample, this one displaces the receptor from the surface, triggering modifications in the electrode—solution interface, which could be sensitively recorded by faradaic impedance spectroscopy (FIS).

This laboratory experiment is designed for three laboratory sessions of 3—4 h and it is directed to Master students. It introduces the measuring setup in FIS, explaining how to analyze the results. In addition, it provides to the students an insight of EIS and its applications in bioanalysis.

11.2 Chemicals and supplies

Reagents:

- Fully 2′-O-methylated RNA aptamer purified by high-performance liquid chromatography.
- (5′-GGCCUGGGCGAGAAGUUUAGGCC-3′).
- Neomycin B sulfate and paromomycin sulfate.
- Mercaptopropionic acid and ethanolamine.
- N-(3-dimethylaminopropyl)-N′-ethylcarbodiimide hydrochloride (EDC).
- N-hydroxysuccinimide (NHS).
  Potassium ferricyanide and ferrocyanide.
- 4-(2-Hydroxyethyl)piperazine-1-ethanesulfonic acid (HEPES).
- 1 M Tris/HCl pH 7.4 solution (RNase-free).
- KCl, NaCl, MgCl$_2$ (all of them RNase-free).

- RNase-free water to prepare solutions in contact to the RNA aptamer.
- Alumina (1, 0.3, and 0.05 μm diameter particles).
- Sulfuric acid, hydrogen peroxide, ethanol, Milli-Q, and RNase-free water.

Instrumentation and materials:

- A computer-controlled potentiostat equipped to perform FIS and cyclic voltammetry.
- A conventional three-electrode potentiostatic system comprised of a Ag|AgCl|KCl(sat)| KNO$_3$(sat) reference electrode, a Pt counter electrode, and a polycrystalline gold working electrode.
- A magnetic stirrer.
- Microcloth pads.
- Pipettes, tips, timer, magnetic bars, and aluminum foil.

Buffer solutions for developing the impedimetric aptasensor:

- *Measurement solution*: 10 mM Tris/HCl pH 7.4, 100 mM KCl, 5 mM Fe(CN)$_6^{3-/4-}$.
- *Affinity solution*: 50 mM Tris/HCl pH 7.4, 250 mM NaCl, 5 mM MgCl$_2$.
- *Washing solution*: 50 mM Tris/HCl pH 7.4, 2 M NaCl, 5 mM MgCl$_2$.

11.3 Hazards

Piranha solution is strongly oxidizing so it should be handled with extreme caution. For its preparation and manipulation, safety glasses and protective gloves are required.

11.4 Experimental procedure

11.4.1 Electrode cleaning and pretreatment

1. Remove organic matter from the gold surface by immersion in piranha solution (3:1 mixture of concentrated sulfuric acid with 30% hydrogen peroxide prepared in glass or Pyrex container) for 10 min and then rinse thoroughly with Milli-Q water.
2. Polish gold electrodes on a microcloth pad using alumina slurries of 1, 0.3, and 0.05-μm particle size, successively. For that, place the electrode face down on the microcloth pad and move it over the slurry by using a figure-eight motion to get a homogeneous polishing.
3. Sonicate the polished surfaces in Milli-Q water for 5 min to remove the exfoliated gold particles.
4. Subject the gold electrodes to an electrochemical polishing via oxidation and reduction under acidic conditions (scan rate 100 mV/s in 0.5 M H$_2$SO$_4$) in the potential range from +0.2 to +1.6 V (vs. a Ag|AgCl|KCl(sat)|KNO$_3$(sat) reference electrode). The scan is repeated until the cyclic voltammogram becomes stable (approximately 10 cycles).

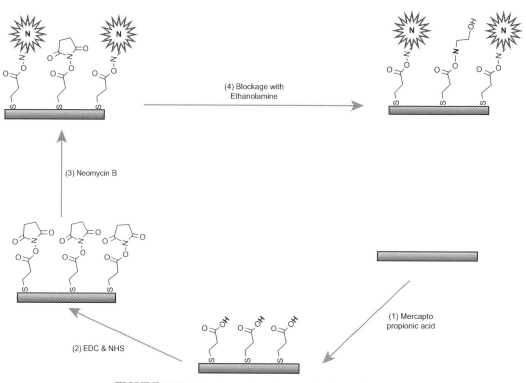

FIGURE 11.3   Reaction scheme for sensing phase formation.

11.4.2 Sensing phase preparation

The procedure is divided into several stages (see Fig. 11.3):

11.4.2.1 Formation of mercaptopropionic acid self-assembled monolayer on the gold surface

1. Prepare an ethanol:water mixture (75:25 v/v) containing 40 mM mercaptopropionic acid in a vial.
2. Immerse the freshly cleaned gold electrode into the mercaptopropionic acid ethanolic solution overnight (until next session) to obtain a compact self-assembled monolayer (SAM). Seal the electrode with parafilm to avoid evaporation.
3. Rinse off mercaptopropionic acid physically adsorbed on the gold surface by using Milli-Q water.

11.4.2.2 Covalent binding of neomycin B to the mercaptopropionic acid self-assembled monolayer

1. Prepare 25 mL of 0.1 M HEPES solution pH 8.64.
2. Prepare 1 mL of 40 mM EDC stock solution in Milli-Q water.
3. Prepare 1 mL of 100 mM NHS stock solution in Milli-Q water.

4. Mix the appropriate volumes of EDC and NHS stock solutions to obtain 1 mL of 2 mM EDC and 5 mM NHS aqueous solution.
5. Transfer the modified electrode to a vial containing the previous solution and incubate under stirring for 1 h to activate the carboxylic groups from the mercaptopropionic acid SAM.
6. Wash the gold surface with Milli-Q water.
7. Prepare 1 mL of 5 mM neomycin B solution in 0.1 M HEPES pH 8.64.
8. Place the gold electrode in a vial containing the neomycin B solution, seal with parafilm, and incubate 1 h under stirring conditions to covalently bind the antibiotic to the modified gold surface.
9. Rinse the surface with 0.1 M HEPES solution pH 8.64.
10. Incubate the gold electrode in a 1 M ethanolamine solution prepared in 0.1 M HEPES pH 8.64 under gentle agitation for 15 min to block the remaining carboxylic groups in the monolayer.
11. Wash the gold surface with 0.1 M HEPES pH 8.64.
12. Store the electrode immersed in this buffer solution in the refrigerator at 4°C until next session.

11.4.2.3 Immobilization of the antineomycin receptor on the antibiotic-modified electrode surface
13. Preequilibrate the modified gold surface for 2 min with the affinity solution.
14. Immerse the electrode into 1 mL of the affinity solution containing 5 μM antineomycin receptor and incubate under agitation for 30 min.

11.4.3 Sensing phase evaluation

FIS allows controlling the suitable construction of the sensing phase. With this aim:

1. Prepare 100 mL of 10 mM Tris/HCl solution pH 7.4 containing 100 mM KCl.
2. Prepare 1 mL of 0.4 M $K_4Fe(CN)_6$ stock solution in Milli-Q water and protect the vial from light with aluminum foil.
3. Prepare 1 mL of 0.4 M $K_3Fe(CN)_6$ stock solution in Milli-Q water and protect the vial from light with aluminum foil.
4. Dispose the electrochemical cell in its stand.
5. Add to the cell 40 mL of the measurement solution prepared from solutions described in steps 1–3. Cover the cell with aluminum foil.
6. Put reference and counter electrodes in the cell and connect them to the potentiostat.
7. Upon each stage of the sensing phase preparation (as well as for the bare gold electrode), immerse the modified gold electrode into the electrochemical cell, connect it to the potentiostat, and evaluate the electron transfer resistance R_{et} by FIS measurements, fixing the following parameters: formal potential of $[Fe(CN)_6]^{4-/3-}$ (previously determined by cyclic voltammetry at 50 mV/s by sweeping the potential between 0 and + 0.5 V), frequency range: 10,000–0.01 Hz, and alternating current amplitude: 5 mV.
8. Measure the R_{et} value.

11.4.4 Displacement assay

1. Expose the aptamer saturated gold surface to increasing concentrations of neomycin B (0.75–500 μM) in the affinity solution for 5 min under gentle agitation.
2. Rinse the surface briefly with RNase-free water.
3. Immerse the modified gold electrode into the electrochemical cell containing the measurement solution, connect it to the potentiostat, and record the impedance spectrum.
4. Before interrogating a higher antibiotic concentration, rinse the electrode surface with the washing solution (with high salt concentration) to remove $Fe(CN)_6^{4-/3-}$ traces from FIS measurement.

11.4.5 Data collection and analysis

With the aim of comparing different sensing phases, changes in R_{et} are expressed as a percentage of the R_{et} decrease ($S_\%$) according to the expression:

$$S_\% = \left(\frac{R_i - R_g}{R_a - R_g}\right) \times 100$$

where R_i is the R_{et} after incubating the sensing phase with an i concentration of neomycin B, R_g is the R_{et} value for neomycin-modified electrode without aptamer (the lowest value), and R_a corresponds to R_{et} for antibiotic-modified electrode saturated with aptamer (the highest value).

When plotting $S_\%$ versus the logarithm of neomycin B concentration in μM, a sigmoidal curve is expected. In this type of plot, the dynamic range of the sensor is comprised between the concentrations corresponding to 20% and 80% of the maximum signal; and the detection limit is defined as the target concentration corresponding to 90% of the maximum recorded signal.

Determine the dynamic response range and the detection limit of the sensor developed considering the previous information.

11.4.6 Regeneration of the sensing phase

The sensing phase can be easily regenerated by incubation in a concentrated neomycin B solution for 1 h, which is capable of completely displacing all the receptor molecules remaining on the surface. The regeneration step is concluded by resaturating the surface with the RNA receptor (30-min incubation in 5 μM antineomycin aptamer).

11.4.7 Selectivity evaluation

To assess the selectivity of the faradaic impedimetric sensor, challenge the aptamer saturated gold surface with 500 μM paromomycin in the affinity solution under stirring for 5 min. Rinse the electrode with RNase-free water and record the impedance spectrum as explained above. This test should be performed in triplicate.

11.5 Lab report

At the end of the experiment, write a lab report that include an introduction with the relevance and the basis of this technique. Comment and discuss the results and finish with main conclusions. Include all the protocols, graphs and data required for facilitating the comprehension ot the methodology. The following points should be considered:

1. Explain in detail the basis of the modification approach.
2. Describe, with the help of schematics, the basis of the assay.
3. Indicate which is the analytical signal and how it could be related to the concentration. Plot the calibration curve.
4. List the main analytical properties of the methodology: sensitivity, selectivity, reusability, simplicity... commenting them in a critical way.

11.6 Additional notes

1. Solutions for sensing phase construction should be prepared immediately before use.
2. Coupling through carbodiimide reaction requires that the amine group is not charged. This is the reason why a pH of 8.64 was selected, which corresponds to the pI of neomycin B. Although at more alkaline solutions all amine groups would be neutral, the yield of the coupling reaction decreases.
3. When washing the surface of modified electrodes, the stream of washing solution should flow over the surface, without hitting it directly.
4. For controlling the potential of the gold working electrode, a silver/silver chloride reference electrode with an additional potassium nitrate salt bridge is used to prevent chloride leakage to the working solution.
5. As EIS is sensitive to changes in the electrode surface, measurable signals could arise from nonspecific interactions. To assess the specificity of the measurements, control experiments without neomycin B attached to the electrode surface are particularly important.
6. Computer fitting of the experimental data to a theoretical model (the equivalent circuit) is used for evaluating the impedance spectra displayed in a Nyquist plot. In this way, in addition to the electron transfer resistance, used as analytical signal, other important parameters can be calculated, such as the double-layer capacitance, C_{dl}. Thus, EIS is not only a suitable transduction technique but also provides a powerful tool for the characterization of the sensing phase.
7. In this experiment, the term aptamer is introduced; however, it is recommended to revise Chapter 22 in this book to obtain more details.
8. This experiment can be easily adapted to a label-based electrochemical bioassay to compare the advantages of each approach. The same chemical reaction scheme used for sensing phase construction can be implemented onto magnetic microparticles modified with carboxylic groups (commercial available), thus circumventing the formation of mercaptopropionic acid SAM. Likewise, RNA sequence should be modified with biotin in its 5'-end for subsequent incorporation of the enzyme conjugate streptavidin—

horseradish peroxidase and finally detect the immobilized enzyme activity by chronoamperometry in the presence of H_2O_2 and tetramethylbenzidine (see Chapter 22).

11.7 Assessment and discussion questions

1. Why is the presence of $Fe(CN)_6^{4-/3-}$ in the measurement solution necessary to record faradaic impedance spectra?
2. Explain the sign of the R_{et} change at each step of the sensor construction and in the displacement assay.
3. If $Fe(CN)_6^{4-/3-}$ is replaced by $Ru(NH_3)^{3+/2+}$, would the spectra corresponding to the sensing phase construction be modified?

References

[1] E. Barsonkov, I.R. Macdonald (Eds.), Impedance Spectroscopy; Theory, Experiment, and Applications, second ed., Wiley Interscience Publications, 2005.
[2] Commission regulation (EU) No 37/2010 of 22 December 2009, Off. J. Eur. Union L15 (2010) 1−72.
[3] M.G. Wallis, U. von Ahsen, R. Schroeder, M. Famulok, A novel RNA motif for neomycin recognition, Chem. Biol. 2 (1995) 543−552.
[4] N. de-los-Santos-Álvarez, M.J. Lobo-Castañón, A.J. Miranda-Ordieres, P. Tuñón-Blanco, Modified −RNA aptamer-based sensor for competitive impedimetric assay of neomycin B, J. Am. Chem. Soc. 129 (2007) 3808−3809.

Electrochemical impedance spectroscopy for characterization of electrode surfaces: carbon nanotubes on gold electrodes

Raquel García-González, M. Teresa Fernández Abedul

Departamento de Química Física y Analítica, Universidad de Oviedo, Oviedo, Spain

12.1 Background

Electrochemical impedance spectroscopy (EIS) is an electrochemical technique that is employed in different aspects of electroanalysis, mainly as a label-free detection technique for (bio)sensors and as a characterization technique of the electrode surface. In Chapter 11, the first aspect is covered with an experiment that illustrates its use in the transduction of an aptasensor. In this chapter, the technique is applied to the characterization of the surface of a gold electrode that suffers different modifications, with the polymer Nafion and with carbon nanotubes (CNTs). The experiment consists in measuring the faradaic impedance (hence, it is also named faradaic impedance spectroscopy, FIS) in such a way that the cell contains a solution with both forms of a redox couple (the pair ferro/ferricyanide in this case) so that the potential of the working electrode is fixed [1]. Then, the mean potential of the working electrode (DC potential) is the equilibrium potential determined by the ratio of oxidized and reduced forms of the couple. A sinusoidal component (E_{AC}) with small peak-to-peak amplitude is overimposed. Therefore, instead of imposing potential sweeps or steps, which drive the electrode to a condition far from equilibrium, the cell is perturbed with an alternating signal of small magnitude. Then, the way in which the system follows the perturbation, at steady state, is measured. As one usually works close to equilibrium, there is no need to know the behavior of the i-E response curve over great ranges of overpotential (additional potential beyond equilibrium needed to drive a reaction at a certain rate).

A variation of the faradaic impedance measurement is ac voltammetry (see Chapter 5). In this case, the potential (E_{DC}) is scanned slowly with time, with addition of a sinusoidal component (E_{AC}). The measured responses are the magnitude of the AC component of the current at the frequency of E_{AC} and its phase angle with respect to E_{AC} [1]. In EIS, the cell or electrode impedance (Z) is measured and represented against frequency (ω, angular frequency, which is 2π the conventional frequency in Hz). Measurements over a wide time or frequency range can be made (10^4–10^{-6} s or 10^{-4}–10^6 Hz).

If we consider the sinusoidal voltage, E, as

$$E = E_0 \sin \omega t$$

where E_0 is the amplitude, then the sinusoidal current, I, can be described as

$$I = I_0 \sin(\omega t + \theta)$$

with θ the phase angle. Both magnitudes, E and I, are not in phase. The current I is phase-shifted an angle θ compared to E (taken as a reference). This potential can be applied through different circuits. Then, if the circuit consists of a pure resistance, R, the phase angle is zero:

$$I = (E_0/R) \sin \omega t$$

and if the circuit consists only of a pure capacitance, then the current is phase-shifted $\pi/2$ because:

$$I = C\,(dE/dt) = \omega C E_0 \cos \omega t = \omega C E_0 \sin\left(\omega t + \frac{\Pi}{2}\right)$$

In an RC circuit, both have to be considered. Then, potential and current are linked by the impedance, the resistance of a circuit to the flow of an alternating current, with a component "real" (R) and another "imaginary" ($1/\omega C$) that changes with frequency, Z_{re} and Z_{im}, respectively.

The variation of the impedance with frequency can be displayed in different ways, mainly Bode and Nyquist plots. In the first one, log Z is plotted against log ω (also the phase angle, θ, is represented against log ω). Nyquist plot displays the imaginary component of the impedance (Z_{im}) against the real component (Z_{re}) for different values of ω.

As commented in Chapter 11, an electrochemical cell can be considered as an impedance to a small sinusoidal excitation [1] and can be represented with an equivalent circuit of resistors and capacitors. One frequently employed is the Randles equivalent circuit (Fig. 12.1). The total current is the sum of two contributions: faradaic and capacitive currents. Both flow together in a circuit where R_Ω is the ohmic solution resistance but in the interface they divide to flow separately through different elements: the double-layer capacitance, C_{dl}, and the faradaic impedance, Z_f, circuit components where capacitive and faradaic currents flow, respectively. The first one is nearly a pure capacitance, but the second one can be considered as the sum of the contributions of some elements in series. The most common is to consider a pure resistance to the charge transfer (R_{ct} or R_{et} for electron transfer) and a resistance to the mass transfer (Z_w, Warburg impedance), although it can also be considered as a resistance and capacitance in series (R_s and C_s).

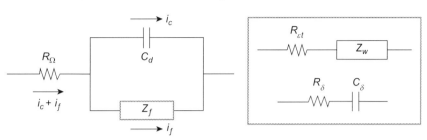

FIGURE 12.1   Equivalent circuit of an electrochemical cell. In blue (light gray in print version) are two of the main representations of the faradaic impedance. *Reprinted with modifications from A.J. Bard, L.R. Faulkner, Electrochemical Methods: Fundamentals and Applications, second ed., John Wiley and Sons, NY, 2001.*

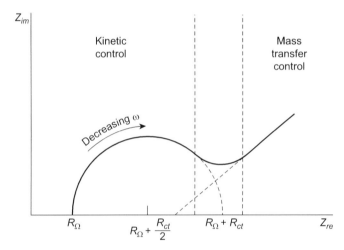

FIGURE 12.2   Scheme of the Nyquist plot and the two main regions for low (mass transfer control, straight line with unit slope) and high frequencies (kinetic control, semicircle with diameter equal to R_{ct}). *Reprinted with permission from A.J. Bard, L.R. Faulkner, Electrochemical Methods: Fundamentals and Applications, second ed., John Wiley and Sons, NY, 2001.*

In this context, the Nyquist plot can be employed to characterize the electrode surface. The presence of new layers of reagents in the interface electrode–electrolyte, conductive or nonconductive, will lead to a different curve in the Nyquist plot, with different values for R_{et}, C_{dl}, etc. In this plot we can differentiate two regions (Fig. 12.2): a semicircle (kinetic control, charge transfer) and a line (mass transfer control). It can be deduced [1] that at low frequencies the plot of Z_{im} versus Z_{re} should be linear with unit slope. The line intersects the real axis at $R + R_{et} - 2\,C_{dl}$. The linear correlation is characteristic of a diffusion-controlled electrode process. As the frequency rises, the charge transfer resistance, R_{et}, and C_{dl} become more important and the relationship between Z_m versus Z_{re} gives a circular plot centered at $R_\Omega + R_{et}/2$ and $Z_{im} = 0$ with a diameter of R_{et}.

In this experiment, adapted from Ref. [2], a screen-printed gold electrode is going to be modified with different layers and to observe what happens with the impedance. Two different modifiers are going to be employed: Nafion and CNTs. Nafion is a cationic exchange polymer that has been extensively employed as a membrane on electrode surfaces, as it allows the passage of cationic species and blocks anionic species from reaching the electrode surface. The increasing popularity of Nafion for the fabrication of modified electrodes arises from easy fabrication and many advantageous chemical and physical properties as good electrical conductivity, high partition coefficients of many redox compounds, and good biocompatibility. A very thin film of Nafion is able to offer minimal obstruction to the diffusion of the analyte to the electrode surface, while at the same time preventing adsorption/desorption processes of organic species in biological fluids. However, when the thickness is very high, it can block diffusion and electron transfer. On the other hand, CNTs have attracted the interest of many researchers belonging to different disciplines (see Chapter 24). Their physical and chemical properties make them very promising materials and composite materials based on solubilizing CNTs with various polymers, especially Nafion that has a polar chain [3], have been employed in (bio)electroanalysis [4] and in the generation of fuel cells [5].

This laboratory experiment can be developed in two laboratory sessions of 3 h and it is directed to Master students of Analytical Chemistry or other related disciplines. It introduces the technique electrochemical impedance spectroscopy as a useful means of characterization of electrodic surfaces.

12.2 Electrochemical cell

Screen-printed gold electrodes (see also Chapter 24) include a traditional three-electrode system printed on the same strip. The configuration includes a gold disk as working electrode, a silver pseudo-reference electrode, and a gold counter electrode using the same ink of the working electrode. All of them were screen-printed on a ceramic substrate and subjected to high-temperature curing. An insulating layer serves to delimit the working area and electric contacts. A specific connector allows their connection to the potentiostat.

12.3 Chemicals and supplies

Reagents:

- *Redox probes*: Potassium ferro/ferricyanide, dopamine, methylene blue.
- *Nanomaterials*: Amine-functionalized multiwalled carbon nanotubes (MWCNTs-NH$_2$), carboxylated carbon nanofibers (CNFs−COOH).
- *Preparation of the polymer*: Nafion (perfluorinated ion-exchange resin, 5 wt.% solution in a mixture of lower aliphatic alcohols and water), isopropyl alcohol.
- *Preparation of background electrolyte*: Potassium chloride, sulfuric acid.
- Purified Milli-Q water is employed throughout the work.

Materials and equipment:

- Microcentrifuge tubes and micropipettes.
- Ultrasonic bath and centrifuge.
- Potentiostat interfaced to a computer with frequency response analysis module, which allows to perform cyclic voltammetry (CV) and electrochemical impedance spectroscopy.

12.4 Hazards

As in a general Analytical Chemistry laboratory, safety glasses, protective gloves, and lab coat are required. Potential hazardous properties of carbon nanotubes are a matter of ongoing research but there are indications that they can be hazardous if workers or users are exposed through inhalation pathways [6]. Then, precautionary measures should be warranted and an appropriate mask should be used when handling carbon nanostructures. Potassium ferrocyanide releases a very toxic gas in contact with acids.

12.5 Electrochemical procedure

12.5.1 Modification of electrodes

Two modifiers are going to be employed: Nafion and MWCNTs (amine-functionalized). Most of the applications of CNTs require a presolubilization step for obtaining a homogeneous suspension. Two important factors have to be taken into account: the CNTs weight/volume of solvent ratio and the dispersion procedure. A fixed amount of CNTs (e.g., 1 mg) is dispersed in a known volume of solvent (1 or 5 mL) and is subjected to cycles, 2 h in the ultrasonic bath and 10 min of centrifugation. After each cycle, the precipitate is discarded and the supernatant is subjected to a new cycle until finally a uniform solution is achieved. Nafion—ethanol was employed for the dispersion of aminated CNTs. In brief, four cycles of ultrasonic bath and centrifugation (5000 rpm) were employed, discarding the supernatant after each cycle. Finally, this dispersion was diluted 1:1 with purified water.

Dispersion of CNFs can be made by simply putting a small amount of CNFs (1 mg) together with a small volume of water (1 mL) and subjecting the mixture to an ultrasonic bath for 5 min. In this case, water was replaced by a solution of Nafion in ethanol, to compare the studies with those of CNTs.

Electrode modification with the dispersing agent (Nafion) or with carbon nanostructures was carried out by evaporation of a 2—5 μL drop of Nafion or CNTs/CNFs dispersion on the working electrode.

12.5.2 Cyclic voltammetry and electrochemical impedance spectroscopy measurements

1. Prepare a 0.1 M KCl solution.
2. Prepare solutions of ferro and ferricyanide (2 mM in each, 10 mM in each) in 0.1 M KCl and protect them from light.

I. Dynamic electroanalytical techniques

3. Deposit 40 μL of a solution of 0.1 M KCl on the screen-printed card, covering well all the three electrodes.
4. Record the CV by scanning the potential between −0.3 and +0.5 V at 50 mV/s.
5. Take off the drop of the background electrolyte and wipe slightly with a soft tissue.
6. Deposit 40 μL of a solution 1 mM in each, ferro and ferricyanide, in 0.1 M KCl, on the screen-printed card.
7. Record the CV by scanning the potential between −0.3 and +0.5 V at 50 mV/s and determine the formal potential of the pair ferro/ferricyanide. Discard the drop.
8. Deposit 40 μL of a solution 5 mM in each, ferro and ferricyanide in 0.1 M KCl, on the screen-printed card.
9. Apply the equilibrium potential and perform EIS in the frequency range comprised between 100,000 and 0.01 Hz and 10 mV of alternating current amplitude.
10. Perform steps 8 and 9 but on two different modified electrodes: (i) Nafion-modified and (ii) CNTs-modified screen-printed gold electrodes (following the procedure stated in Section 12.5.1).
11. Discard the drop of the solution of ferro/ferricyanide that is on the Nafion-modified electrode and rinse slightly with 0.1 M KCl.
12. Deposit 2 μL of the MWCNTs-NH$_2$ dispersion and let dry.
13. Perform again steps 8 and 9.
14. Compare the results by plotting the corresponding Nyquist plot for all the four electrodes: bare, Nafion-modified, CNT-modified, and CNT/Nafion-modified electrode.
15. Determine the values of R_{et} and C_{dl} and discuss the results.

In Fig. 12.3, an example of the Nyquist plots that can be obtained for the different modifications is shown.

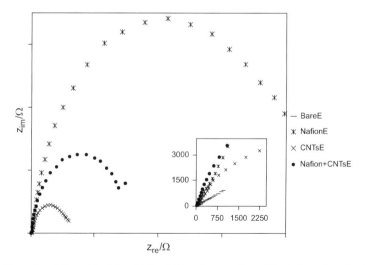

FIGURE 12.3   Example of Nyquist plots obtained from bare and modified electrodes (with Nafion, carbon nanotubes [CNTs], or CNT/Nafion-modified electrodes), E, electrode. *Reprinted with modifications from R. García-González, M.T. Fernández-Abedul, A. Costa-García, Nafion® modified screen-printed gold electrodes and their carbon nanostructuration for electrochemical sensors applications, Talanta 107 (2013) 376−381.*

12.6 Lab report

At the end of the experiment, write a lab report that includes an introduction with the current importance of electrochemical impedance spectroscopy, the basis, and the main information that can be extracted from the spectra. Indicate also, very briefly, the solutions and instrumentation employed as well as important notes on the protocols. Comment and discuss the results regarding the differences observed between the different modifications. Include also graphical representation of raw data. Indicate in a table the values obtained for R_Ω, R_{et}, and C_{dl} and discuss the differences. Include schematics or drawings when necessary.

12.7 Additional notes

1. When washing the surface of modified electrodes, the stream of washing solution should flow over the surface, without hitting it directly.
2. A discussion on the differences observed for the three electrodes (bare, Nafion-, and CNTs-modified electrodes) should be made. Parameters of R_Ω, R_{et}, and C_{dl} can be calculated from the Nyquist plot. The resistance to the electron transfer increases notoriously when passing from bare (mass transport control) to Nafion-modified electrode (kinetic control). In this case is important to note also that Nafion is a cationic exchanger and then ferro/ferri ions will be repelled. Then, the Nafion membrane is not only a "physical" impediment to the electron transfer but also an element that produces charge repulsion. This is represented by a semicircle with a high diameter (R_{et}).
3. When CNTs are added over the Nafion layer, R_{et} decreases again (and then the dimension of the semicircle) because conductive CNTs favor electron transfer.
4. Comparison with other CNT-modified electrodes can be made. The behavior observed with EIS can be corroborated with CV, another important characterization technique. Then, CVs can be recorded on the different electrodes and the extension of the electron transfer can be compared with data obtained with EIS.
5. It is interesting to evaluate different nanostructures. For example, in this case, amino-functionalized CNTs are employed, but if carboxylated CNFs are added instead, reversion of the effect is not observed.
6. In this case, the redox pair ferro/ferricyanide is employed for the studies. The data on the reversibility of its behavior are always important to obtain valuable conclusions. Then, the difference between anodic and cathodic peaks, the ratio between anodic and cathodic current, and the variation of the peak potential with the scan rate should be studied.
7. Different analytes have different behaviors. Regarding CV studies, it could be interesting to compare the behavior of the pair ferro/ferricyanide with an organic molecule such as dopamine or methylene blue. The behavior is quite different when all the four (bare and three modified) electrodes are compared [2].

12.8 Assessment and discussion questions

1. Why is the presence of $Fe(CN)_6^{4-/3-}$ in the measurement solution for recording faradaic impedance spectra necessary?
2. Why the equilibrium potential of the pair ferro/ferricyanide is set for performing FIS?
3. Explain why R_{et} increases or decreases for the different electrode modifications.
4. Which behavior should be expected if the pair ferro/ferricyanide is changed by a cationic molecule?
5. How can R_{et} be measured graphically?
6. Explain the two zones that can be commonly observed in the Nyquist plot.
7. Explain why the current disappears in the CV of ferro/ferricyanide recorded in an electrode modified with Nafion.

References

[1] A.J. Bard, L.R. Faulkner, Electrochemical Methods: Fundamentals and Applications, second ed., John Wiley and Sons, NY, 2001.
[2] R. García-González, M.T. Fernández-Abedul, A. Costa-García, Nafion® modified screen-printed gold electrodes and their carbon nanostructuration for electrochemical sensors applications, Talanta 107 (2013) 376–381.
[3] J. Wang, M. Musameh, Y. Lin, Solubilization of carbon nanotubes by Nafion toward the preparation of amperometric biosensors, J. Am. Chem. Soc. 125 (2003) 2408–2409.
[4] N.F. Atta, Y.M. Ahmed, A. Galal, Layered-designed composite sensor based on crown ether/Nafion/polymer/carbon nanotubes for determination of norepinephrine, paracetamol, tyrosine and ascorbic acid in biological fluids, J. Electroanal. Chem. 828 (2018) 11–23.
[5] K. Janghorban, P. Molla-Abbasi, Modified CNTs/Nafion composite: the role of sulfonate groups on the performance fo prepared proton exchange methanol fuel cell's membrane, J. Part. Sci. Technol. 3 (2018) 211–218.
[6] C.A. Poland, R. Duffin, I. Kinloch, A. Maynard, W.A.H. Wallace, A. Seaton, V. Stone, S. Brown, W. Macnee, K. Donaldson, Carbon nanotubes introduced into the abdominal cavity of mice show asbestoslike pathogenicity in a pilot study, Nat. Nanotechnol. 3 (2008) 423–428.

PART II

Electroanalysis and microfluidics

CHAPTER

13

Single- and dual-channel hybrid PDMS/glass microchip electrophoresis device with amperometric detection

Andrea González-López[1], Mario Castaño-Álvarez[2],
M. Teresa Fernández Abedul[1]

[1]Departamento de Química Física y Analítica, Universidad de Oviedo, Oviedo, Spain;
[2]MicruX Technologies, Gijón, Asturias, Spain

13.1 Background

Undoubtedly, miniaturization continues being one of the main scientific trends in Analytical Chemistry, after Manz et al. made a relevant contribution in 1990 [1] introducing the first microchip for electrophoresis. Microfluidics is the field where the fluids are manipulated in channels with dimensions of tens of micrometers [2] and is the basis of many analytical applications [3,4]. Real analysis usually requires a separation step, and capillary electrophoresis (CE) is a well-known and established technique. As capillaries are made of fused silica, this technique (CE) has been "transferred" to the microchip format employing glass as the first material for manufacturing the miniaturized devices.

One of the main advantages of the chip versus the capillary format is that microchips can be manufactured in different materials. There are some works with ceramic materials, but polymers have gained terrain to facilitate the manufacture process and reduce the cost of the devices. Among them, thermostable and thermoplastic materials have been employed. Elastomers are a special case of non–cross-linked thermostable polymers, which can be flexible and elastic. Polydimethylsiloxane (PDMS), first introduced by Whitesides' group [5], is one of the most employed elastomers, which can also be sealed to a wide variety of substrates. Different materials will have different electroosmotic flow (EOF). This is one of the electrokinetic forces that move species in the fluid (and the fluid itself). Therefore, a species will move with a velocity that will

Laboratory Methods in Dynamic Electroanalysis
https://doi.org/10.1016/B978-0-12-815932-3.00013-9

be the sum of the electrophoretic and electroosmotic velocities, both being proportional to the electric field. The electrophoretic (μ_{ep}) or electroosmotic (μ_{eo}) mobilities (in cm^2 V/s) are, respectively, the factors of proportionality. The electrophoretic mobility will depend on the charge/ radius ratio and also on the viscosity of the running buffer. The EOF will mainly depend on the charges on the microchannel wall; this is the reason why the material becomes so important. In some cases, it has to be even reversed (see Chapter 15).

On the other hand, the elasticity of the PDMS is an advantage that can be used for different purposes: among them, the integration of different components (e.g., detectors) in the microchip. Ideally, a whole integration is intended. The adequate fitting of the electrochemical detection (ED) with miniaturized devices has produced devices with different configurations and for a great variety of applications. The lack of dependence on sample turbidity and on the optical path length as well as the use of a simple instrumentation makes ED a promising alternative. Conductometry and amperometry are the most common detection modes. While the first one is a universal method of detection that provides response to all species, amperometry is a selective principle with useful advantages, such as (i) the low sample volumes required, (ii) the type of data, that do not need final conversion, (iii) the high number of electroactive molecules, and (iv) the possibility of derivatization with easily reducible or oxidizable species.

Among the possibilities, off-chip and on-chip modes relate to the detection/device integration. Depending on the location of the detection in regard to the channel, it could be in-, end-, or off-channel. In the last case, an electrode is located before those employed for detection to avoid interferences from the high voltage applied. Gold wires (100 μm in diameter) have been aligned at the end of the microchannel of glass (single- and dual channel), PMMA or Topas (thermostable polymer) microchips [6,7], with a reservoir integrated for the detection. As PDMS can be reversibly sealed to glass, a polymeric plate with microchannel and reservoirs (made by mixing and curing the polymer on a master mold) can be slightly removed for situating and aligning the wire acting as working electrode [8]. The rest of electrodes could be located in the detection reservoir.

Single-channel microchips are the most common. However, complicate architectures can be found in the bibliography. A benefit of ME (microchip electrophoresis) compared with conventional capillary electrophoresis, apart from the different materials that can be employed, is the possibility to design structures that are more complex. In this chapter, a dual-channel hybrid PDMS/glass microchip with amperometric detection is proposed for simultaneous measurements [8]. In this case only one injection channel is employed and detection is performed at the same potential. Dopamine (DA), *p*-aminophenol (*p*AP), hydroquinone (HQ), epinephrine (EP), and catechol (CA) are the analytes that will be injected, as molecules of interest that have been traditionally determined by electrochemical methodologies.

13.2 Chemicals and supplies

— *Analytes*: DA, EP, *p*AP, HQ, and CA. Solutions are daily prepared in the running buffer.
— *Buffer preparation*: 2-(N-morpholino)ethanesulfonic acid (MES), L-histidine (His), sodium hydroxide, and potassium chloride. The running buffer is 25 mM MES-His pH 5.95.

- *Master mold and PDMS/glass microchip fabrication:* Positive photoresist AZ4562, developer AZ400K, Sylgard 184 silicon elastomer Kit, and soda lime microscope slides. Isopropyl alcohol, 10% hydrofluoric acid, and N_2 gas are also employed.
- *General materials:* Nylon syringe filters (0.1 μm, 30 mm), micropipettes and corresponding tips, syringe rubber piston, soda lime microscope slides.
- *Electrodes for applying high voltage:* 0.3 mm in diameter, 1-cm long platinum wires.
- *Electrodes for ED:* 100-μm diameter gold wire (working electrode), 0.3-mm diameter platinum wire (counter electrode) and 1-mm diameter silver wire (reference electrode). Saturated KCl is used as electrolyte of the reference electrode. Copper cable, conductive silver epoxy resin, and insulating tape are employed for connection.
- *Apparatus and instruments:* High-voltage power supply (HVPS), vacuum pump, (bi) potentiostat interfaced to a computer system, pH meter, spin coater, hotplate, microscope, and weighing scale.
- Milli-Q purified water was employed for preparing the solutions and washing.

13.3 Microchip fabrication

Microchannels are fabricated on PDMS by replication of a master mold made using soft lithography. Different microchannel designs can be fabricated on a 10-cm diameter silicon wafer.

1. After cleaning it with 10% hydrofluoric acid, deposit the positive photoresist AZ4562.
2. Place the wafer in a spin-coating machine. Around 20-μm thick layer is obtained using a spin rate of 2000 rpm during 3 s.
3. Soft bake on a hotplate at 100°C for 50 s.
4. To transfer the pattern with the microchannels, expose it to UV light through a mask using an energy density of 400 mJ/cm^2.
5. Solve the exposed photoresist by immersing it into the AZ400K developer that has been 1:4 diluted.
6. Hard bake it at 115°C during 50 s on a hotplate or 60 min in a heater. After this, the master mold is ready for replication.
7. Prepare a mixture of the elastomer precursor and the curing agent (in a 10:1 ratio) contained in the Sylgard 184 silicon elastomer Kit.
8. Pour the mixture over the master mold to replicate the pattern on the PDMS.
9. Cure it overnight at 35°C.
10. Peel off the PDMS plate (containing the microchannels) from the master mold.
11. Drill holes of 2-mm diameter at the end of the PDMS channels.
12. Wash the glass cover plate (soda lime microscope slides) and the PDMS plate with isopropyl alcohol and dry them with N_2.
13. Put together both plates to obtain a reversible-sealed hybrid microchip. The adherence between PDMS and glass is strong enough to allow liquids to run through the microchannels without leakage.

14. To obtain reservoirs of approximately 100 μL of volume, cut micropipette tips in 0.5-cm long pieces with a diameter of 0.3 cm at the top and stick them concentrically to the chip holes with an epoxy adhesive.

13.4 Microchip designs

Two different microchips designs can be evaluated: single- and dual channel. With this methodology, the microchannels have a depth of around 20 μm and a width of 50 μm (Fig. 13.1). The single-channel microchip consists of a 45-mm long separation channel (between buffer reservoir A and detection/waste reservoir B) and a 10-mm long injection channel (from sample reservoir C to sample waste reservoir D). The two channels crossed each other halfway between C and D reservoirs, at 5 mm from the separation channel buffer reservoir (A).

The dual-channel microchip (Fig. 13.2) consists of two 45-mm long parallel channels, separated 2 mm from each other. The injection channel is 10-mm long and crosses each separation channel at 5 mm from the buffer reservoir A. Separation channels cross the injection channel at 4 mm from C and D reservoirs.

13.5 Electrochemical detector design

The amperometric detector is placed in the waste reservoir B with a three-electrode (for the single-channel microchip) or four-electrode (for the dual-channel microchip) configuration. Then,

1. As working electrode, a 100-μm diameter gold wire (two in the case of the dual-channel design, see Fig. 13.2) with an end-channel configuration is used. The electrode alignment is easy, thanks to the reversibility of the PDMS fixed to the glass plate.

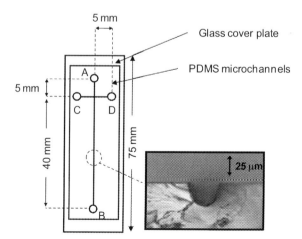

FIGURE 13.1   Scheme of the single-channel microchip including a picture of a polydimethylsiloxane (PDMS) microchannel section taken by optical microscopy.

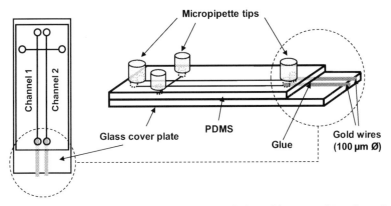

FIGURE 13.2 Schemes of the dual-channel microchip with the gold wire working electrodes inserted.

Once the electrode is accurately aligned with the aid of a microscope, both surfaces (PDMS and glass) are sealed again. A copper cable is fixed to the gold wire with conducting silver epoxy resin for electrical connection.

2. The reference electrode consists of a 1-mm diameter silver wire anodized in saturated KCl solution and introduced in a 250-µL micropipette tip through a syringe rubber piston.

3. The auxiliary (or counter) electrode consists of a 0.3-mm diameter platinum wire that is externally fixed to the tip of the reference electrode with insulating tape. The tip is then introduced in the detection reservoir (B) of the microchip.

13.6 Hazards

High voltage is required and then appropriate precautions should be taken to avoid an electrical shock. Specifically, it has to be ensured that the power supplies are off before touching any wire or high voltage cable.

Hydrofluoric acid is corrosive and causes severe burns. Always wear safety glasses, protective gloves, and lab coat. It is toxic if inhaled, and always handle in a fume hood.

13.7 Experimental procedure

13.7.1 Electrophoretic separation in a single-channel microchip

1. Rinse the PDMS/glass microchip with 0.1 M NaOH for 15 min. Washing could be made with a simple vacuum system.

2. Fill the microchip with running buffer. Wash it for 10 min and make sure all reservoirs have the same level (50 µL).

3. Insert the high-voltage Pt wires in the corresponding reservoirs (A, B, C, D) and connect them to the power supplies by means of alligator clips (those in A and B to one HVPS and those in C and D to the other).
4. Apply a high voltage (e.g., +1000 V) between A and B reservoirs as well as the detection potential (e.g., +1.2 V), meanwhile baseline current is recorded with time. The value of the detection potential has to be chosen after previous studies on the electrochemical behavior of the analytes.
5. Remove the running buffer of the sample reservoir C (or D) when a stable baseline is obtained. Fill it with the standard or sample solution (25 μL of a mixture of, e.g., DA, *p*AP, and HQ, 100 μM in each).
6. Apply the injection voltage (e.g., +750 V) for a fixed time (5 s) between C and D.
7. Apply the separation voltage between the running buffer reservoir (A) and the detection reservoir (B) recording the corresponding electropherogram (i-t curve).

To see the effect of different parameters, record electropherograms with

1. Separation voltages varying from +500 to +2500 V.
2. Injection times decreasing from 5 to 1 s.
3. Injection voltages varying from +500 to +1000 V.

Observe, in all the cases, what happens with the migration time and the peak current. Choose the best parameters and record seven successive electropherograms to know the precision of the measurements (through the value of the relative standard deviation, RSD) performed with the same microchip.

PDMS microchips are considered single-day devices. To know the precision of the measurements obtained in different microchips, electropherograms are similarly recorded in several devices. The RSD is calculated to estimate the precision.

13.7.2 Electrochemical detection in a dual-channel microchip

This configuration allows studying different types of injection, using times from 1 to 10 s. Then, after performing steps 1−3 from Section 13.7.1, different formats can be followed:

1. *InjC* (sample situated in reservoir C, V_{inj} applied at C with D and B grounded),
2. *InjD* (sample situated in reservoir D, V_{inj} applied at D with C and B grounded),
3. *InjCD* (sample situated in reservoirs C and D, V_{inj} applied at C and D with B grounded), and
4. *InjACD* (same than InjCD with a high-voltage electrode introduced in A).

A 100-μM DA solution could be used this time using the different injection formats with a separation voltage of +1000 V, an injection voltage of +750 V, and a detection potential of +1.2 V. Simultaneous analytical signals will be achieved in both channels with all the injection formats, even using an injection time for 1 s. As an example, in Fig. 13.3, electropherograms for C or D injection formats are shown.

Apart from simultaneous measurements (Fig. 13.4A), dual-channel microchips also allow performing simultaneous separations of different analytes in both channels and no

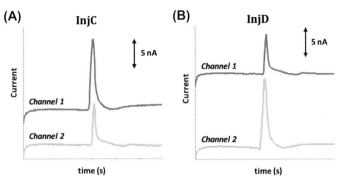

FIGURE 13.3   Influence of the injection format on the dual-channel polydimethylsiloxane/glass microchip.

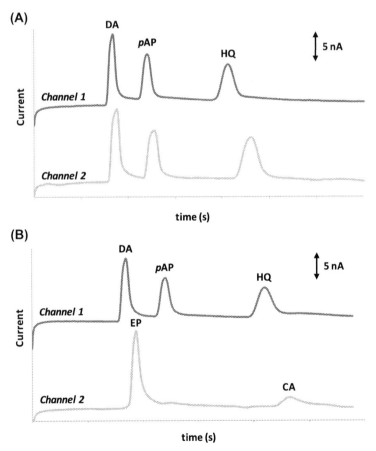

FIGURE 13.4   Simultaneous separation on the dual-channel microchip for: (A) the same sample mixture of dopamine (DA), *p*-aminophenol (*p*AP), and hydroquinone (HQ) and (B) different sample mixtures of DA, *p*AP, and HQ (channel 1) and EP and catechol (CA) (channel 2).

cross-separation is obtained (Fig. 13.4B). So, two different assays can be achieved in a same experiment with an important reduction in analysis time.

13.8 Lab report

Write a lab report which includes an abstract, a brief introduction explaining the purpose of the experiment, a detailed experimental section, results and discussion, and conclusions. Include tables, graphics, or figures wherever necessary. The following points should be considered:

1. In the introduction, include a revision of reported works with examples of microchips made of different materials and with different designs, increasing in complexity. Applications from different fields can be considered: clinical, food, environmental, etc.
2. In the experimental section, explain briefly the different steps required for the fabrication of the microchip and draw schematics with the sequence for injecting, separating, and detecting.
3. In the results and discussion section, show the most representative electropherograms, including the incidences during the course of the experiment, for both single- and dual-channel experiments. Comment optimizations performed and the effect of the different variables on the migration time and the peak current.
4. In the conclusions, highlight the most representative results and a critical opinion on the future of these devices.

13.9 Additional notes

1. The procedure for the creation of the master mold must be carried out in a clean room.
2. A methacrylate holder ($18 \times 13 \times 2$ cm) is fabricated for accommodating the chips. Once the chip is inserted on the holder, a small rectangular piece ($5 \times 1 \times 0.8$ cm) is fixed with screws with the aim of securing the chips.
3. All experiments are performed at room temperature.
4. The volume in the sample reservoir has to be lower than the other reservoirs to avoid hydrodynamic effects (in reservoirs C or D when no sample analysis is performed, the buffer volume has to be 25 μL, too).
5. When the separations of DA, pAP, and HQ are performed in successive days with the same microchip, the analytical response of the different compounds worsened (it could be checked). This implies that PDMS microchips must be used as single-day devices. This is not an inconvenience because of their low cost, ease, and low time of fabrication that allow obtaining several microchips in the same manufacturing process.
6. The pretreatment of the microchip is very important to obtain precise results. Comparison between measurements made with or without pretreatment could be also evaluated.

13.10 Assessment and discussion questions

1. Indicate the fabrication process of a hybrid PDMS/glass microchip.
2. Explain how a mixture of three different analytes can be injected, separated, and detected using a microchip electrophoresis.
3. If a new microchip design wants to be evaluated, which are the steps that should be followed?
4. A neutral molecule migrates with the EOF. How could EOF be determined from the recorded electropherograms?
5. Calculate the resolution and efficiency of the separation (see Chapter 14).

References

[1] A. Manz, N. Graber, H.M. Widmer, Miniaturized total chemical analysis systems: a novel concept for chemical sensing, Sens. Actuators B 1 (1990) 244—248.
[2] G.M. Whitesides, The origin and the future of microfluidics, Nature 442 (2006) 368.
[3] J.C. Jokerts, J.M. Emory, C.S. Henry, Advances in microfluidics for environmental analysis, Analyst 137 (2012) 24—34.
[4] A. González Crevillén, M. Hervás, M.A. López, M.C. González, A. Escarpa, Real sample analysis on microfluidic devices, Talanta 74 (2007) 342—357.
[5] A. Kumar, H.A. Biebuyck, G.M. Whitesides, Patterning self-assembling monolayers: applications in materials science, Langmuir 10 (1994) 1498—1511.
[6] D.F. Pozo-Ayuso, M. Castaño-Álvarez, A. Fernández-la-Villa, M. García-Granda, M.T. Fernández-Abedul, A. Costa-García, J. Rodríguez-García, Fabrication and evaluation of single- and dual-channel (π-design) microchip elecrophoresis with electrochemical detection, J. Chromatogr. A 1180 (2008) 193—202.
[7] M. Castaño-Álvarez, M.T. Fernández-Abedul, A. Costa-García, Poly(methylmethacrylate) and Topas capillary electrophoresis microchips performance with electrochemical detection, Electrophoresis 26 (2005) 3160—3168.
[8] M. Castaño-Álvarez, M.T. Fernández-Abedul, A. Costa- García, A., Multiple-point electrochemical detection for a dual-channel hybrid poly(dimethylsiloxane)-glass microchip electrophoresis device, Electrophoresis 30 (2009), 3372—2280.

CHAPTER

14

Analysis of uric acid and related compounds in urine samples by electrophoresis in microfluidic chips

Diego F. Pozo-Ayuso, Mario Castaño-Álvarez, Ana Fernández-la-Villa

MicruX Technologies, Gijón, Asturias, Spain

14.1 Background

Clinical analysis is demanding new and competitive analytical methods. Conventional systems require a large amount of time and money for reagents and instruments, which is not compatible with the enormous number of analyses that are needed in daily life.

Clinical analysis of urine is one of the most important inspections for medical diagnosis. Uric acid (UA) is the final breakdown product of dietary or endogenous purines and is generated by the enzyme xanthine oxidase. Clinical studies have shown that monitoring UA levels in urine and blood serum can be used to diagnose several diseases. Sustained elevations of UA in the blood usually result from an increase in the endogenous production of UA or a reduction in renal urate excretion, or a combination of both. Such elevations can cause hyperuricemia and gout [1,2].

UA can be determined in serum or plasma and 24-h urine. Typical reference values for UA [1] are:

— Urine:	250–750 mg/24 h	≈ 1.49–4.5 mmol/24 h
— Serum or plasma:		
＊ Women	2.5–6.8 mg/dL	≈ 149–405 μmol/L
＊ Men	3.6–7.7 mg/dL	≈ 214–458 μmol/L
＊ Children	2.0–5.5 mg/dL	≈ 120–330 μmol/L

Conventional methods for the measurement of UA are based on the conversion of urate to allantoin via uricase. Colorimetric techniques based on this reaction are, however, temperature-dependent, expensive, and require labile reagents. An additional obstacle in monitoring UA levels is the interference from other compounds such as epinephrine (EP), DOPA, ascorbic acid (AA), paracetamol, xanthine, theophylline, or caffeine. Thus, capillary electrophoresis (CE) has become a powerful tool for the separation and determination of UA and interferences.

CE is a powerful analytical tool for the separation of different analytes. CE has proved to be useful for separations of compounds such as amino acids, chiral drugs, vitamins, pesticides, inorganic ions, organic acids, dyes, surfactants, peptides and proteins, carbohydrates, oligonucleotides and DNA restriction fragments, and even whole cells and virus particles. The miniaturized CE approach, microfluidic electrophoresis chips (see also Chapters 13 and 15), is a novel technology considered as the first stage to get a "true" lab-on-a-chip [3].

Microfluidic electrophoresis platforms enable the possibility of monitoring UA levels and the separation of interference compounds such as EP, L-DOPA, AA, acetaminophen (APAP), xanthine, theophylline, or caffeine coming from endogenous or exogenous sources. They bring novel methodologies for fast, inexpensive, and high throughput urine analysis. In combination with an appropriate instrumentation, they offer high speed, great versatility, high throughput, low cost, performance of parallel assays, and negligible consumption of reagents/sample and waste generation. Thus, microfluidic chips provide a great selectivity for the separation of compounds with similar chemical structures (Fig. 14.1).

Separation by CE is obtained by differential migration of solutes in an electric field. Different separation modes are available by CE depending on the electrophoretic mobility (based on analyte charge and size) and electroosmotic flow (EOF) (based on capillary/channel charge). In the most typical electrophoresis mode (normal mode), all the analytes (cations, anions, and neutral species) are driven toward the cathode Fig. 14.2) by the action of the EOF. However, as EOF can be suppressed or reversed (reverse mode), the analytes can also migrate toward the anode (see Chapter 15).

Electropherograms provide key analytical information for the identification and quantification of the compounds of a complex sample (Fig. 14.3). Thus, the migration time (t_m) is used for the identification of a compound. The peak height (h) enables the determination of the concentration of each compound. Finally, other parameters such as peak width (w) or half-peak width ($w_{1/2}$) are also significant for the determination of the efficiency and resolution of separations.

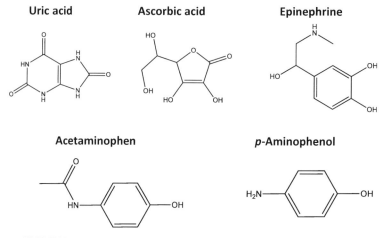

FIGURE 14.1   Chemical structure of uric acid and related compounds.

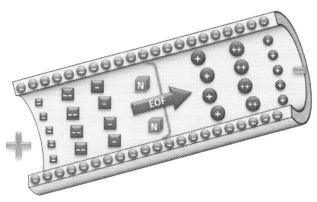

FIGURE 14.2 Ion migration in capillary electrophoresis. *EOF*, electroosmotic flow.

The separation efficiency is expressed as the theoretical plate number (N):

$$N = 5.54 \times \left(\frac{t_m}{w_{1/2}}\right)^2 \tag{14.1}$$

The separation resolution (R) between two peaks with a Gaussian distribution:

$$R = 1.18 \times \left(\frac{t_{m2} - t_{m1}}{w_{1/2(1)} - w_{1/2(2)}}\right) \tag{14.2}$$

On the other hand, these microdevices require a miniaturized and sensitive detection system. Electrochemical detection has proven to be very effective because of characteristics such as inherent miniaturization, sensitivity, low cost, portability, and compatibility with microfabrication technology. Electrochemical approaches also enable the direct detection of UA and related compounds. So, conventional methodologies for UA determination, which are temperature-dependent, expensive, and require labile reagents (enzymes), are improved. Moreover, the establishment of a simple, economical, and accurate analytical method for the determination of UA and related compounds would be useful in point-of-care settings [4].

This lab experiment is focused on undergraduate and postgraduate students of different areas (Chemistry, Pharmacy, Biology, etc.) and can be performed in two/three sessions of

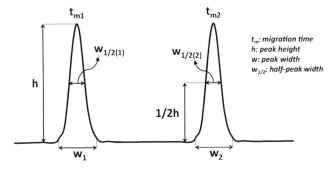

FIGURE 14.3 Typical electropherogram and main separation parameters.

3 h. The students will be able to acquire practical skills in the use of microfluidic electrophoresis devices and miniaturized analytical instrumentation.

14.2 Electrophoresis system setup

The setup of the complete system, including the HVStat instrument and chip holder, is shown in the next picture (Fig. 14.4). In case iHVStat instrument is used, external cables will not be necessary. Thus, microchip holder is integrated in the socket of the instrument (see iHVStat manual).

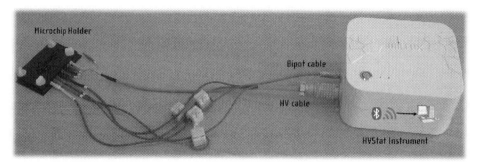

FIGURE 14.4   Electrophoresis system setup. *HV*, high voltage.

SU-8/glass microchips integrate platinum electrodes in an on-chip end-channel approach (Fig. 14.5, see also Chapter 15) for amperometric measurements.

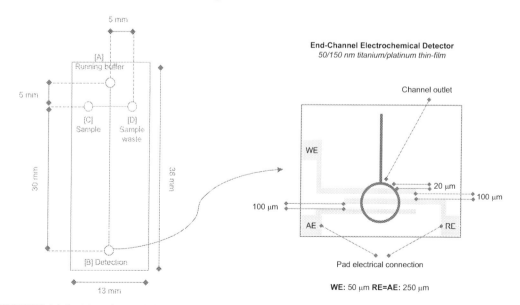

FIGURE 14.5   Microchip and electrodes design. *AE*, auxiliary electrode; *RE*, reference electrode; *WE*, working electrode.

14.3 Chemicals and supplies

Reagents:

— *Analytes*: APAP, AA, EP, *p*-aminophenol (*p*AP), and UA.
— *Preparation of the running buffer*: 2-(N-morpholino)-ethanesulfonic acid (MES)–NaOH–sodium dodecyl sulfate (SDS) buffer pH 6.0 (it can be provided already prepared by Micrux Fluidic, hereinafter referred to as MES buffer), sodium hydroxide (NaOH), hydrochloric acid (HCl).

Instrumentation and materials:

— MicruX HVStat (or iHVStat) instrument.
— Microfluidic platform.
— SU-8/glass microfluidic chips with integrated platinum electrodes.
— Volumetric material (flasks, pipettes, vessels, micropipettes, etc.) and all the materials necessary for the preparation of the solutions, which should be of analytical reagent grade.
— Syringes and syringe filters (0.1–0.45 µm) are used for removing particles of the working solutions and samples.

14.4 Hazards

HVStat and iHVStat instruments use high voltages (up to 3 kV). The high-voltage power supply has a current limit set for protection, but appropriate precautions should be taken to avoid an electrical shock. Specifically, it has to be ensured that the power supplies are off before touching any wire or high-voltage cable.

14.5 Experimental procedure

14.5.1 Preparation of solutions

Running buffer:
The buffer solution used in the determination of UA and related compounds is a mixture of MES, sodium hydroxide, and SDS, mainly. The MES–NaOH–SDS buffer is provided as single dose for dissolving in 100 mL distilled water to get a final concentration of 20 mM MES, pH 6.0.

Stock and working standard solutions:

— 10 mM stock solutions of APAP, EP, and *p*AP have to be prepared in 10 mM HCl. Prepare 1 mL of each stock solution.
— 10 mM stock solution of UA has to be prepared in 20 mM NaOH, and 10 mM stock solution of AA has to be prepared in deionized water. Prepare 1 mL of each stock solution.

Stock and working standard solutions:

- Standard solutions (1 mL) of AA (500 µM), APAP (200 µM), EP (100 µM), pAP (100 µM), and UA (250 µM).
- A mixture (1 mL) of EP (100 µM), APAP (200 µM), pAP (100 µM), AA (500 µM), and UA (250 µM).
- Other working solutions indicated along the experimental section.

14.5.2 Sample preparation

UA is determined in urine samples. Fresh urine samples should be daily collected in a sterilized container. Samples should have a pH higher than 8.0. In case the pH is lower than 8.0, it has to be adjusted with 1 M NaOH to pH 8.0–8.5. If urine is cloudy, warm the specimen to 60°C for 10 min to dissolve precipitated urates and UA. Do not refrigerate.

The urine sample is directly diluted 1:10 or 1:20 in the running buffer. The solutions have to be filtered using a syringe filter (0.1–0.45 µm) and injected directly to the microchip for analysis.

14.5.3 Electrophoretic procedure

14.5.3.1 Microchip pretreatment

Rinse the SU-8/glass microchip with 0.1 M NaOH for 20–30 min, deionized water for 15 min, and then with the running buffer for 10 min. Washing could be made with the aid of a simple vacuum system. After washing, fill the microchip with running buffer and be sure all reservoirs have the same level (50 µL).

14.5.3.2 Baseline stabilization

After filling the microchip and connecting the system (see Section 14.2), the baseline should be stabilized. Thus, in the main window of MicruX Manager (see MicruX Manager User Manual), apply a high voltage between A (also possible C or D) and B reservoirs and the detection potential, meanwhile baseline current is recorded with time. The value of the detection potential has to be selected according to previous studies on the electrochemical behavior of the analytes and the hydrodynamic curve (see Section 14.5.4.2).

When a stable baseline is obtained, remove the running buffer of the sample reservoir C (or D) and fill it with the standard or sample solution (25 µL). The sample volume has to be lower than the other reservoirs to avoid hydrodynamic effects (in reservoirs C or D when no sample analysis is performed, the buffer volume has to be 25 µL, too).

14.5.3.3 Unpinched injection

The system has been designed for unpinched approaches. Thus, in the experiment window of MicruX Manager (see MicruX Manager User Manual), select the injection voltage between sample (C or D) and sample waste (D or C) reservoirs with a short injection time (e.g., 4–5 s).

14.5.3.4 Separation and detection

In the experiment window of MicruX Manager (see MicruX Manager User Manual), select also the separation voltage between the running buffer reservoir (A) and the detection reservoir (B). You can select a separation time of 200–300 s. Finally, in the same window, select the amperometric mode (DC) and the detection potential for WE1. Then, start the experiment and record the electropherogram for the compounds mixture.

14.5.4 General electrophoretic behavior

14.5.4.1 Buffer solution

A good performance for the separation of AA, APAP, EP, pAP, and UA is observed using MES buffer. To obtain an adequate electropherogram for all the five compounds:

1. In the experiment window of the MicruX Manager configure an injection voltage of +750 V for 3 s (configuration C-D or D-C), a separation voltage of +1000 V (configuration A-B), an interval time of 50 ms, and DC amperometry with a detection potential of +0.8 V (WE1).
2. Inject a mixture containing 500 μM AA, 200 μM APAP, 100 μM EP, 100 μM pAP, and 250 μM UA to test if all the system is working adequately.
3. In the same way, inject individually the standard solution of each compound to identify these analytes. Under these conditions, the separation of the five compounds is achieved Fig. 14.6).

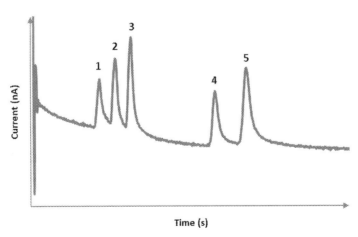

FIGURE 14.6   Typical electropherogram for the separation and detection of the five compounds.

14.5.4.2 Electrochemical detection

A hydrodynamic voltammogram (HDV, i vs. E curve) has to be performed in the microchip with amperometric detection to get the optimal detection potential where the best signal-to-noise ratio (S/N) is achieved. To obtain it,

1. Fix a separation and injection voltage for recording the HDV as in the previous section.
2. Study the detection potential between 0.0 and +1.2 V for a mixture containing 500 μM AA, 200 μM APAP, 100 μM EP, 100 μM pAP, and 250 μM UA.

The complete experiment can be programmed in the experiment window (Advanced Options) of the MicruX Manager (see MicruX Manager User Manual).

14.5.4.3 Separation and injection performance

The influence of the separation and injection voltages as well as of the injection time and configuration have on the analytical has to be evaluated. Then, study the effect of:

1. The separation voltage between +500 and +1500 V for a mixture of 250 μM AA, 100 μM APAP, 50 μM EP, 50 μM pAP, and 130 μM UA.
2. The injection voltage between +500 and +1500 V for the mixture of UA and related compounds used in point 1.
3. The injection configuration (C-D, C-A, or C-B) for the mixture of UA and related compounds used in point 1.
4. The injection time from 1 to 10 s for the mixture of UA and related compounds used in point 1.

Other conditions for the studies are the same that were employed in previous experiments. The complete experiment for each study can be programmed in the experiment window (Advanced Options) of the MicruX Manager (see MicruX Manager User Manual).

14.5.5 Analytical parameters

Using the optimized separation and injection voltage, injection time, and detection potential, to:

1. Study the precision (repeatability) for successive measurements of a sample mixture of 250 μM AA, 100 μM APAP, 50 μM EP, 50 μM pAP, and 130 μM UA.
2. Study the effect of the compounds concentration on the analytical signals (calibration plot). Prepare different mixtures of the compounds for studying the concentration (Fig. 14.7):
   - AA: from 25 μM to 1 mM
   - APAP: from 10 to 600 μM
   - EP: from 5 to 500 μM
   - pAP: from 5 to 500 μM
   - UA: from 10 to 750 μM

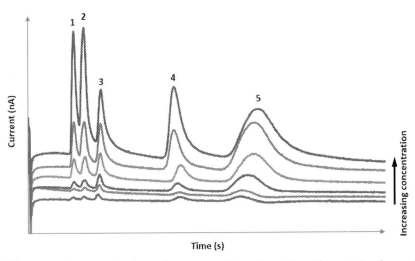

FIGURE 14.7 Electropherograms for increasing concentrations of ascorbic acid, acetaminophen, epinephrine, *p*-aminophenol, and uric acid.

14.5.6 Real sample analysis

The analysis of urine samples is performed in the microchips using the optimal conditions. UA determination in urine samples is carried out using the standard addition method.

Prepare four 100-μL aliquots of the urine sample in separated microtubes. One of the aliquots is directly diluted with the buffer solution (1:10 or 1:20). Spike the other three aliquots with increased amounts of UA, from a standard solution (solution of known concentration of analyte) (Fig. 14.8).

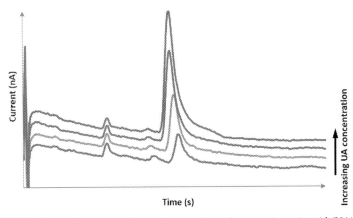

FIGURE 14.8 Electropherograms for a urine sample with successive uric acid (UA) additions.

14.6 Lab report

At the end of the experiment, write a lab report including an introduction, experimental (materials, equipment, and protocols), results and discussion, and conclusions sections. The following points should be considered in the report:

1. Sketch the HDVs for UA and related compounds and determine the optimal potential for each individual compound and for the sample mixture.
2. Include the electropherograms for the compound mixtures for the separation and injection studies. Determine the main parameters for the separation and detection of UA and related compounds in each performed study:
   - Migration time (t_m)
   - Peak current (i_p)
   - Half-peak width ($w_{1/2}$)
   - Theoretical plate number (N) Eq. (14.1)
   - Resolution (R) Eq. (14.2).

3. Determine the precision (% RSD- Relative Standard Deviation) of the migration time and peak current for the compound mixture.
4. Draft the calibration plots for each compound of the sample mixture and determine the main analytical parameters (linear range, sensitivity, limit of detection, and limit of quantification) for the compounds in the optimal conditions.
5. Sketch the calibration curve of UA obtained by the standard addition method taking into account the UA additions. Determine the UA amount in the urine sample. The amount of the UA in urine is expressed as mmol UA/24 h or mg UA/24 h.

14.7 Additional notes

1. The buffer substance (for the running buffer) in the bottle should be initially dissolved in 2–3 mL of distilled water and transferred to a beaker. Then, approximately 80 mL distilled water is added to the beaker and the pH is measured. If the pH is lower than 6.0, it has to be adjusted using 1 M NaOH. Then, the solution is transferred to a flask and filled to 100 mL with distilled water.
2. Stock solutions (except UA solution) should be kept on dark and stored at 4°C.
3. Working solutions are prepared daily from stock solutions by diluting these to the appropriate concentration in the running buffer.
4. All working solutions have to be filtered using a syringe filter of 0.1–0.45 μm to remove small particles, which can block the microchannels.
5. In all the experiments, typically, three consecutive injections are performed and the first injection is always discarded. The first injection is used to preload the sample in the injection channel.
6. It is recommended to perform an injection of buffer solution between each standard solution to clean the injection channel.

7. In this experiment, as in those regarding microfluidic electrophoresis, the symbol E refers to the electric field (V/cm). However, in the rest of chapters, E is simply the potential applied to the working electrode (V). This should be taken into account to avoid misunderstandings.

14.8 Assessment and discussion questions

1. Taking into account the properties of the compounds and the buffer solution, what is the theoretical migration order of the compounds? Is it in accordance with the experimental data?
2. What are the electrochemical processes involved for each compound?
3. Taking into account the results of the experiments, what are the optimal conditions for the separation and detection of UA and related compounds?

References

[1] A. Schultz, Uric acid, in: A. Kaplan, et al. (Eds.), Clinical Chemistry. The C.V. Mosby Co. St Louis. Toronto. Princeton, 1984, pp. 1261–1266, 418.
[2] N.W. Tietz, et al., Clinical Guide to Laboratory Tests, third ed., AACC, 1995.
[3] J.P. Landers (Ed.), Handbook of Capillary and Microchip Electrophoresis and Associated Microtechniques, third ed., CRC Press, Taylor & Francis Group, 2008.
[4] A. Fernández-la-Villa, V. Bertrand-Serrador, D.F. Pozo-Ayuso, M. Castaño-Álvarez, Fast and reliable analysis using a portable platform based on microfluidic electrophoresis chips with electrochemical detection, Anal. Methods 5 (2013) 1494–1501.

Microchannel modifications in microchip reverse electrophoresis for ferrocene carboxylic acid determination

Rebeca Alonso-Bartolomé, Andrea González-López, M. Teresa Fernández Abedul

Departamento de Química Física y Analítica, Universidad de Oviedo, Oviedo, Spain

15.1 Background

Detection of molecules can be approached using an intrinsic property such as e.g., electroactivity, absorbance of radiation, or fluorescence. In the case of molecules that do not have groups providing these properties, derivatization with a marker (indicator molecule) allows their detection. Ferrocene, in the form of monocarboxylic acid (ferrocene carboxylic acid [FCA], Fig. 15.1A), has been employed as mediator of electron transfer reactions, usually modifying the working electrode [1] or, more scarcely, as covalent electrochemical label of biomolecules [2], such as antibodies, DNA, peptide nucleic acids, or aptamers. In this context, it is of paramount relevance to have appropriate techniques to monitor the bioconjugation process. Then, measurement of the label, either free or conjugated to biomolecules for following the bioconjugation procedure to determine analytes indirectly, is required in many cases.

Electrophoresis is a powerful separation technique based on the differential mobility of charged species in an electric field. It can be performed in submillimeter diameter capillaries (capillary electrophoresis [3]) or in microchannels (even nanochannels), in devices named microchips (ME for microchip electrophoresis or MCE for microchip capillary electrophoresis [4]). The field is advancing with developments not only in materials but also in surface modifications as well as in detection systems and associated instrumentation [4–7].

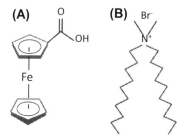

FIGURE 15.1   Structures of (A) ferrocene monocarboxylic acid and (B) didecyldimethylammonium bromide.

Different detection principles have been integrated with MEs. Electrochemical detection schemes are gaining acceptance and maturity, mainly because of their simplicity, high sensitivity, and possibility of electrode integration into the chip for different applications [8–10]. In this experiment, a microchip based on the photoresist EPON SU-8 (SU-8) [11], a negative tone epoxy photopatternable resist, mechanically reliable, optically transparent, chemically resistant, and hydrophilic, is employed with fully integrated electrochemical detection system.

Regarding the electrophoresis, there are two different electrokinetic forces to move species in the fluid: electrophoretic and electroosmotic [3]. The electrophoretic velocity (v_{ep}, cm/s) of an analyte toward the oppositely charged electrode depends on the electric field (E, V/cm), with the electrophoretic mobility (μ_{ep}, cm^2/Vs) being the proportionality constant:

$$v_{ep} = \mu_{ep}E \tag{15.1}$$

Two forces act on a particle inside the microchannel: the electrical force, F_{el}, that depends on the charge and electric field ($F_{el} = q\,E$), and the frictional force, $F_r = -6\pi\eta\,rv$ for a spherical ion. During electrophoresis, a steady state is attained, where the two forces are equal, but in opposite directions. Then, equalizing and solving for velocity ($v = (q/6\pi\eta\,r)\,E$) results that the electrophoretic mobility is:

$$\mu_{eo} = q/6\pi\eta r \tag{15.2}$$

Apart from the electrophoretic flow (EF), there is an electroosmotic flow (EOF), which depends on the charge of the wall of the capillary/microchannel. The EOF corresponds to the velocity of uncharged species, which are not attracted by any of the electrodes. Usually the wall surface is negatively charged, e.g., in silica capillaries or glass microchips as in most of the microchip materials, silanol groups in the surface are deprotonated above pH 3. Anions can be also adsorbed on the surface and then charges negatives are present. Then, cations from the solution are attracted forming both fixed and mobile layers and are positively charged. In this scenario, when the voltage is applied, positive charges are attracted to the cathode (negative electrode) dragging the solution in this direction, with neutral and negative charges included. This uniform force produces the EOF and acts as an "electric field" – driven pump. The electroosmotic velocity (v_{eo}) depends also on the electric field:

$$v_{eo} = \mu_{eo}E \tag{15.3}$$

with μ_{eo} being the electroosmotic mobility ($\mu_{eo} = \varepsilon\,\zeta/\eta$) where ε is the relative permittivity of the buffer, ζ is the zeta potential of the wall of the microchannel, and η is the viscosity. The global velocity of the analyte is the sum of both, $v = v_{ef} + v_{eo}$, with an apparent mobility (μ_{app}) that includes both contributions:

$$v = \mu_{app}E = (\mu_{ef} + \mu_{eo})E \qquad (15.4)$$

The mobility can be calculated from known values of electric field and velocity, as $v = L_{eff}/t_m$ (with L_{eff} the distance from the injection point to the detector and t_m the migration time) and $E = V/L_t$ (with V the voltage and L_t the total length). Then,

$$\mu_{app} = (L_{eff}/t_m)(L_t/V) \qquad (15.5)$$

The μ_{eo} can be calculated similarly but using the migration time of a neutral marker, a species that migrates only with EOF. As, usually EOF is higher than EF, all the analytes go in this direction (in normal mode electrophoresis), even anions.

The order of the analytes depends on the charge to size ratio (q/r), as can be seen in Fig. 15.2A. With the microchannel wall negatively charged, anions migrate after cations and neutral molecules in between. Then, separation of relatively small anions is often a challenging task under cathodic EOF. A high charge-to-size ratio gives some inorganic anions extremely high electrophoretic mobility, which is generally comparable to or is even higher than EOF. As a result and because of the net migration toward the anode, many anions can be detected only if polarity is changed. If the wall is positively charged (see Fig. 15.2B), both EF and EOF have the same direction for anions and they migrate before EOF and cations.

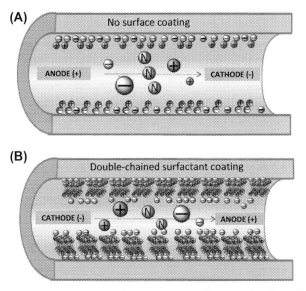

FIGURE 15.2 Order of flow of ions and neutral species in a capillary or microchannel when an electric field is applied for: (A) normal, and (B) reverse modes of electrophoresis.

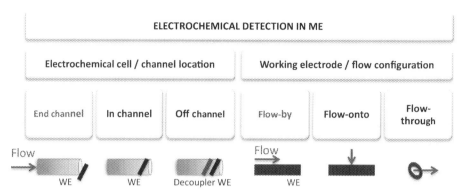

FIGURE 15.3 Classification of amperometric detection approaches following different criteria. *WE*, working electrode.

Then, flow reversal is desirable in most of anion determinations, as happens in this experiment with FCA. With this aim, modification of the microchannel wall to generate positive surface charges has to be performed. A dynamic (vs. covalent) approach has been chosen with a surfactant as additive that provides a semipermanent noncovalent coating: didecyldimethylammonium bromide (DDAB, Fig. 15.1B). This is a double-chained surfactant with two hydrophobic chains that interact in solution forming bilayers or vesicles, creating more stable coatings. It has been employed to obtain reversed EOF [12]. In this experiment it will be employed for generating an appropriate surface for FCA determination.

Microchip electrophoresis (see also Chapters 13 and 14) is one of the first examples of μTAS (micro total analysis systems), developed by Manz in 1990 [13], aimed to integrate in one device the different steps of the analytical process. Later on, they were considered for the same reason a good example of lab-on-a-chip devices. Here, we use one of the simplest designs of microchips (two crossed microchannels) where injection, separation, and detection steps can be performed. Detection is performed amperometrically. Similarly to what is explained in Chapter 10, detection is based on an electron transfer from an electroactive compound at a solid electrode (here, a platinum thin film). There are several possibilities (see Fig. 15.3), but in this experiment, adapted from Ref. [14], on-chip end-channel detection with a flow-by configuration is employed [15].

Platinum electrodes are inserted in the reservoirs situated at the end of the two microchannels for performing injection and separation. Injection is done by applying a voltage during a fixed time between the deposits at the end of the injection channel, one of them filled with sample (see Fig. 15.4). Similarly, separation is performed by applying a voltage between the deposits located at the end of the longer separation channel (both filled with the running buffer [RB]).

15.2 Electrophoresis microchip

The SU-8/Pyrex chips used in this study were purchased from MicruX Fluidic (Oviedo, Spain). They consist of 35- and 10-mm long separation and injection channels, respectively.

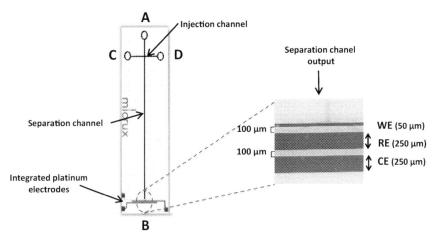

FIGURE 15.4 Electrophoresis microchip with crossed injection and separation channels. Electrodes for amperometric detection are shown in detail. *CE*, counter electrode; *RE*, reference electrode; *WE*, working electrode.

The last one is situated 5 mm from the beginning of the separation channel. Both are 50-μm wide and 20-μm high. Holes of 2-mm diameter that act as reservoirs are situated at the end of the channels: A and B in the separation channel and C and D in the injection channel (Fig. 15.4). One of the reservoirs of the injection channel (C) is filled with sample solution, meanwhile reservoirs A, B, and D are filled with the RB. Reservoir B is employed for detection and contains the electrodes of the potentiostatic system. The working electrode was a 50-μm wide platinum film and was located at 20 μm from the end of the separation channel. The other two electrodes (reference and auxiliary) were 250-μm wide platinum films and were separated 100 μm from each other. The distance between working and reference electrode was also 100 μm.

15.3 Chemicals and supplies

- *Analyte:* Ferrocene monocarboxylic acid.
- *Surfactant:* DDAB.
- *Components of* RB: 4-(2-Hydroxyethyl) piperazine-1-ethanesulfonic acid (HEPES) and sodium hydroxide.
- *General materials, apparatus, and instruments:* 1.5-mL microcentrifuge tubes, 10-μL, 100-μL, and 1-mL micropipettes with corresponding tips, Nylon syringe filters (0.45 μm, 25 mm), weighing scale, pH meter, oven, ultrasound bath, and HVStat (high-voltage power supply and bipotentiostat).
- Milli-Q water is employed for preparing the solutions and washing.

15.4 Hazards

HVStat instrument uses high voltages (up to 3 kV). The high-voltage power supply has a current limit set for protection, and appropriate precautions should be taken to avoid an electrical shock. Specifically, check the power supplies are off before touching any wire or high-voltage cable. Ferrocene carboxylic acid is harmful if swallowed, and is irritant to eyes, respiratory system and skin. Students are required to wear lab coat, appropriate gloves, mask when required, and safety glasses.

15.5 Experimental procedure

15.5.1 Solutions and sample preparation

The following solutions are needed (the amount of solution required has to be previously estimated):

— A 25 mM HEPES—NaOH buffer (from now HEPES buffer) of pH 7.0 (all working solutions used along this experiment will be prepared in this buffer solution).
— A 0.1 mM DDAB in 25 mM HEPES buffer of pH 7.0.
— A 0.1 M NaOH solution.
— A 1 mM FCA stock solution. From this one, FCA solutions of concentrations ranging between 5 and 150 μM are prepared. FCA solutions were prepared daily in the RB.
— All solutions were filtered through Nylon syringe filters.

15.5.2 Dynamic coating of microchannels

For the modification of the wall microchannel with DDAB:

1. Rinse the microchannel with 0.1 M NaOH for 10 min.
2. Rinse then with surfactant-containing buffer (0.1 mM DDAB in 25 mM HEPES buffer pH 7.0) for 20 min.
3. Let stabilize the microchannel in 0.1 mM DDAB in RB for 15 min.
4. Wash the microchannel with the RB for 5 min.

For the determination of FCA with DDAB semipermanent coating, RB does not contain DDAB.

15.5.3 Electrophoresis and electrochemical detection

Microchip zone electrophoresis was carried out using a high-voltage power supply integrated with the potentiostat (HVStat, Micrux Fluidic). A microchip is included in a commercial holder for performing electrophoresis with electrochemical detection as follows:

1. Precondition the microchannels, before use, as commented in Section 15.5.2.
2. Fill reservoirs C and D with 50 μL of sample and RB solutions, respectively.
3. Fill A and B reservoirs with 50 μL of RB.

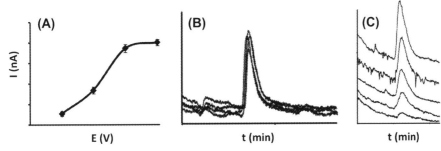

FIGURE 15.5 Example of (A) hydrodynamic curve, (B) precision evaluation, and (C) calibration curve for successive injections of ferrocene carboxylic acid. *From R. Alonso-Bartolomé, A. González-López, M.T. Fernández-Abedul, Double-chained cationic surfactant modification of SU-8/Pyrex microchips for electrochemical sensing of carboxylic ferrocene after reverse electrophoresis, Sens. Actuator. B Chem. 255 (2018) 490e497.*

4. Apply the appropriate detection potential to the working electrode for amperometric detection.
5. Before analyte injection, apply the separation voltage between A and B for baseline stabilization.
6. Record the electropherogram (I *vs.* t curve)
7. Perform unpinched sample injection by applying −700 V between C and B for 30 s.
8. Apply a voltage of −900 V between A and B for separation.
9. Record the electropherogram and measure the migration time and intensity of the peak current.

15.5.4 Evaluation of the detection potential

Perform a hydrodynamic curve to determine the most adequate detection potential. Record electropherograms corresponding to injections of RB and 200 µM FCA solutions in RB for +0.3, +0.4, +0.5, +0.6, and +0.7 V. Select the most appropriate potential (highest analytical signal with maximum precision, usually at the plateau, Fig. 15.5A).

15.5.5 Calibration curve

After preconditioning the microchip to generate a stable DDAB semipermanent coating, FCA solutions are injected sequentially with concentrations ranging between 5 and 150 µM in RB. Washing is not required in between if solutions are injected from diluted to concentrated.

Measure each concentration in triplicate to estimate the precision of the methodology.

Plot the intensity of the current at the maximum versus the concentration of FCA. Specify the linear range, the sensitivity (as the slope of the calibration curve), and the limit of detection (as the concentration corresponding to a signal that is 3 times the standard deviation of the intercept (or of the estimate) in the lower range of concentrations).

15.6 Lab report

At the end of the experiment, write a lab report that includes an introduction, experimental procedures, results obtained and discussion, finishing with main conclusions. In a more detailed manner, the following points should be considered in the report:

1. Discuss in the introduction the advantages of the miniaturization and the evolution of electrophoresis (conventional, capillary, and microchip electrophoresis).
2. Indicate the different operations performed in the microchip and how are they approached: injection (unpinched vs. pinched), separation (zone electrophoresis vs. other modes; normal vs. reverse), detection (electrochemical vs. other principles; amperometric vs. other modes).
3. Comment on the possible modifiers that can be employed in electrophoresis and the two main approaches (static and dynamic modifications) with arguments for and against.
4. Indicate the effect, in general, of using positive, neutral, or negative surfactants as modifiers.
5. Include pictures and schematics of the device (dimensions, steps of fabrication, etc.) as well as the procedure followed for injection and separation.
6. Indicate the basis of the analytical signal.
7. Include graphical representation of the data obtained along the experiment. The motivation for doing these studies should be commented first. The reasons for selecting different parameters (separation and injection voltage, injection time, detection time, or buffer composition) should be also included.
8. Represent the calibration curve and include the linear range, sensitivity, and limit of detection.

15.7 Additional notes

1. The semipermanent DDAB coating has to be prepared at the beginning of each working day.
2. At the end of the day, the microchip is rinsed with RB and kept dried. The following day, the preconditioning procedure (Section 15.5.2) is performed before recording electropherograms.
3. Modification with surfactant is made to record the signal of FCA. However, electrophoresis could be performed also in the normal mode to confirm that the signal for FCA, an anionic relatively small probe, is not obtained.
4. It is interesting to comment the effect of the modification with a single-chained positive surfactant (cetyltrimethylammonium bromide), in terms of precision and double-peak generation, as explained in Ref. [10].
5. The procedure for optimizing the detection potential is commented. Similarly, the influence of the separation and injection voltages (as well as the injection time) can also be studied.

6. Calibration curves can be performed in different microchips or different days to evaluate the robustness of the methodology. Precision can be evaluated by measuring the relative standard deviation of the migration time and the intensity of the peak current.
7. In this experiment, as in those regarding MEs, the symbol E refers to the electric field (V/cm). However, in the rest of chapters, E is simply the potential applied to the working electrode (V). This should be taken into account to avoid misunderstandings.

15.8 Assessment and discussion questions

1. Microchips for electrophoresis are included in the μTAS or lab-on-a-chip systems, explain the reason.
2. Explain the design of the microchip employed and how injection, separation, and detection is performed, indicating the types (see the point 2 in Section 15.6).
3. Explain the two modes of electrophoresis, normal and reverse and how the use of different surfactants can affect.
4. Indicate the origin of the two electrokinetic phenomena, electrophoretic flow and EOF.
5. Indicate the order of anions, cations, and neutral species.
6. What are the main parameters in an electropherogram?
7. What are the main parameters that should be optimized in ME?

References

[1] Z. Sun, L. Deng, H. Gan, R. Shen, M. Yang, Y. Zhang, Sensitive immunosensor for tumor necrosis factor based on dual signal amplification of ferrocene modified self-assembled peptide nanowire and glucose oxidase functionalized gold nanorod, Biosens. Bioelectron. 39 (2013) 215–219.
[2] W. Lai, J. Zhuang, J. Tang, G. Chen, D. Tang, One-step electrochemical immunosensing for simultaneous detection of two biomarkers using thionine and ferrocene as distinguishable signal tags, Microchim. Acta 178 (2012) 357–365.
[3] J.P. Landers (Ed.), Handbook of Capillary Electrophoresis, second ed., CRC Press, Boca Ratón (Florida, USA), 1997.
[4] C.S. Henry (Ed.), Microchip Capillary Electrophoresis, Methods and Protocols, Humana Press Inc., Totowa (New Jersey, USA), 2006.
[5] E.R. Castro, A. Manz, Present state of microchip electrophoresis: state of the art and routine applications, J.Chromatogr. A 1382 (2015) 66–85.
[6] M.T. Fernández-Abedul, I. Álvarez-Martos, F.J. García Alonso, A. Costa García, Improving the separation in microchip electrophoresis by surface modification (Chapter 6), in: C.D. García, K.Y. Chumbimuni-Torres, E. Carrilho (Eds.), Capillary Electrophoresis and Microchip Capillary Electrophoresis: Principles, Applications, and Limitations, J.Wiley & Sons, Inc., 2013, pp. 95–125.
[7] I. Álvarez-Martos, R. Alonso-Bartolomé, V. Mulas Hernández, A. Anillo, A. Costa-García, F.J. García Alonso, M.T. Fernández-Abedul, Poly(glycidyl methacrylate) as a tunable platform of modifiers for microfluidic devices, React. Funct. Polym. 100 (2016) 89–96.

[8] M.P. Godoy-Caballero, M.I. Acedo-Valenzuela, T. Galeano-Díaz, A. Costa-García, M.T. Fernández-Abedul, Microchip electrophoresis with amperometric detection for a novel determination of phenolic compounds in olive oil, Analyst 137 (2012) 5153–5160.

[9] M. Ávila, M.C. González, M. Zougagh, A. Escarpa, A. Ríos, Rapid sample screening methdo for authenticity controlling vanilla flavors using a CE microchip approach with electrochemical detection, Electrophoresis 28 (2007) 4233–4239.

[10] R. Kikura-Hanajiri, R.S. Martin, S.M. Lunte, Indirect measurement of nitric oxide production by monitoring nitrate and nitrite using microchip electrophoresis with electrochemcial detection, Anal. Chem. 74 (2002) 6370–6377.

[11] M. Castaño-Álvarez, M.T. Fernández-Abedul, A. Costa-García, M. Agirregabiria, L.J. Fernández, J.M. Ruano-López, B. Barredo-Presa, Fabrication of SU-8 based microchip electrophoresis with integrated electrochemical detection for neurotransmitters, Talanta 80 (2009) 24–30.

[12] M.M. Yassine, C.A. Lucy, Factors affecting the temporal stability of semipermanent bilayer coatings in capillary electrophoresis prepared using double-chained surfactants, Anal. Chem. 76 (2004) 2983–2990.

[13] D.J. Harrison, A. Manz, Z.H. Fan, H. Ludi, H.M. Widmer, Capillary electrophoresis and sample injection systems integrated on a planar glass chip, Anal. Chem. 64 (1992) 1926–1932.

[14] R. Alonso-Bartolomé, A. González-López, M.T. Fernández-Abedul, Double-chained cationic surfactant modification of SU-8/Pyrex® microchips for electrochemical sensing of carboxylic ferrocene after reverse electrophoresis, Sens. Actuator. B Chem. 255 (2018) 490–497.

[15] J. Wang, Electrochemical detection for microscale analytical systems: a review, Talanta 56 (2002) 223–231.

Integrated microfluidic electrochemical sensors to enhance automated flow analysis systems

Mario Castaño-Álvarez, Diego F. Pozo-Ayuso, Ana Fernández-la-Villa

MicruX Technologies, Gijón, Asturias, Spain

16.1 Background

Nowadays, new analytical devices that fulfill features such us portability, automation, user-friendliness, low cost, low power requirements, etc., are required. Therefore, a current trend in Analytical Chemistry is focused on the development of micro total analysis systems (μTAS) or lab-on-a-chip (LOC) platforms [1].

Microfluidics is the engineering or use of devices that applies fluid flow to channels smaller than 1 mm in at least one dimension. Microfluidic devices can reduce reagent consumption, allow well-controlled mixing and particle manipulation, integrate and automate multiple assays, and facilitate imaging and tracking.

Electrochemical transducers offer also multiple advantages such as low cost, portability, and low power requirements. Moreover, they can be easily integrated on microfluidic devices because of their compatibility with the microfabrication technologies. Thus, the possibility of integrating both systems, electrochemistry and microfluidics, contributes to the development of true LOC platforms [2,3].

Microfluidic devices and electrochemical sensors can be coupled into a flow injection analysis (FIA) system to improve the automation and high throughput of the platforms (see also Chapters 5, 9 and 28 with other electrochemical FIA systems). Flow cells facilitate analyte delivery to electrode surfaces for a range of applications such as clinical diagnostics or food and pharmaceutical analysis. The most common flow cells used in an FIA system are wall-jet and thin-layer arrangements (Fig. 16.1) [4].

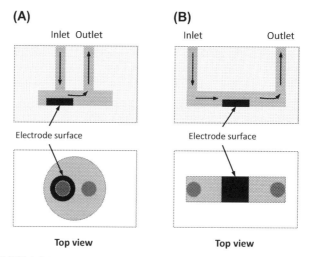

FIGURE 16.1   Sketch of typical (A) wall-jet and (B) thin-layer flow cells.

In a wall-jet cell, the axis of the solution stream is normal to the electrode surface and the solution is drained away completely from the electrode vicinity after contacting the electrode.

In a thin-layer cell, the solution flows through a thin flat channel parallel to the electrode surface, which is embedded in the channel wall. The use of microfluidics improves the design and performance of thin-layer flow cells. Thus, the solution layer thickness is easily controlled through the dimensions of the microfluidic channels. Microfluidics enable a perfect control of the dimensions of microchannels and the positioning of electrodes, decreasing the cell volume, the sample, and reagents requirements and enhancing the efficiency and sensitivity of the system.

Therefore, integration of microfluidics and electrochemical detection on a single chip allow enhancing the control of fluids over the electrode surface. The use of a thin-layer-based flow cell in FIA systems is ideal for (bio)chemical sensors development.

Acetaminophen, N-acetyl-p-aminophenol, or paracetamol (APAP) is a common analgesic and antipyretic drug formulated in a variety of dosage forms. APAP is used for the relief of fever, headaches, and other minor aches and pains. Their determination in pharmaceuticals is of paramount importance because an overdose of APAP can cause fulminating hepatic or renal necrosis and other toxic effects. Hepatic toxicity begins with plasma levels of APAP in the $120\,\mu g/mL$ range 4 h after the ingestion and an acute damage is presented with plasmatic levels up to $200\,\mu g/mL$ 4 h after the ingestion.

p-aminophenol (pAP), the primary hydrolytic degradation product of APAP, can be present in pharmaceutical preparations of APAP as a synthetic intermediate or as a degradation product. pAP is limited to the low level of $50\,\mu g/mL$ (0.005% w/w) in the drug raw material and 0.1% w/w in tablet formulations by the European [5], United States [6], and Chinese [7] pharmacopoeias. The low level ensures APAP drug safety because pAP has significant nephrotoxicity and teratogenic effects.

Therefore, establishment of simple, economical, and accurate analytical methods for determination of APAP and *p*AP would be useful to medical manufacturers, etc., for investigation of the stability of APAP, pharmaceutical analysis, and quality control.

This lab experiment is focused on undergraduate and postgraduate students of different areas (Chemistry, Pharmacy, Biology, etc.) and can be performed in two/three sessions of 3 h. The students will be able to acquire practical skills in the use of microfluidic devices integrating electrochemical sensors in an FIA system.

16.2 Flow injection analysis system setup

The FIA system setup for using the microfluidic electrochemical sensor and thin-film electrodes is shown in Fig. 16.2.

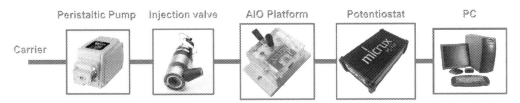

FIGURE 16.2 Flow injection analysis system setup for microfluidic sensors and thin-film electrodes. *AIO*, all-in-one; *PC*, personal computer.

Microfluidic electrochemical sensors (Fig. 16.3A) consist of a three-electrode system fabricated on gold (150 nm) or platinum (150 nm) deposited on a glass substrate (10 × 6 mm). The working electrode has a geometrical area of 0.3 mm^2. An SU-8 resin layer is used for building the microfluidic channel on the glass substrate with the electrodes. The channel is 40 μm height with a width of 250 μm and 1 mm for the electrochemical cell. The single channel

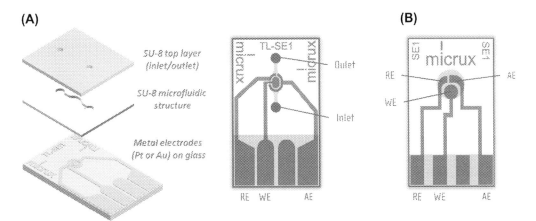

FIGURE 16.3 (A) Microfluidic electrochemical sensor and (B) thin-film single-electrode layouts. *AE*, auxiliary electrode; *RE*, reference electrode; *WE*, working electrode.

(A)   **(B)**

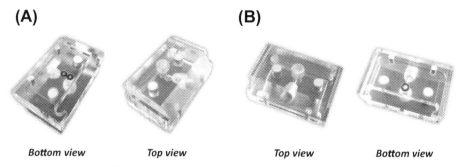

Bottom view   *Top view*   *Top view*   *Bottom view*

FIGURE 16.4   (A) Thin-layer and (B) wall-jet flow cell add-ons.

has a length of 3 mm. Finally, an SU-8 film containing the inlet/outlet is used as cover for closing the microfluidic channel. The basic thin-film single electrodes (Fig. 16.3B) also consist of a three-electrode system fabricated on gold (150 nm) or platinum (150 nm) deposited on a glass substrate (10 × 6 mm). In this case, an SU-8 layer is simply used for delimiting the electrochemical cell (2 mm diameter). The working electrode has a diameter of 1 mm with a geometrical area of 0.8 mm^2.

Two different flow cell add-ons are used with the all-in-one (AIO) platform depending on the sensor. For the microfluidic sensors, the flow cell add-on (Fig. 16.4A) consists of two independent inlet and outlet isolated with two O-rings enabling a thin-layer approach. For the thin-film single electrodes, the flow cell add-on (Fig. 16.4B) consists of two inlets/outlets embedded in the same O-ring enabling a wall-jet approach. Both approaches integrate standard fluidic ports ($^1/_4$ "-28 UNF) with inlet channel of 0.5 mm ID.

16.3 Chemicals and supplies

Reagents:

— *Analytes:* APAP and *p*AP.
— *Preparation of electrolytes:* Sulfuric acid (H_2SO_4), phosphoric acid (H_3PO_4), hydrochloric acid (HCl), and sodium hydroxide (NaOH).

Solutions:

— *Background electrolyte:* The carrier used in the FIA system is 0.1 M H_2SO_4 and 0.1 M phosphate buffer (PB) pH 7.4. Prepare 1000 mL of each solution.
— *Stock and standard solutions:* 10 mM stock solutions of APAP and *p*AP are prepared in 10 mM HCl. Prepare 100 mL of each stock solution.

For the FIA system, standard solutions of APAP and *p*AP (1, 10, 50, 100, 250, and 500 μM) are prepared by diluting the stock solutions in 0.1 M H_2SO_4 or 0.1 M PB pH 7.4. Prepare 25 mL of each standard solution.

Instrumentation and materials:

— Potentiostat
— AIO platform (drop-cell and thin-layer/wall-jet flow cell).
— Thin-layer microfluidic single electrodes (TL-SE1-Pt/Au).
— Thin-film single electrodes (ED-SE1-Pt/Au).

Additional instrumentation for the FIA system:

— Peristaltic pump or syringe pump.
— Six-port injection valve (20 μL sample loop).
— Tubing and fittings.
— Volumetric material (flasks, pipettes, vessels, micropipettes, etc.) and all the materials necessary for the preparation of solutions, which should be of analytical reagent grade.
— Syringes and syringe filters (0.1–0.45 μm), used for removing particles of working solutions and samples.

16.4 Hazards

Acid and alkaline solution preparations should be carried out under a fume hood. Protective garment and gloves should be worn at all times.

16.5 Experimental procedure

16.5.1 Electrochemical procedures

16.5.1.1 Electrode precleaning

The metal surface of thin-film electrodes integrated into microfluidic chips should be cleaned to get the best electrochemical signals. The electrode surface is cleaned by a simple electrochemical pretreatment. With this aim:

1. Using the AIO cell with thin-layer (for TL-SE1 sensors) or wall-jet (for ED-SE1 sensors) flow cell add-on, drive the carrier throughout the FIA system (be sure the solution covers all the electrode surfaces in the cell without any bubble).
2. Stop the flow and perform a cyclic voltammetry experiment between −1.5 and +1.5 V with a scan rate of 0.1 V/s (at least 10 cycles) for platinum-based electrodes and between −1.0 and +1.0 V with a scan rate of 0.1 V/s (at least 12 cycles) for gold-based electrodes.

16.5.1.2 Amperometric measurements

A flow cell (thin-layer or wall-jet) is used in the FIA system for obtaining the amperometric measurements in dynamic conditions. To get the signals:

1. Connect the flow cell to the FIA system and to the potentiostat.
2. Perform the electrode precleaning, as commented in Section 16.5.1.

3. Drive the carrier continually throughout the system using the peristaltic pump (be sure no bubbles are formed in the tubing or flow cell).
4. Apply a constant detection potential to the electrode and record the baseline on the corresponding software window.
5. When the current of the baseline is stable, inject the standard solution (or sample) by using the six-port injection valve of the FIA system. This valve enables the injection of a sample volume in a very fast and reproducible way without disturbing the flow of carrier. In all the experiments, a sample volume of 20 μL is used. Record the signals on the software window.

16.5.2 Influence of the electrode material, carrier solution, and detection potential

The electrode material, carrier (composition, ionic strength, pH, etc.), and detection potential affect the analytical signals obtained in the FIA system.

The optimal detection potential is determined by performing the hydrodynamic voltammogram (HDV, see Chapters 5, 9 and 28). Using the microfluidic sensors in the FIA system with the thin-layer flow cell,

1. Study the influence of the detection potential for APAP by varying it between 0.0 and +1.0 V, for a 100-μM solution, with microfluidic gold- and platinum-based electrodes using different carrier solutions (0.1 M H_2SO_4 and 0.1 M PB pH 7.4).
2. Study the influence of the detection potential for pAP by varying it between 0.0 and +1.0 V, for a 100-μM solution, with microfluidic gold- and platinum-based electrodes using different carrier solutions (0.1M H_2SO_4 and 0.1 M PB pH 7.4).

For all the experiments, fix a flow rate of 1.0 mL/min. The flow rate in the system can also affect analytical signals (shape and height of the peaks). Then, a study of the flow rate can be also accomplished.

After these studies, the most appropriate detection potential, carrier, and electrode material is selected for the detection of each analyte.

16.5.3 Analytical parameters

Using the optimal detection potential, electrode material, and carrier:

1. Study the precision (repeatability) for successive injections (at least 15 measurements) of standard samples of 10 and 100 μM APAP solutions as well as 10 and 100 μM pAP solutions. Evaluate the intrachip (in the same chip) and interchip (in different chips) precision (Fig. 16.5).
2. Study the effect of the compound concentration on the analytical signals (calibration plot). Prepare solutions of different concentrations of APAP and pAP (values ranging from 1 μM to 1 mM) to be injected in the flow system.
   Perform the calibration plot, starting first by the lowest concentration to the highest. Then, repeat injections from the highest concentration to the lowest one. This study also allows checking the durability of the electrodes without spotting an electrode fouling effect (Fig. 16.6).

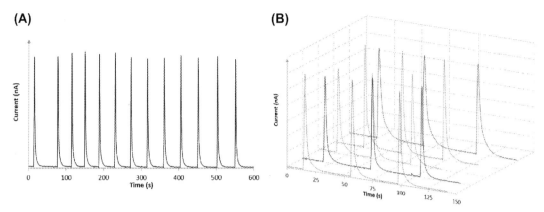

FIGURE 16.5   Fiagrams for successive injections in (A) the same and (B) different microfluidic chips.

16.5.4 Comparison of flow injection analysis systems: wall-jet versus microfluidic thin-layer flow cells

Microfluidic sensors enable the use of a thin-layer approach in an FIA system. Thus, they enable an excellent control of fluids through the electrode surface, decreasing the volume of sample and reagents as well as enhancing the sensitivity with very low dead volume.

To check the high performance of the thin-layer approach, a comparison study between the microfluidic sensors with the thin-layer approach and a thin-film single electrode with a wall-jet configuration is accomplished.

Perform successive injections of standard samples of $100\,\mu M$ APAP and $100\,\mu M$ pAP (separately) in the FIA system with the microfluidic thin-layer sensors using the optimal

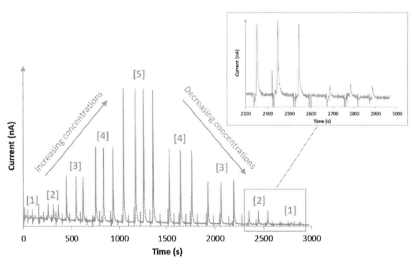

FIGURE 16.6   Amperometric response for increasing/decreasing concentrations of acetaminophen.

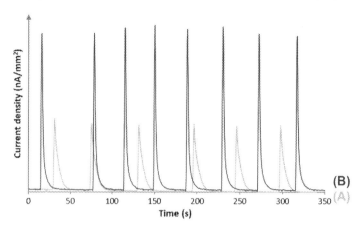

FIGURE 16.7 Comparison of analytical signals for (A) wall-jet and (B) microfluidic thin-layer approaches.

conditions. Repeat the same experiments but using the thin-film single electrodes with the wall-jet approach.

In this case, as the electrode area is different, for a better comparison, the analytical signals are normalized (current density, J) considering the area of the electrodes (Fig. 16.7).

16.5.5 Paracetamol determination

16.5.5.1 Sample preparation
1. Pulverize finely a tablet of paracetamol in a mortar.
2. Weigh accurately three samples. Powder can be directly weighed.
3. Solve each weighed sample directly in deionized water and transfer quantitatively to a 100-mL volumetric flask, making up with deionized water.
4. Filter the solutions using a syringe filter (0.1–0.45 μm).
5. Dilute the sample to the desired concentration with the background electrolyte. Prepare 25 mL of each diluted sample.
6. Inject directly into the FIA system.

16.5.5.2 Sample measurement
The analysis of the paracetamol samples is performed using the microfluidic thin-layer sensors under optimal conditions.

APAP determination in drug samples is carried out using two different methodologies:

1. *Calibration with standard solutions*: The calibration plot accomplished in Section 16.7 is used for the determination of the paracetamol concentration in the sample. Thus, the sample (paracetamol tablet/powder dissolved in 100 mL of deionized water) diluted in the carrier is directly injected into the FIA system using the six-port valve.

   Perform at least three replicates of the sample. The mean current intensity of the peaks is used to calculate the paracetamol using the calibration curve.

2. *Standard addition method*: It is used in instrumental analysis to determine the concentration of a substance (analyte) in an unknown sample by comparison to a set of standards of known concentration included in the sample matrix. Standard addition method can be applied to most analytical techniques and is used instead of a calibration curve to solve matrix effect problems and get a better precision.

Prepare four drug samples (paracetamol tablet/powder dissolved in 100 mL of deionized water) in separated flasks. One of the samples is directly diluted with the carrier to the desired concentration (as in the previous methodology). In the other samples, increased volumes of an APAP standard solution (solution of known concentration of analyte) are added. For obtaining the standard addition curve, the intensity of current is plotted against the concentration of standard added. The intercept in the *x*-axis corresponds to the concentration of APAP in the sample.

16.6 Lab report

At the end of the experiment, write a lab report including an introduction, experimental (materials, equipment, and protocols), results and discussion, and conclusions sections. The following points should be considered in the report:

1. Sketch the HDVs for APAP and *p*AP. Determine the optimal potential for each compound.
2. Determine the precision (% RSD, relative standard deviation) of the peak current for the compounds in the same (intra) and different (inter) chips.
3. Draft the calibration plots and determine the main analytical parameters (linear range, sensitivity, limit of detection, and limit of quantification) for each compound under optimal conditions.
4. Include the fiagrams for APAP and *p*AP using the wall-jet and microfluidic thin-layer approaches. For a better comparison, the peak current must be normalized taking into account the area of the working electrode.
5. Sketch the calibration curve of paracetamol obtained by the standard addition method considering the APAP additions. Determine the APAP concentration in the drug samples. Compare the results with those obtained by using the external calibration curve (using standard solutions of APAP).

16.7 Additional notes

1. Stock solutions should be kept on dark and stored at 4°C.
2. All working solutions/samples have to be filtered using a syringe filter of 0.1—0.45 μm to remove small particles, which can block the tubing/flow cell.
3. Be sure the flow cell add-on is correctly placed on the AIO platform to avoid the leakage of fluids.
4. Be sure there are no bubbles in the system (tubing and flow cell), especially on the electrode surface. Bubbles affect the accuracy, precision, and performance of the system.

5. The electrode surface should be cleaned before starting first experiments. In this pretreatment the different particles adhered to the electrode surface are removed with hydrogen and oxygen gases generated. After the pretreatment, the electrodes could be used for several measurements depending on the sample. Thin-film electrodes could be reused after a new precleaning process.
6. The flow rate employed affect the analytical signals (shape and height of the peaks). A study of the flow rate can be accomplished. The peristaltic pump or syringe pump should be calibrated previously to the experiments performed in the FIA system.
7. In all the experiments, typically, three consecutive injections are performed to improve the precision and obtain better results.
8. Generally, pharmaceuticals state the paracetamol amount in the leaflet. It can be used to calculate the method recovery and also to select the most appropriate dilution of the sample (different dilutions should be performed depending of the pharmaceuticals). Be sure, after the dilution, the APAP concentration is in the lineal range (with and without the additions).

16.8 Assessment and discussion questions

1. What are the electrochemical processes involved for each compound?
2. Draw the wall-jet and thin-layer flow cell configurations.
3. What is the optimal detection potential for each compound? How is it chosen?
4. Which flow cell approach shows a higher throughput for the determination of paracetamol? Why?
5. In real sample analysis, which methodology is more precise, the calibration curve with standard solutions or the standard addition method?

References

[1] J. West, M. Becker, S. Tombrink, A. Manz, Micro total analysis systems: latest achievements, Anal. Chem. 80 (2008) 4403–4419.
[2] Lab-on-a-Chip Technology, in: K.E. Herold, A. Rasooly (Eds.), Fabrication and Microfluidics, Caister Academic Press, 2009.
[3] K.E. Herold, A. Rasooly (Eds.), Lab-on-a-Chip Technology: Biomolecular Separation and Analysis, Caister Academic Press, 2009.
[4] M. Trojanowicz, Flow Injection Analysis: Instrumentation and Application, World Scientific, 2000.
[5] The European Pharmacopeial Convention, the Sixth Edition European Pharmacopoeia, 2007, p. 0049.
[6] The United States Pharmacopeial Convention, USP (the United States pharmacopoeia) 27-NF (The National Formulary) 22, vol. 27, 2004, p. 2494.
[7] Editor Committee of National Pharmacopocia, Chinese Encyclopedia of Medicines, vol. 2, Chemical Industry Press, Beijing, 2000, p. 206.

PART III

Bioelectroanalysis

Bienzymatic amperometric glucose biosensor

*María Carmen Blanco-López, M. Jesús Lobo-Castañón,
M. Teresa Fernández Abedul*

Departamento de Química Física y Analítica, Universidad de Oviedo, Oviedo, Spain

17.1 Background

The development of electrochemical biosensors represents a mature field in Electrochemistry, but it is also an important area of research. This chapter starts a series of six experiments that present various classes of electrochemical biosensors organized according to the nature of their biological component, balancing classical designs and contemporary trends. A biosensor is defined as an analytical device that contains two basic functional units: a biochemical or biomimetic recognition element, which mainly provides selectivity to the device, integrated or in close proximity with a physicochemical transducer that transforms the chemical energy involved in the recognition reaction into an electrical signal. Electrochemical transducers are the most widely used in biosensors design.

Biosensors exploit the capability of natural biomolecules such as enzymes, antibodies, or nucleic acids to selectively recognize a desired target or analyte. In this way, biosensors allow the quantification of biological or environmentally important species after direct contact with the sample, ideally without the need for any sample pretreatment and in a continuous and reversible way. Among the biocomponents, the biological catalysts (enzymes) are especially attractive as they allow a true reversible response. They catalyze the conversion of the analyte into a product, without being altered themselves. They were historically the first type of receptors used in the construction of biosensors.

Enzyme-based amperometric biosensors are constructed by immobilizing enzyme(s) on a highly conductive material. In this type of biosensors, the analytical signal is obtained by monitoring the faradaic current produced at a constant potential as a result of the exchange of electrons between the biological system and the transducer. Therefore, oxidoreductases are

173

of particular interest in the construction of this type of biosensors because in the enzymatic conversion of the substrate, an electron transfer event takes place.

Oxidases are one of the broadest groups of oxidoreductases (see also Chapter 19). These enzymes depend on a cofactor with redox activity, strongly bound to the catalytic site, commonly flavin adenine dinucleotide. The general reaction for an oxidase-catalyzed reaction is

$$S + O_2 + 2\,H^+ \xrightarrow{\text{Oxidase}} P + H_2O_2$$

where S and P are the substrate and the product of the enzymatic reaction, respectively. The substrate is oxidized by the enzyme, with oxygen as the natural electron acceptor from the reduced form of most oxidases. It allows the regeneration of the oxidized enzyme, producing hydrogen peroxide in this process. Thus, the enzymatic transformation of the substrate may be monitored by detecting the biocatalytically produced H_2O_2. This can be carried out by a number of methods. With that aim, the enzyme peroxidase can be immobilized with an oxidase in the sensing layer for constructing a bienzyme biosensor able to detect the corresponding substrate of the oxidase enzyme. This is a general design for the detection of different compounds that act as substrate of oxidases such as glucose, lactate, or ethanol. In this work, we describe the fabrication of a glucose bienzymatic sensor based on the coimmobilization of glucose oxidase (GOx) and horseradish peroxidase (HRP) by simple dispersion into a carbon paste material, covering the resulting sensor by an anion exchange membrane to increase its stability [1].

Peroxidase is a heme-containing enzyme that uses hydrogen peroxide to oxidize a wide variety of organic and inorganic compounds. These molecules are referred to as "electron transfer mediators" (M), which shuttle electrons between the redox center of the enzyme and the electrode, at relatively low overpotentials. In this way, H_2O_2 detection can be performed at a potential determined by the formal potential of the mediator according to the following mechanism:

$$H_2O_2 + HRP_{red} \rightarrow H_2O + HRP_{ox}$$

$$HRP_{ox} + M_{red} \rightarrow HRP_{red} + M_{ox}$$

$$M_{ox} + e^- \rightarrow M_{red}$$

In this particular case, ferrocene (Fc), an organometallic sandwich compound consisting of two cyclopentadienyl rings bound symmetrically to an iron atom, is used as mediator for HRP [2]. Ferrocene is water-insoluble and undergoes reversible oxidation to ferricinium cation (Fc^+) at a potential around +0.25 V (vs. Ag/AgCl). Therefore, by incorporating GOx, HRP, and Fc to the carbon paste, amperometric detection of glucose can be performed at this potential with high sensitivity and selectivity. Fig. 17.1 illustrates the set of reactions taking place to obtain a reduction current, which is related to the concentration of glucose.

To improve the stability of the sensor while hindering the access of potential interferences, the working electrode is coated with a Nafion layer. A thin film of this cation-exchange

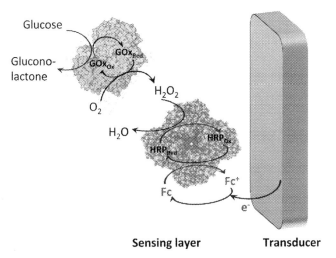

Sensing layer   **Transducer**

FIGURE 17.1 Schematic illustration of the reactions taking place at the biosensor constructed in this experiment.

polymer (Fig. 17.2) is formed on the sensor surface by cast-coating with an isopropanol–water solution of the polysulfonated polymer. The pores of this polymer do not allow the enzyme in the carbon paste to diffuse out of the sensor and, however, it is permeable to small neutral species such as glucose. The negatively charged sulfonic groups in the film electrostatically exclude anions, such as ascorbic acid, from the electrode surface, thus improving the selectivity of the device.

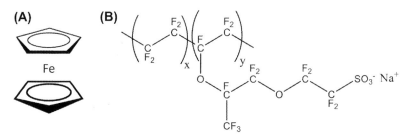

FIGURE 17.2 Chemical structure of (A) the mediator (ferrocene, Fc) and (B) the polymer (Nafion) used in the construction of the sensor.

Glucose is a very important analyte, mainly due to the incidence of diabetes mellitus. Its enzymatic determination is also considered in Chapters 25, 27 and 28, using in those cases paper-based or pin-based carbon ink electrodes as examples of low-cost electroanalysis, with potassium ferrocyanide as electron transfer mediator.

This experiment, adapted from Ref. [1], is intended to be performed by advanced undergraduate Chemistry and Biotechnology students in three sessions of 3 h. It will introduce them to electrochemical sensor technology, providing a view of the usefulness of cyclic voltammetry (CV, see also Chapter 2) for the characterization of the sensor performance and gaining insight into the different steps required for the construction, evaluation, and validation of electrochemical sensors.

17.2 Electrochemical setup

A potentiostat combined with a magnetic stirrer and a conventional three-electrode potentiostatic system is employed. Electrodes used are the following:

- Working electrode: carbon paste electrode with or without modifiers.
- Auxiliary electrode: platinum wire.
- Reference electrode: Ag/AgCl/KCl (3M).

For the working electrode, different carbon pastes are prepared:

- *Paste A*: Mix thoroughly graphite powder (1 g) and paraffin oil (0.360 μL) in a mortar.
- *Paste B*: Weigh 300 mg of the paste A and add ferrocene (Fc) to have a final Fc concentration of 3% (w/w).
- *Paste C*: Weigh 300 mg of paste A and mix with ferrocene (3% (w/w)), GOx, and HRP (5% (w/w)) for each enzyme). These enzyme amounts were optimized for typical enzyme preparations with the following specific activities: 15,500 U/g for GOx and 1000 U/g for HRP.

Once the carbon paste (A, B, or C) is homogenized in the mortar, the necessary amount is introduced into a Teflon cylinder having an electrical contact, polishing the surface in a circular fashion on a sheet of paper. The rest is stored in dark glass bottles at room temperature.

17.3 Chemicals and supplies

Reagents:

- *Analyte*: Glucose standard (0.1 M solutions are prepared in the background electrolyte).
- *Enzymes*: GOx and HRP.
- *Background electrolyte*: 0.1 M phosphate buffer (PB) pH 7.5 (it can be prepared from phosphoric acid adjusting the pH with sodium hydroxide).
- *Electrode preparation*: Graphite powder, paraffin oil, ferrocene (Fc), and Nafion (5% solution; it is diluted to a final concentration of 0.5% (v/v) with a 1:1 mixture of 2-propanol and water).
- Milli-Q water is employed for washing and preparing solutions.

Samples:
Glucose was determined in fruit juices, honey, and cola beverages.

Materials and equipment:

- *CV and amperometry*: 20-mL glass electrochemical cell, carbon paste electrode (Teflon cylindrical holder, carbon paste, and stainless steel rod), Ag/AgCl/KCl (3M) reference electrode, platinum wire, cell stand, potentiostat, and recorder/computer.
- *General materials and apparatus*: Volumetric flasks, 100 and 1000 μL micropipettes and corresponding tips, magnetic bars and stirrer, analytical balance, and pH meter.

17.4 Hazards

Although the quantities employed are very small, ferrocene is harmful if swallowed. Protective gloves and glasses are always highly recommended, especially when handling concentrated acids or bases.

17.5 Experimental procedure

17.5.1 Electrochemical study of ferrocene behavior

1. Prepare the working electrode by filling the hole of the Teflon holder (approximately 2 mm deep) with the carbon paste B.
2. After compacting the paste and polishing the surface circularly on a clean piece of paper, place it together with the RE and the AE in the cell stand containing 20 mL of 0.1 M PB solution pH 7.5.
3. Incorporate a magnet bar in the cell to stir gently in between scans.
4. Record a cyclic voltammogram between −0.2 and +0.8 V at a scan rate of 50 mV/s in 0.1 M PB solution (pH 7.5) and confirm the reversible behavior of the electron transfer mediator (see Chapter 2).
5. Stir the solution for a few seconds and repeat the scan. The absence of reduction peak in this case points out the need for a membrane at the electrode.
6. Prepare a membrane for the working electrode by drop casting 20 μL of a 0.5% Nafion solution. Let it dry.
7. Place the working electrode together with the RE and the AE in the cell stand containing 20 mL of 0.1 M PB solution pH 7.5.
8. Let the membrane hydrating in the background electrolyte for 10 min.
9. Record a new CV to confirm the efficiency of the membrane (in Fig. 17.3, examples of CVs recorded with and without membrane are shown).
10. Determine the optimum potential for amperometric determination from these voltammograms.

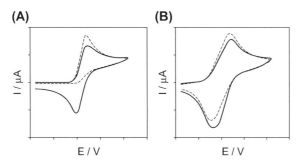

FIGURE 17.3 Example of cyclic voltammograms recorded with a carbon paste electrode modified with the mediator ferrocene: (A) without and (B) with a Nafion membrane (*continuous line*: stationary solution, *dotted line*: after stirring the solution).

17.5.2 Construction of the amperometric glucose sensor. Evaluation of its analytical performance

1. Prepare the working electrode by filling the hole of the Teflon holder (approximately 2 mm deep) with the carbon paste C.
2. After compacting the carbon paste and polishing the surface circularly on a clean piece of paper, coat the working electrode surface with a Nafion membrane and hydrate it as indicated in the steps 5−7 of Section 17.5.1.
3. Incorporate a magnet in the cell and start a gentle stirring. Keep it during the entire experiment.
4. Apply the potential value selected in step 10 of Section 17.5.1 and start the amperometric measurement under hydrodynamic conditions.
5. When a stable baseline is obtained, add the suitable volume of a 0.1 M glucose stock solution to get a final concentration of 5×10^{-5} M in the cell.
6. Once the current intensity is stabilized, i.e., a plateau is reached, proceed with a new addition of glucose, covering a range of concentrations up to 1×10^{-3} M.
7. Construct the I−C (intensity of current vs. glucose concentration) graph, considering the cumulative current for each concentration and correcting the concentration with the volume of glucose solution added in each step.
8. Calculate the value of the apparent K_M and I_{max} kinetic constants for the enzyme reaction, from the Lineweaver−Burk representation (1/I vs. 1/[glucose]).
9. Determine the analytical performance of the electrochemical sensor: linear response range, limit of detection, and response time.
10. Evaluate the reproducibility by repeating the calibration curve with different sensors.

17.5.3 Determination of glucose in real food samples

1. Prepare an enzymatic glucose sensor as commented in Section 17.5.2 and follow steps 1−4.
2. Apply the potential value selected for the amperometric measurement of glucose under hydrodynamic conditions.
3. When a stable baseline is obtained, add the suitable volume of an appropriate dilution of the sample (e.g., fruit juice, honey, or cola beverage) to get a final concentration in the cell located in the middle of the calibration curve.
4. Once the current intensity is stabilized, proceed with additions of glucose standard, taking always into account that the current is included in the linear part of the calibration curve (see Fig. 17.4A).
5. Plot the standard addition curve (current intensity vs. added glucose concentration, Fig. 17.4B) and determine the concentration in the sample.
6. Give the final value as the average of at least three replicates. Use the standard deviation to ascertain the number of significant figures of the result.

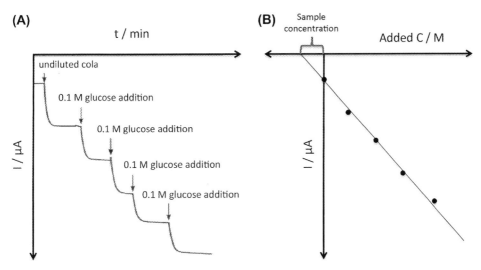

FIGURE 17.4 (A) Example of the amperometric signal recorded with the enzymatic sensor at a fixed potential for sample determination. The first step corresponds to the sample and the rest to additions of glucose standard. (B) Example of a standard addition plot.

17.6 Lab report

Write a lab report following the typical scheme of a scientific article, including a brief introduction, experimental part (materials, equipment, and protocols), results and discussion, and conclusions. The following points should be borne in mind:

1. In the introduction, explain the importance of glucose as analyte not only in the food field but also in clinical samples (e.g. Ref. [3]). Revise the bibliography with the main methodologies. Because of the enormous amount of scientific literature for glucose determination, this could be restricted to a field (e.g., food samples), principle of detection (e.g., electrochemical), biosensor (GOx/HRP-based sensors), electron transfer mediators (ferrocene and derivatives, ferrocyanide), etc. Alternatively, students could choose between different topics, always related to the basis of the determination considered in this experiment.
2. Commercial electrochemical enzymatic glucometers are available from many years. Then, it is also interesting to revise not only the scientific bibliography but also commercial products. In this case, properties such as cost, portability, autonomy, and disposability are of relevance.
3. Protocols must be suitably detailed including schemes where appropriate and the necessary calculations.
4. Include figures with representative raw data (cyclic voltammograms for ferrocene and amperograms for glucose determination) as well as results presented in tables and

graphs (calibration or standard addition curves) paying special attention to the significant figures in each case.

5. Discuss the values obtained considering the expected results and the incidences during the course of the experiment.

17.7 Additional notes

1. The carbon paste is packed in a piston-type Teflon electrode. Alternatively, a polyethylene syringe can be used to prepare the sensor.
2. The response time of the sensor can be considered as the time required to reach 95% of the steady current.
3. The interelectrode precision can be evaluated with the relative standard deviation (RSD) of the intensities of the currents obtained for a glucose concentration or, better and alternatively, with the RSD of the slopes of the calibration curves performed with different sensors.
4. All the students of a group of e.g., 3, can prepare a sensor at the same time, and later sensors can be tested with a calibration curve to compare their performance and precision.
5. Depending on the time available, calibration curves could be repeated in the presence of interferences such as ascorbic acid, glutathione, uric acid, galactose, or mannose to evaluate the selectivity.
6. Enzymes are biological reagents and then the variability in the source and activity is high. Therefore, their concentration can be adjusted after appropriate optimization.
7. As low current intensities are expected to discriminate the faradaic current from the noise, an RC filter could be employed.
8. Biosensors can be applied to the direct determination of glucose in several liquid samples without any preparation step apart from the dilution. These should be made with buffer solution considering the glucose value stated in the label.
9. Glucose could be also determined in clinical samples (e.g., serum), but in this case approval from an ethical committee to warranty adequate sample management is required/encouraged.
10. The methodology could be validated using the Trinder method [4], which is a kinetic spectrophotometric method that uses the same enzymatic system (GOx/HRP).
11. It would be very interesting to discuss about the history of electrochemical glucose biosensors, from the first generation to more recent approaches such as noninvasive or subcutaneous glucose monitoring [5].

17.8 Assessment and discussion questions

1. Explain the basis of the enzymatic sensor for glucose determination.
2. Is it necessary to couple a second reaction to the reaction catalyzed by GOx for determining glucose?

3. What are the advantages of using a mediator of the electron transfer?
4. Why is it stated that catalysts regenerate?
5. How is the detection potential chosen?
6. Indicate how the enzymatic kinetic constants can be obtained.

References

[1] M.C. Blanco-López, M.J. Lobo-Castañón, A.J. Miranda-Ordieres, Homemade bienzymatic amperometric biosensor for beverages analysis, J. Chem. Educ. 84 (2007) 677–680.

[2] P. Domínguez-Sánchez, A.J. Miranda-Ordieres, A. Costa-García, P. Tuñón-Blanco, Peroxidase-ferrocene modified carbon paste electrode as an amperomeric sensor for the hydrogen peroxide assay, Electroanalyisis 3 (1991) 281–285.

[3] G. Reach, G.S. Wilson, Can continuous glucose monitoring be used for the treatment of diabetes? Anal. Chem. 64 (1992) 381A.

[4] D. Barham, P. Trinder, An improved colour reagent for the determination of blood glucose by the oxidase system, Analyst 97 (1972) 142–145.

[5] J. Wang, Electrochemical glucose biosensors, Chem. Rev. 108 (2008) 814–825.

CHAPTER

18

Determination of ethyl alcohol in beverages using an electrochemical enzymatic sensor

Rebeca Miranda-Castro, M. Jesús Lobo-Castañón

Departamento de Química Física y Analítica, Universidad de Oviedo, Oviedo, Spain

18.1 Background

The determination of ethanol in beverages is carried out routinely for the purpose of controlling compliance with the tax regulations that establish the tax regime of these beverages based on their ethyl alcohol content. A second purpose of this determination is to ensure compliance with the food laws that establish maximum alcohol content based on the commercial classification of beverages. For example, in Spain, a "nonalcohol" beer may not contain more than 1% (v/v) ethanol, while a "0.0" beer must contain no more than 0.1% [1]. These standards pose a significant analytical problem because of the high volume of samples to be examined in the laboratories responsible for these controls and the sensitivity needed to accurately examine samples of low alcohol content, such as "0.0" beers. Enzymatic sensors offer an interesting alternative to solve this problem, because of the great saving of high-priced reagents, which also allow simplification and time saving in measuring procedures. In this chapter, we describe the construction and evaluation of an amperometric ethanol sensor and its application to the analytical determination of ethanol in beer.

Dehydrogenase enzymes are a subclass of oxidoreductases, which together with oxidases are very useful for the construction of electrochemical enzymatic sensors. Many dehydrogenase enzymes use the nicotinamide adenine dinucleotide $(NAD(P)^+/NAD(P)H)$ redox system as coenzyme, catalyzing the following general reaction:

$$SH_2 + NAD^+ \underset{}{\overset{\text{Dehydrogenase}}{\rightleftarrows}} S + NADH + H^+$$

There is a great variety of dehydrogenase enzymes, catalyzing the transformation of important substrates such as lactate, glutamate, glucose, glycerol, etc., which can be used in the design of electrochemical biosensors [2]. Alcohol dehydrogenase (ADH), isolated from yeast, is a tetrameric metalloenzyme that contains zinc in the active site. It catalyzes the oxidation of ethanol (and to a less extent other short-chain aliphatic alcohols except methanol) by the oxidized coenzyme NAD^+, yielding β-nicotinamide adenine dinucleotide (NADH) and acetaldehyde according to the following reaction:

$$CH_3CH_2OH + NAD^+ \overset{ADH}{\rightleftarrows} CH_3CHO + NADH + H^+$$

The immobilization of ADH and NAD^+ on a conductive surface leads to a sensor for ethanol. To visualize the enzymatic reaction that occurs on the electrode surface when it interacts with an ethanol solution, amperometric transduction is employed. The signal is obtained by measuring the current generated in the oxidation of the NADH enzymatically generated at a constant applied potential. However, the direct electrochemical oxidation of NADH to NAD^+ poses two serious difficulties [2]:

(1) The electrode reaction is highly irreversible, that is, of very slow kinetics. This forces the sensor to operate at potentials greater than $+0.4$ V, a potential region where different species in the sample may be electroactive, causing possible interference.
(2) The direct oxidation of NADH does not exclusively produce NAD^+ but there are also secondary reactions that lead to the formation of dimers and other enzymatically inactive compounds. This causes the progressive and irreversible decrease of the amount of NAD^+ immobilized in the sensing phase, as well as the formation of deposits of these by-products on the electrode surface leading to limited sensor stability.

As explained in Chapter 7, a way to decrease the high overpotential required for NADH oxidation, and at the same time to obtain the quantitative regeneration of the oxidized coenzyme, NAD^+, is using a catalyst of the electrochemical oxidation of NADH (electrocatalyst or mediator) [3]. A catalyst of this type is a compound involved in a reversible redox couple; its oxidized form chemically reacts with NADH, oxidizing it to NAD^+, while the catalyst is reduced. The reduced form of the catalyst is then electrochemically oxidized at the potential applied to the working electrode. Thus, the overall reaction is the transfer of electrons from the NADH to the electrode surface at a substantially decreased potential. The catalyst, therefore, is an electron transfer mediator, which shuttles electrons from NADH to the electrode, reducing the overpotential needed for the direct NADH oxidation.

In our case, 2,8-dioxoadenosine, a product generated on the electrode by electrochemical oxidation of adenosine at high potentials ($+1.2$ V), is used as the catalyst [3]. The generation of the catalyst is described in Chapter 7. Fig. 18.1 shows the reversible redox reaction of the adsorbed catalyst and the sequence of reactions of the electrocatalytic process used for the detection of the reduced coenzyme obtained as product of the enzymatic dehydrogenation of ethanol.

With this catalytic system, the oxidation of NADH is achieved at the potential of the catalyst pair, which is much lower than the anodic peak potential of NADH on carbon electrodes and, in parallel, the process is free of secondary reactions that derive from the mechanism of the direct oxidation of NADH.

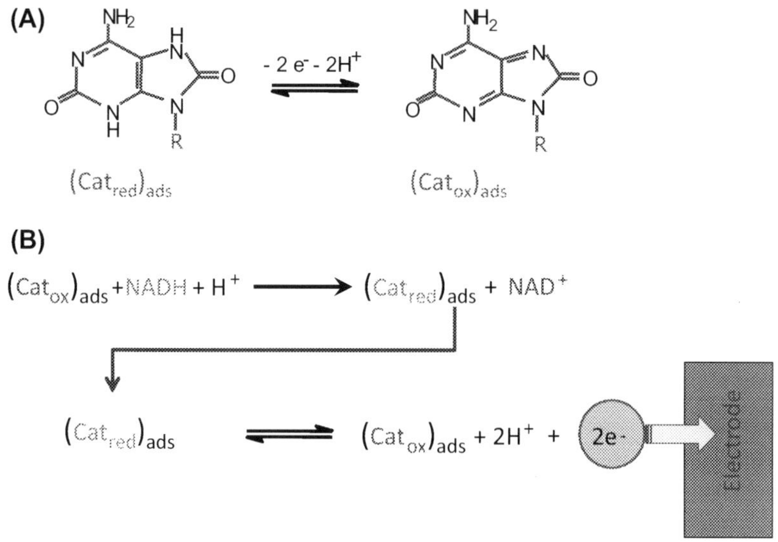

FIGURE 18.1 (A) Electrochemical process associated to 2,8-dioxoadenosine, electrocatalyst for NADH oxidation. The adenine moiety is the electroactive part of the nucleoside, and R represents the ribofuranose moiety. (B) The sequence of reactions involved in the amperometric detection of NADH enzymatically obtained.

The design of the sensing phase used is very simple. The bioreagents (ADH, adenosine, and the oxidized coenzyme NAD$^+$) are incorporated into a carbon paste electrode by simple dispersion in the paste. This type of electrode is formed by pyrolytic graphite powder agglomerated with silicone oil, which produces a highly conductive moldable paste. This paste is placed in a small cavity of the electrode provided with an electrical contact. Various reactive materials, such as enzymes, ion exchangers, etc., can be easily incorporated to the paste as modifiers, maintaining their chemical activity, with the possibility of reusing them after analytical measurement, which is not possible (or it is too expensive) when the same enzymatic reagents are used in solution. As enzymes and NAD$^+$ coenzyme are modifiers highly soluble in water, it is necessary to coat the electrode surface with a membrane that acts as a diffusional barrier, preventing the loss of the bioreagents of the sensing phase by solubilization in the sample solution. The membrane to be used is formed on the surface of the electrode by electrochemical polymerization of o-phenylenediamine. This aromatic amine is oxidized on carbon electrodes to a cation which initiates a radical polymerization process, resulting in a layer of entangled strands of a linear polymer, poly(o-phenylenediamine) (PPD), insoluble and strongly adhered to the surface of the electrode. This layer is permeable to small molecules, such as ethanol, but totally prevents the enzyme and coenzyme NAD$^+$ to pass through, which are large molecules [4]. The PPD membrane is generated by subjecting the electrode modified with the enzyme and NAD$^+$ to a cyclic potential sweep in a solution of the o-phenylenediamine monomer.

The calibration of the sensor consists of the representation of the current (I) obtained at a constant applied potential versus the ethanol concentration ([EtOH]) in the test solution. The corresponding graph is not linear but has the hyperbolic form characteristic of the enzymatic kinetics because the signal (current intensity) is directly proportional to the rate of enzymatic conversion of ethanol. Taking into account that for a given enzyme concentration if the concentration of NAD^+ is kept constant at a saturating level and the concentration of ethanol is varied, the initial rate of the enzymatic reaction will vary with the initial concentration of ethanol according to the Michaelis–Menten equation. The representation of I versus [EtOH] will give rise to an equilateral hyperbola of general equation:

$$I = \frac{I_{max} \, [\text{EtOH}]}{[\text{EtOH}] + K_M}$$

Two limiting situations can be considered:

(1) If [EtOH] $<<K_M$, under these conditions $K_M +$ [EtOH] $\approx K_M$ and, therefore, at low concentrations of ethanol, the enzymatic reaction will be pseudo first order with respect to this substrate.

(2) If [EtOH]$>>K_M$, then $K_M +$ [EtOH] $\approx$ [EtOH] and therefore, at high concentrations of ethanol (saturating concentrations), the enzymatic reaction is of order zero with respect to this substrate. Hence, K_M is an empirical constant that corresponds to the substrate concentration at which the initial reaction rate is half of the maximum possible speed under given experimental conditions. This constant gives us valuable information about the range of concentrations at which the enzyme responds to changes in substrate concentration (analytically useful range of concentrations). To calculate the kinetic constant K_M, a linearized form of the Michaelis–Menten equation, known as Lineweaver–Burk representation, can be used. According to it, a plot of 1/I versus 1/[EtOH] yields a straight line with slope K_M/I_{max} and intercept $1/I_{max}$.

This experiment can be included in an advanced Analytical Chemistry course for undergraduate students in Chemistry or Biotechnology. They will realize the wide potential of coupling enzymatic and electrochemical reactions by constructing and evaluating the analytical performance of the sensor. In addition, they may apply it to the analytical determination of ethanol in beer. They can also perform the study by cyclic voltammetry (see also Chapter 2) of the electrocatalytic process responsible for the sensor signal. This study should clarify the nature of the electrode reaction that allows visualizing the enzymatic dehydrogenation of the ethanol and, in addition, will provide the criterion of selection of the main parameter of sensor operation, which is the detection potential to be applied.

18.2 Electrochemical setup

A potentiostat combined with a magnetic stirrer and a conventional three-electrode potentiostatic system is employed. Electrodes used are the following:

— Auxiliary electrode: platinum wire.
— Reference electrode: Ag| AgCl| KCl(3M).
— Working electrode: carbon electrode with or without modifiers.

For the working electrode, different carbon pastes are prepared:

- *Unmodified carbon paste*: Mixture of 1 g of graphite and 0.323 g of silicone oil.
- *Carbon paste modified with adenosine*: The necessary amount of adenosine is added to the above carbon paste so that the mixture contains 5% of this nucleoside.
- *Carbon paste modified with adenosine, NAD^+, and ADH*: The required amount of these reagents is added to the unmodified carbon paste so that the mixture contains 5% adenosine, 10% NAD^+, and 8% ADH.

Once the carbon paste is homogenized in the mortar, the necessary amount is introduced into a Teflon cylinder having an electrical contact, polishing the surface in a circular fashion on a sheet of paper.

18.3 Chemicals and supplies

Materials and reagents:

- Graphite powder and silicone oil.
- 2,8-Dioxoadenosine, NADH, O-phenylenediamine (PPD), and ADH.
- Phosphoric acid and sodium hydroxide.
- Ethanol.
- Milli-Q water is employed for preparing solutions and washing.
- Micropipettes with corresponding tips.

Solutions:

- 0.1 M phosphate buffer pH 8.5.
- 0.1% (v/v) ethanol solution, prepared by dilution with the above buffer solution.
- 1 mL of 0.1 M NADH solution in the phosphate buffer pH 8.5.
- 1 mL of 0.1 M o-phenylenediamine in the phosphate buffer pH 8.5.

18.4 Hazards

Protective gloves and glasses are recommended, especially when handling o-phenylenedi-amine (a potential dermal hazard).

18.5 Experimental procedure

18.5.1 Electrocatalytic detection of NADH on carbon paste electrodes modified with 2,8-dioxoadenosine

The noncatalyzed direct oxidation process of NADH on unmodified carbon paste electrodes is firstly studied by cyclic voltammetry. For this purpose:

1. Dispose the electrochemical cell in its stand on the magnetic stirrer.
2. Add to the cell 20 mL of 0.1 M phosphate buffer solution pH 8.5.

3. Immerse the three electrodes in the previous solution and connect them to the potentiostat.
4. Record the cyclic voltammogram between -0.2 and $+1.0$ V at five different scan rates ranging from 10 to 100 mV/s (background voltammograms).
5. Add the suitable volume of 0.1 M NADH stock solution into the electrochemical cell to obtain a final NADH concentration of 5×10^{-4} M.
6. Repeat step 4 but, before each voltammogram, stir the solution for 30 s with the help of a magnet to regenerate the diffusion layer. Select the same scan rates as in step 4.
7. For each scan rate evaluated, measure the peak potential and the peak current corresponding to the NADH oxidation process after subtraction of the background voltammogram from that in the presence of NADH. Plot the NADH oxidation peak current versus the scan rate to determine the nature of the redox process (diffusion- or adsorption-controlled redox process).

Next, the same experiment is repeated on a carbon paste electrode modified with 2,8-dioxoadenosine, an NADH oxidation catalyst generated on the electrode by electrochemical oxidation of its precursor. Then,

8. Prepare the working electrode using carbon paste modified with 5% adenosine.
9. Add 20 mL of clean phosphate buffer pH 8.5 to the electrochemical cell, immerse the three electrodes, and connect them to the potentiostat.
10. Subject the working electrode to five potential cycles between -0.2 and $+1.2$ V at 50 mV/s to generate the catalyst 2,8-dioxoadenosine.
11. To check the formation of an oxidation product strongly adsorbed on the electrode surface (2,8-dioxoadenosine or $(Cat_{ox})_{ads}$ in Fig. 18.1), wash the working electrode surface, renew the phosphate buffer solution in the electrochemical cell, and record a cyclic voltammogram between -0.2 and $+0.7$ V at 50 mV/s. Measure the formal potential of the resulting reversible redox process and determine the amount of catalyst 2,8-dioxoadenosine generated in mol/cm^2, assuming a two-electron process.
12. Repeat the previous cyclic voltammogram varying the scan rate between 10 and 100 mV/s and plot the peak current versus the scan rate to confirm the adsorption nature of the reversible redox process.
13. Using the working electrode modified with 2,8-dioxoadenosine, repeat the voltammetric study of NADH oxidation (steps 5 and 6). Discuss and explain the observed differences between the two experiments.
14. Select the potential to be applied in the amperometric sensor.

Voltammetric behavior of NADH in the absence/presence of the electrocatalyst 2,8-dioxoadenosine is illustrated in Fig. 18.2.

18.5.2 Construction of an amperometric ethanol sensor. Evaluation of its analytical performance

The previously studied NADH oxidation catalyst, 2,8-dioxoadenosine, is used in the design of an amperometric ethanol sensor whose sensing layer is obtained as follows:

1. Fill the Teflon cylinder with carbon paste modified with 5% adenosine, 10% NAD$^+$, and 8% ADH and polish the surface on a piece of paper.

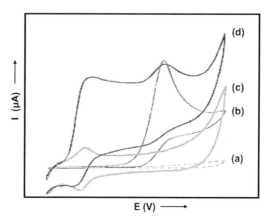

FIGURE 18.2 Cyclic voltammograms recorded onto carbon paste electrodes (scan rate: 50 mV/s) by using unmodified electrodes: (a), (- - -) background electrolyte; (b), (—) NADH; by using electrodes modified with 2,8-dioxoadenosine; (c), (--) background electrolyte; (d), (–) NADH.

2. Add to the electrochemical cell 20 mL of 0.1 M phosphate buffer solution pH 8.5 as well as the appropriate volume of 0.1 M o-phenylenediamine stock solution to get a final concentration of 5×10^{-4} M and stir the solution.
3. Perform five potential cycles between +0.7 and 0.0 V at 50 mV/s to electrogenerate a poly(o-phenylenediamine) layer with the aim of retaining the hydrophilic bioreagents within the carbon paste.
4. Wash thoroughly the electrode surface and the electrochemical cell with distilled water.
5. Pour fresh phosphate buffer solution pH 8.5 in the clean electrochemical cell and repeat step 10 (Section 18.5.1) to generate the NADH oxidation catalyst in the sensing layer.
6. Verify the generation of 2,8-dioxoadenosine by sweeping the potential between −0.2 and +0.7 V in clean phosphate buffer solution pH 8.5.

At this point, the amperometric sensor of ethanol is ready to use. The reaction scheme that occurs at the electrode surface when this sensor is brought into contact with an ethanol-containing solution is summarized in Fig. 18.3.

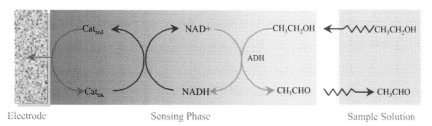

FIGURE 18.3 Reaction scheme for the determination of ethanol.

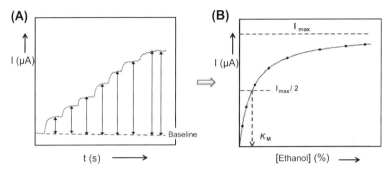

FIGURE 18.4 (A) Representative amperometric signal recorded with the enzymatic sensor at a fixed potential. Each step corresponds to a new addition of ethanol and the arrows indicate the catalytic current corresponding to each evaluated ethanol concentration. (B) Sensor calibration.

Firstly, the value of the Michaelis–Menten constant (K_M) for the immobilized enzyme is determined by recording the oxidation current obtained for increasing concentrations of ethanol in the range from 5×10^{-4} to 5×10^{-3}% (v/v):

7. Add 20 mL of fresh phosphate buffer solution pH 8.5 to the electrochemical cell, place the three electrodes, and connect them to the potentiostat.
8. Incorporate a magnet in the cell, start a gentle stirring, and keep it during the entire experiment.
9. Set the potential value selected in step 14 (Section 18.5.1) and start the amperometric measurement under hydrodynamic conditions.
10 As soon as the background current is stable, add the suitable volume of 0.1% (v/v) ethanol stock solution to get a final concentration of 5×10^{-4}% (v/v) in the cell.
11. Once the current intensity is stabilized, i.e., a plateau is reached, proceed with a new addition of ethanol until the stated ethanol concentration range is assessed (Fig. 18.4).
12. Plot the amperometric current recorded versus the ethanol concentration in the cell. This latter should be corrected by taking into account the volume of ethanol added into the cell in each case.
13. Calculate the value of K_M from Lineweaver–Burk representation.
14. Determine the analytical performance of the electrochemical sensor: limit of detection, linear response range, response time, and reproducibility.

18.5.3 Determination of ethanol content in beer

The ethanol content in a beer is determined by using the standard additions method. The sample is processed in triplicate:

1. Pour the sample in a beaker and degas its content by magnetic stirring.
2. Perform steps 1–4 (Section 18.5.2) to obtain a fresh ethanol sensor.
3. Repeat steps 7–9 (Section 18.5.2) to prepare the amperometric setup.
4. Considering the amount of ethanol stated in the beer label and the change in the oxidation current measured for the standard solutions of ethanol, add an appropriate volume

of beer to the 20-mL aliquot of phosphate buffer solution in the cell and record the amperometric signal.

5. Then, once a plateau current is achieved, make small additions of 0.1% (v/v) ethanol solution to the sample in the cell and record the changes in signal.
6. Calculate the ethanol content in the beer according to the standard additions method.

18.6 Lab report

Write a lab report following the typical scheme of a scientific article, including a brief introduction, experimental part (materials, equipment, and protocols), results and discussion, and conclusions. The following points should be beared in mind:

1. In the introduction, explain the purpose of the experiment and the importance thereof in food industry. Do a short review of the described methods to tackle this problem.
2. Protocols must be suitably detailed including schemes where appropriate and the necessary calculations.
3. Include figures with representative raw data (cyclic voltammograms and amperograms) as well as results presented in tables and graphs (calibration curves), paying special attention to the significant figures in each case.
4. Discuss the values obtained considering the expected results and the incidences during the course of the experiment.

18.7 Additional notes

1. NADH and o-phenylenediamine stock solutions should be freshly prepared and protected from light.
2. Thickness of PPD films obtained by electropolymerization of o-phenylenediamine on the carbon paste electrode is very sensitive to the electrolysis conditions. Control carefully the time that the electrode is poised at +0.7 V.
3. Although a Michaelis—Menten response is expected when plotting the amperometric signal against the ethanol concentration, a linear response could be also obtained if PPD film is thick, meaning that sensor response is controlled by diffusion of ethanol through the membrane instead of by the enzymatic reaction.
4. As current intensities are in the nA range, to discriminate the faradaic current from the noise, an RC filter could be employed.

18.8 Assessment and discussion questions

1. How can you determine the adsorption/diffusion nature of a redox process?
2. Explain the working conditions that allow an enzymatic bisubstrate reaction to be described by the Michaelis—Menten equation with respect to one of the substrates.
3. Explain the reaction scheme involved in the designed sensor as well as the function of the various sensor components.

References

[1] Real Decreto 678/2016, Boletín Oficial del Estado (BOE), 17 de diciembre, Spain, 2016. http://www.boe.es.

[2] N. de-los-Santos-Álvarez, P. de-los-Santos-Álvarez, M.J. Lobo-Castañón, A.J. Miranda-Ordieres, P. Tuñón-Blanco, in: C.A. Grimes, E.C. Dickey, M.V. Pishko. (Eds.), NADH-based Electrochemical Sensors, Encyclopedia of Sensors. American Scientific Publisher, vol. 6, 2006, pp. 349–378.

[3] M.I. Álvarez-González, S.B. Saidman, M.J. Lobo-Castañón, A.J. Miranda-Ordieres, P. Tuñón-Blanco, Electrocatalytic detection of NADH and glycerol by NAD^+-modified carbon electrodes, Anal. Chem. 72 (2000) 520–527.

[4] M.J. Lobo-Castañón, A.J. Miranda-Ordieres, P. Tuñón-Blanco, Amperometric detection of ethanol with poly-(o-phenylenediamine)-modified enzyme electrodes, Biosens. Bioelectron. 12 (1997) 511–520.

CHAPTER

19

Enzymatic determination of ethanol on screen-printed cobalt phthalocyanine/carbon electrodes

Julien Biscay[1], Estefanía Costa-Rama[2, 3]

[1]Department of Pure and Applied Chemistry, Technology and Innovation Centre, University of Strathclyde, Glasgow, United Kingdom; [2]REQUIMTE/LAQV, Instituto Superior de Engenharia do Porto, Instituto Politécnico do Porto, Porto, Portugal; [3]Departamento de Química Física y Analítica, Universidad de Oviedo, Oviedo, Spain

19.1 Background

In the age of miniaturization and simplification of analytical devices, screen printing has become a well-established technique to fabricate electrochemical sensors because it allows the manufacturing of electrodes with small size and good analytical features [1–3]. Briefly, screen-printed electrodes (SPEs) are fabricated depositing an ink layer onto a substrate employing a screen or mesh that controls the thickness of the film layer and the shape of the final electrode [4]. These electrodes are very versatile, not only in their design or material (of substrate and electrodes), but also in their facility to be modified (with mediators, proteins, polymers, nanomaterials, etc.) [4–7]. This great versatility, together with their low cost and possibility of mass production, made SPEs a very trendy device for cost-effective measurements in assorted fields such as environment [8], food [9], or clinical and pharmaceutical [4,10].

As mentioned, SPEs can be made of different conductor materials, e.g., gold, silver, platinum, etc., with carbon in different varieties (e.g., graphite, graphene, carbon nanotubes, etc.) being the most employed [4,10]. The substrate where the electrodes are screen-printed is also varied: ceramics are widely employed but alumina, glass, or rigid polymeric materials are also commonly used. Recently, flexible materials, such as paper or plastics, have been also successfully employed obtaining paper-based SPEs [11] or highly innovative glove-based electrodes [12]. Regarding design, nowadays commercial SPEs with different electrode configurations can be found; for example, SPEs with two or more working electrodes or with

Laboratory Methods in Dynamic Electroanalysis
https://doi.org/10.1016/B978-0-12-815932-3.00019-X

eight different electrochemical cells in the same card [13] (see Chapters 30 and 31). In respect of their modification, it can also be very diverse [4]. SPEs can be modified with different materials such as nanomaterials (metallic nanoparticles, carbon nanotubes, graphene, etc.), polymers or biomolecules. The SPEs can be modified, not only by the addition of substances on the printing inks or deposition of the substance onto their surface, but also by other procedures (especially for biomolecules) such as entrapment, crosslinking, or covalent attachment.

The determination of ethanol is required in many different fields because of its toxicological and psychological effects. For example, accurate quantification of ethanol in the clinical field is very important for the analysis of human body fluids such as blood or urine. In the case of food industry, simple and low-cost ethanol determination is necessary to monitor fermentation processes and control food quality/safety. However, in both cases, miniaturized, portable, and fast-response devices for its analysis are very helpful because they allow on-site determinations with a small amount of sample (especially important in clinical samples).

There are several methods and strategies for the centralized determination of ethanol such as gas or liquid chromatography or spectrophotometry. In contrast with these conventional methods, which require expensive and complex equipment and qualified operators to handle them, electrochemical biosensors arise as advantageous tools because they are inexpensive and easy to use. Moreover, electrochemical detection owns a particular interest for developing simple and portable analytical devices because of its low cost, ease of miniaturization, low sample requirement, and high accuracy [14].

In this chapter, the construction of an amperometric enzymatic sensor, based on SPEs, for determination of ethanol and its application in alcoholic beverages are described. Enzymes are well-acclaimed biorecognition elements because of their availability and high selectivity. The enzymes alcohol dehydrogenase (see Chapter 18) and alcohol oxidase (AOx) are among the most employed for developing alcohol biosensors [15,16]. Oxidase enzymes catalyze the oxidation of the substrate employing molecular oxygen as electron acceptor. Although there are oxidases that do not contain or require any cofactor, the most common ones have a cofactor, necessary for the electron transfer, strongly bound inside their structure. This cofactor can be flavin-based (the flavin adenine dinucleotide [FAD] or the flavin mononucleotide [FMN]) or based on a metallic ion. AOx is an oligomeric enzyme consisting of eight identical subunits, each one with a strongly bound FAD cofactor. This enzyme allows the irreversible catalytic conversion of low molecular weight alcohols to the corresponding aldehyde, employing the oxygen as electron acceptor and producing hydrogen peroxide [15]. Therefore, the oxidation of alcohol by AOx can be monitored by measuring either the decrease of oxygen or the increase in H_2O_2 concentration. The direct detection of H_2O_2 requires significantly high potentials; this may cause interferences in the current intensity because other species present in the sample can suffer redox process at those potentials. This problem can be substantially overcome employing an electron transfer mediator with lower redox potential. A mediator is a low molecular weight redox couple that participates in the redox reaction with the biological component facilitating the electron transfer between the enzyme and the electrode surface [17]. Phthalocyanines are aromatic macrocyclic compounds with an intense blue-green color widely used in dyeing [18]. They show the capability of coordinating metallic cations in the center of their structure forming metallic complexes as the Co-phthalocyanine (Fig. 19.1A). These phthalocyanine metal complexes have been widely used in catalysis [19].

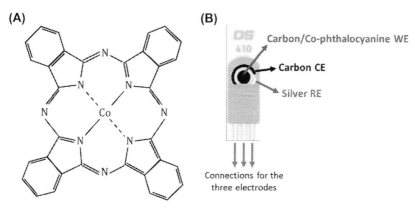

FIGURE 19.1 (A) Structure of Co-phthalocyanine. (B) Picture of a screen-printed Co-phthalocyanine/carbon electrode. *CE*, counter electrode; *RE*, reference electrode; *WE*, working electrode.

In this work, adapted from Ref. [20], a sensor for ethanol determination is developed using the enzyme alcohol oxidase (AOx) as biorecognition element and Co-phthalocyanine as redox mediator. In this case, the mediator is already included in the commercial SPE since the working electrode is made of carbon ink containing Co-phthalocyanine (Fig. 19.1B). The enzyme AOx is simply immobilized onto the surface of the working electrode by adsorption. Thus, AOx catalyzes the oxidation of ethanol in the presence of oxygen, producing H_2O_2. Then, H_2O_2 is oxidized while the mediator is reduced. Hence, the concentration of reduced mediator, which is determined by chronoamperometry (see Chapter 8), measuring the intensity of the current at a fixed time and at a potential where it is oxidized, is proportional to the initial concentration of ethanol (Fig. 19.2A). Therefore, the ethanol detection is performed through the measurement of the current intensity due to the oxidation of the mediator previously reduced by the enzymatic reaction. This intensity of current is the analytical signal (Fig. 19.2B).

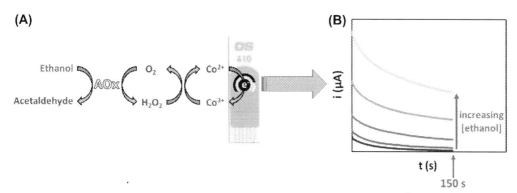

FIGURE 19.2 (A) Scheme of the enzymatic reactions and the electrochemical oxidation of Co-phthalocyanine enzymatically reduced. (B) Schematic representation of chronoamperograms obtained for different ethanol concentrations by applying +0.4 V for 150 s.

Along this experiment, students will optimize the biosensor, perform a calibration plot, and finally, to test its usefulness, determine the ethanol concentration in alcoholic beverages. Thus, they will not only learn about the development of enzymatic sensors and their application for real samples but also become familiar with SPEs that are an inexpensive and miniaturized analytical devices widely employed nowadays. This work can be completed in two laboratory sessions of 3-4 h, and it is appropriate for undergraduate students of advanced Analytical Chemistry or Master students from different fields where analysis is required.

19.2 Electrochemical cell

The SPEs employed in this work (Fig. 19.1B) are commercial cards with a conventional three-electrode electrochemical cell screen-printed on a ceramic strip with dimensions $3.4 \times 1.0 \times 0.05$ cm. This screen-printed carbon electrode platform allows working with microvolumes ($\approx 40 \, \mu L$). The counter electrode (CE) is made of carbon ink and the reference electrode (RE) of silver ink. The working electrode (WE) is made of carbon ink including Co-phthalocyanine.

19.3 Chemical and supplies

- *Enzymes:* Alcohol oxidase from *Hansenula polymorpha* (AOx).
- *Components of background electrolyte:* Orthophosphoric acid and sodium hydroxide.
- *Analyte:* Ethanol.
- *Samples:* Red wine and tequila.
- *Electrochemical cell:* Commercial SPEs with a silver RE, a carbon CE and a carbon/Co-phthalocyanine WE.
- *General materials, apparatus, and instruments:* 1.5-mL microcentrifuge tubes, 10-µL, 100-µL, and 1-mL micropipettes and corresponding tips, weighing scale, pH meter, edge connector, and potentiostat.
- Milli-Q water is employed for preparing solutions.

19.4 Hazards

Phosphoric acid, used for the preparation of phosphate buffer (PB), is corrosive and causes serious burns. A lab coat, gloves, and safety glasses should be worn during the experiment, especially while phosphoric acid is employed.

19.5 Experimental procedure

19.5.1 Solutions and sample preparation

Prepare the following solutions estimating the required volume for each one:

- 0.1 M PB solution pH 6 (all working solutions required along this work will be prepared in this buffer solution).

- AOx enzyme solutions with the following concentrations: 0.05, 0.10, 0.15, and 0.20 U/μL.
- 0.1 M ethanol stock solution. From this one, ethanol solution of concentrations 0.10, 0.25, 0.50, 0.75, 1.0, 5.0, and 10.0 mM should be prepared by dilution.

Regarding the samples, the only pretreatment required for both (red wine and tequila) is a dilution step (about 1:10,000) in 0.1 M PB pH 6 to achieve an ethanol concentration within the linear range of the biosensor.

19.5.2 Construction of the sensor

The construction of the biosensor is based on the simple adsorption of the enzyme AOx on the surface of the working electrode of the SPE. The necessary steps for the biosensor preparation are the following:

1. The screen-printed carbon/Co-phthalocyanine electrode (SPCCPE) is washed with Milli-Q water.
2. A 10-μL aliquot of AOx solution is deposited on the working electrode and is left at room temperature until dried (about 45 min). Then, the biosensor is ready to perform the measurements.

As the constructed biosensor is of single use, each measurement will be performed employing a different sensor.

19.5.3 Electrochemical measurements

To perform electrochemical measurements with the SPCCPE, a 40-μL aliquot of working solution (ethanol solution or sample) is deposited covering the three electrodes.

The analytical signal correlated with ethanol concentration is obtained by recording chronoamperograms at +0.4 V for 150 s (see Fig. 19.2B). The analytical signal is the intensity of the current measured at this fixed time (150 s). This current intensity is due to the electrochemical oxidation of the Co-phthalocyanine (mediator) previously reduced enzymatically. It is related to ethanol concentration because the higher the concentration of ethanol, the higher the concentration of mediator reduced available to be electrochemically oxidized.

19.5.4 Optimization of the biosensor

With the aim of optimizing the enzyme concentration, different biosensors are fabricated employing the AOx solutions previously prepared (with concentrations 0.05, 0.10, 0.15, and 0.20 U/μL). These biosensors are tested recording chronoamperograms in 0.5 mM ethanol solution. Thus, the concentration of enzyme that provides the best analytical response would be selected for further experiments.

19.5.5 Calibration and determination of ethanol in alcoholic beverages

A calibration is performed with biosensors prepared employing the optimized concentration of AOx. Chronoamperograms are recorded for ethanol solutions of concentrations comprised between 0.1 and 10 mM A different biosensor is employed for each measurement.

A calibration curve is obtained representing the intensity of current measured versus ethanol concentration. Thus, the linear range, the sensitivity, the limit of detection, and the limit of quantification can be obtained. It is convenient to measure each concentration in triplicate to evaluate the precision. On the other hand, to evaluate the robustness of the biosensor, several calibration curves could be performed in the same day, or even in different days.

To analyze ethanol concentration in samples of alcoholic beverages (red wine and tequila), biosensors are prepared and chronoamperograms are recorded in the sample dilutions previously prepared. Three measurements for each sample should be performed to evaluate the precision. Then, results are presented as the average value, indicating the standard deviation, and paying attention to the number of the significant figures.

19.6 Lab report

Students should write a lab report following the typical scheme of a scientific article including an introduction, experimental part, results and discussion, and conclusions. Diagrams, tables, and images should be included and the following points should be considered:

1. In the introduction, include the importance and advantages of SPEs. Search and discuss recent publications on (i) other available designs of SPEs, (ii) SPEs made of other materials (specially modified carbon inks or metallic inks) and printing in other substrates (polymers, glass, paper, etc.), (iii) other low-cost and miniaturized approaches employed for electrochemical determination of ethanol, and (iv) other interesting analytes that can be determined employing SPEs.
2. Include graphs representing the data obtained along the experiment: (i) diagrams representing the current intensities achieved for the different concentrations of enzyme, (ii) chronoamperograms recorded in the presence of different concentrations of ethanol and calibration curve obtained (with the equation and the coefficient of determination, R^2).
3. Discuss the values obtained for the linear range, sensitivity, limit of detection, and limit of quantification of the biosensor, and compare them with other found in the literature.
4. Report the ethanol concentration in the samples including the standard deviation. Perform a statistical t-student test to compare the values obtained with those indicated in the package (considering them as real values). The results obtained by different (groups of) students can be also compared to calculate the accuracy of the results.

19.7 Additional notes

1. The PB solution and the enzyme stock solutions can be stored at 4°C for 1 week.
2. This experiment can be combined with other in which ethanol concentration is also determined but employing other techniques/methods. For example, chromatographic techniques or enzymatic kits with colorimetric detection could result very interesting for students. Thus, they could perform the determination of ethanol employing, e.g., gas

chromatographic methodologies and then compare the results with those obtained using the electrochemical biosensor. Then, students can not only validate their methodology but also compare, critically, their advantages and disadvantages.

3. Chapter 18 also deals with the development of an electrochemical sensor for ethanol determination. However, in that chapter, a carbon paste electrode is employed as working electrode and the enzyme used is alcohol dehydrogenase. Thus, performing both experiments could be interesting for working with and comparing both conventional and miniaturized electrochemical cells. Moreover, they can work with systems based on dehydrogenase and oxidase enzymes.

4. A Michaelis–Menten kinetics behavior is expected. Therefore, the apparent Michaelis–Menten constant (K_M) could be estimated by the Lineweaver–Burk linearization (see Chapter 18) Discussion about the value obtained and comparison with other found in the literature is encouraged.

19.8 Assessment and discussion questions

1. The described ethanol sensor is based on alcohol oxidase enzyme and Co-phthalocyanine as mediator (already included in the working ink). Performing a bibliographic search indicates other possible approaches for the same purpose: (i) different mediators, (ii) alternative enzymes, (iii) bienzymatic sensors, (iv) other approaches for enzyme immobilization, (v) employment of nanomaterials, etc.
2. Why +0.4 V is the detection potential chosen? What problems would involve the employment of higher potentials?
3. Explain the differences between oxidase and dehydrogenase-based approaches.

References

[1] J.P. Metters, R.O. Kadara, C.E. Banks, New directions in screen printed electroanalytical sensors: an overview of recent developments, Analyst 136 (2011) 1067–1076.
[2] Z. Taleat, A. Khoshroo, M. Mazloum-Ardakani, Screen-printed electrodes for biosensing: a review (2008-2013), Microchim. Acta 181 (2014) 865–891.
[3] E.C. Rama, A. Costa-García, Screen-printed electrochemical immunosensors for the detection of cancer and cardiovascular biomarkers, Electroanal. 28 (2016).
[4] R.A.S. Couto, J.L.F.C. Lima, M.B. Quinaz, Recent developments, characteristics and potential applications of screen-printed electrodes in pharmaceutical and biological analysis, Talanta 146 (2016) 801–814.
[5] Z. Chu, J. Peng, W. Jin, Advanced nanomaterial inks for screen-printed chemical sensors, Sensor. Actuator. B Chem. 243 (2017) 919–926.
[6] M. Albareda-Sirvent, A. Merkoçi, S. Alegret, Configurations used in the design of screen-printed enzymatic biosensors. A review, Sensor. Actuator. B Chem. 69 (2000) 153–163.
[7] O.D. Renedo, M.A. Alonso-Lomillo, M.J.A. Martínez, Recent developments in the field of screen-printed electrodes and their related applications, Talanta 73 (2007) 202–219.
[8] M. Li, D.W. Li, G. Xiu, Y.T. Long, Applications of screen-printed electrodes in current environmental analysis, Curr. Opin. Electrochem. 3 (2017) 137–143.
[9] A. Vasilescu, G. Nunes, A. Hayat, U. Latif, J.L. Marty, Electrochemical affinity biosensors based on disposable screen-printed electrodes for detection of food allergens, Sensors 16 (2016) 1863.

[10] H.M. Mohamed, Screen-printed disposable electrodes: pharmaceutical applications and recent developments, TrAC Trends Anal. Chem. 82 (2016) 1−11.

[11] J. Mettakoonpitak, K. Boehle, S. Nantaphol, P. Teengam, J.A. Adkins, M. Srisa-Art, C.S. Henry, Electrochemistry on paper-based analytical devices: a review, Electroanal. 28 (2016) 1420−1436.

[12] R.K. Mishra, L.J. Hubble, A. Martín, R. Kumar, A. Barfidokht, J. Kim, M.M. Musameh, I.L. Kyratzis, J. Wang, Wearable flexible and stretchable glove biosensor for on-site detection of organophosphorus chemical threats, ACS Sens. 2 (2017) 553−561.

[13] Metrohm DropSens. http://www.dropsens.com/, July 2018.

[14] D.G. Rackus, M.H. Shamsi, A.R. Wheeler, Electrochemistry, biosensors and microfluidics: a convergence of fields, Chem. Soc. Rev. 44 (2015) 5320−5340.

[15] A.M. Azevedo, D.M.F. Prazeres, J.M.S. Cabral, L.P. Fonseca, Ethanol biosensors based on alcohol oxidase, Biosens. Bioelectron. 21 (2005) 235−247.

[16] P. Goswami, S.S.R. Chinnadayyala, M. Chakraborty, A.K. Kumar, A. Kakoti, An overview on alcohol oxidases and their potential applications, Appl. Microbiol. Biotechnol. 97 (2013) 4259−4275.

[17] A. Chaubey, B.D. Malhotra, Mediated biosensors, Biosens. Bioelectron. 17 (2002) 441−456.

[18] J.K.F. Van Staden, Application of phthalocyanines in flow- and sequential-injection analysis and microfluidics systems: a review, Talanta 139 (2015) 75−88.

[19] A.B. Sorokin, Phthalocyanine metal complexes in catalysis, Chem. Rev. 113 (2013) 8152−8191.

[20] E.C. Rama, J. Biscay, M.B. González García, A.J. Reviejo, J.M. Pingarrón Carrazón, A. Costa García, Comparative study of different alcohol sensors based on screen-printed carbon electrodes, Anal. Chim. Acta 728 (2012) 69−76.

CHAPTER

20

Immunoelectroanalytical assay based on the electrocatalytic effect of gold labels on silver electrodeposition

Olaya Amor-Gutiérrez[1], Estefanía Costa-Rama[1,2],
M. Teresa Fernández Abedul[1]

[1]Departamento de Química Física y Analítica, Universidad de Oviedo, Oviedo, Spain;
[2]REQUIMTE/LAQV, Instituto Superior de Engenharia do Porto, Instituto Politécnico do
Porto, Porto, Portugal

20.1 Background

Human serum albumin (HSA) is a significant indicator for an early diagnosis of renal disease in diabetic patients, so its quantification is commonly performed in clinical laboratories [1]. The concentration of HSA in the urine of a healthy person is approximately 20 μg/mL, whereas it can increase up to 300 μg/mL in patients suffering from very bad renal disease, also called microalbuminuria. There are many reports on the determination of HSA using antigen—antibody reactions, and different immunoassay formats and strategies have been employed to detect the immunological event [1—4].

Electrochemical immunosensors include a wide variety of devices based on the combination of immunological reactions with electrochemical transduction [5—7]. They involve the immobilization of an immunoreagent (antigen or antibody) on the surface of the electrode. The detection is carried out recording an electrodic process which results from the antigen—antibody interaction.

Regarding the electrochemical transducers, there are many possible different platforms that can be combined with a wide variety of immobilization procedures. Many of the devices rely on the use of metallic (mainly gold) or carbon electrodes (mainly reusable glassy carbon electrodes (GCEs) or disposable screen-printed carbon electrodes). The sensing phase can be constructed onto the electrode surface by, e.g., covalent linkage, physical adsorption or membrane entrapping of the specific immunoreagent [6,8]. Although there are label-free approaches (e.g., using electrochemical impedance spectroscopy detection), most of the devices require a label linked to either the antigen or the antibody, which is responsible for the electroanalytical signal [7]. Enzymes, that catalyze the conversion of a substrate into an electroactive product, or electroactive species can be employed as labels.

In the last case, an antibody is labeled with a group that renders it electroactive. It has to be possible to reduce or oxidize the label conjugated to the antibody (or antigen) in a potential window at which other species are not electroactive, thereby allowing unlabeled reagents to be distinguished without interferences. Apart from cyclic voltammetry, differential pulse voltammetry (DPV), square wave voltammetry and their combination with anodic stripping analysis are extremely sensitive electrochemical methods that could be used for the trace determination of electroactive labels. These can be classified into different groups, depending on whether they are metallic ions [9] or colloids [10], inorganic complexes [11], organometallic tracers [12] or organic redox compounds [13]. Metal colloids, such as colloidal gold (also known as gold nanoparticles, AuNPs), are very advantageous for electrochemical detection [14]. The behavior of colloidal gold-labeled antihuman IgG has been studied on carbon paste electrodes (CPEs) [15]. It is strongly adsorbed on the electrode surface in open circuit and it is oxidized at a high potential in a hydrochloric acid medium where the $AuCl_4^-$ complex is formed. Afterward, gold can be reduced at around $+0.4$ V (versus Ag/AgCl). Using this methodology and DPV as electrochemical technique, low detection limits can be achieved.

However, a higher sensitivity is always required and then amplification approaches are continuously being developed. One of the most common is based on the catalytic effect of gold on the electrodeposition of silver and its subsequent stripping [16]. Silver reduction process occurs at a less negative potential when colloidal gold is attached to a CPE surface. This fact generates a range of potentials in which silver reduction is only due to the colloidal gold attached to the electrode surface. Thus, the direct reduction of silver ions on the carbon electrode surface can be avoided. Therefore, they are reduced to metallic silver only on the surface of colloidal gold particles, forming a shell around them. When an anodic scan is performed, this shell of silver is oxidized and this process constitutes the analytical signal.

Moreover, it has been found that ionic gold electrodeposited on carbon paste or GCEs has the same catalytic effect on silver electrodeposition, especially when gold is first oxidized [17,18]. Again, reduction of silver occurs at more positive potentials when ionic gold is electrodeposited on the electrode surface. This gives rise to a range of potentials at which direct reduction of silver does not occur. Then, if an appropriate potential is chosen, electrodeposition of silver occurs only on the electrodeposited gold surface.

In this experiment, adapted from reference [19], ionic gold is used as electroactive label for the determination of HSA. A simple labeling procedure is used, without the need of chelating agents. The immunological reaction between gold-labeled anti-HSA (anti-HSA/Au) and its antigen (HSA) is performed on GCEs using two different formats: noncompetitive and competitive. HSA is adsorbed on the electrode surface and the immunoreaction is monitored by measuring voltammetrically the concentration of gold conjugated to the antibody through its catalytic effect on silver electrodeposition.

In the noncompetitive immunoassay format, HSA (analyte) is previously adsorbed on the electrode surface and the immunological reaction with anti-HSA/Au is conducted. In this case, the higher the concentration of HSA, the higher the amount of anti-HSA/Au that is bound to the HSA immobilized on the electrode surface. As a result, an increased analytical signal is obtained, proportional to HSA concentration.

In the competitive format, a fixed amount of HSA adsorbed on the electrode surface competes with this present in solution, in a reaction with a limited amount of anti-HSA/Au. The higher the concentration of HSA in solution, the lower the amount of free anti-HSA/Au available to react with the HSA immobilized on the electrode surface, and consequently, a decrease in the analytical signal is obtained.

In this experiment, some of the main parameters affecting the assay are studied and the analytical characteristics of the methodology, including linear range, limit of detection, sensitivity and reproducibility, will be discussed. Along this work, students will learn, in two laboratory sessions of 3-4 hours, to work with a conventional cell with a glassy carbon working electrode (WE) to develop immunoassays with electrochemical detection. Finally, the students are encouraged to discuss the main advantages and drawbacks of the label used, the immobilization procedure and the assay formats.

20.2 Electrochemical cells

All measurements in this experiment are made at room temperature in a 20-mL cell (protected from light) with a three-electrode configuration: a GCE as WE, a platinum wire as CE and a Ag/AgCl as RE. Five homemade GCEs are used all along the work. They consist of a 3-mm diameter glassy carbon rod sealed into a Teflon holder with Spurr resin for electron microscopy. Electric contact is a brass rod welded to glassy carbon with a silver-loaded conductive epoxy resin.

Preparation and renewal of the glassy carbon surfaces is achieved by following the steps below:

1. Polish the surface of the GCEs first with 1.0 μm α-alumina particles (adding water to generate a suspension) on an 8-inch microcloth polishing sheet.
2. Put them in a water-filled beaker and place it in an ultrasonic bath. Sonicate for 5 min.
3. Polish again the surface with 0.3-μm α-alumina particles (adding a small volume of water to prepare a suspension) on an 8-inch microcloth polishing sheet and repeat sonication. Finally, wash it with water.

When the WE is ready to use, place it together with the platinum wire (CE) and the Ag/AgCl electrode (RE) into the electrochemical cell. A magnetic stirrer is used when necessary.

20.3 Chemicals and supplies

20.2.1 Chemicals:

Labeling of anti-HSA with sodium aurothiomalate:

- NaCl (0.15 M unbuffered solution, with pH 7.5 adjusted with 0.1 M NaOH).
- Anti-HSA solution (1.0×10^{-4} M) in 0.15 M NaCl pH 7.5.
- Sodium aurothiomalate solution (2.0×10^{-3} M) in 0.15 M NaCl pH 7.5.

Electrode pretreatment:

- HCl (0.1 M).
- Tris-HCl buffer (0.1 M, pH 7.2).

Immunoassays:

- Tris-HCl buffer (0.1 M, pH 7.2).
- HSA solutions (varying from 1.0×10^{-10} to 1.0×10^{-8} M) prepared in Tris-HCl buffer.
- BSA (bovine serum albumin).
- Anti-HSA/Au solution (5.0×10^{-8} M).

Recording of the analytical signal:

- HCl (0.1 M).
- H_2SO_4 (0.1 M).
- Silver nitrate ($AgNO_3$, 2.0×10^{-4} M) in 1.0 M NH_3 solution.
- KCN (0.1 M) prepared in 0.1 M NaOH.

Materials:

- Glass electrochemical cell (20 mL).
- Teflon holder.
- Spurr resin.
- Glassy carbon rods (3-mm diameter).
- α-Alumina powder (1.0 and 0.3 μm).
- Microcloth polishing sheets (8 in.).
- Ultrasonic bath.
- Platinum wire (counter electrode [CE]).
- Ag/AgCl reference electrode (RE).
- Magnetic stirrer.
- Dialysis cassettes (10,000 MWCO).
- Potentiostat and computer system.
- pH-meter.
- 1.5-mL microcentrifuge tubes, 100-μL and 1-mL micropipettes with corresponding tips.

Ultrapure water is employed throughout the work to prepare solutions and to clean electrodes and glassware.

20.4 Hazards

Wear protective gloves and glasses, especially when handling concentrated acids and bases. Ammonia is irritating, corrosive to exposed tissues, and toxic if inhaled. It causes severe skin burns and eye damage. Potassium cyanide is fatal if inhaled, swallowed or in contact with skin. In both cases, revise carefully the corresponding safety sheets. Prepare solutions in a fume hood and work in well-ventilated areas.

20.5 Experimental procedures

20.5.1 Labeling of anti-HSA with sodium aurothiomalate

1. Mix an aliquot of 20 μL of 1.0×10^{-4} M anti-HSA solution with 1000 μL of 2.0×10^{-3} M sodium aurothiomalate solution (1:1000 ratio) and leave it to react at 37°C for 24 h.
2. Purify the conjugate (anti-HSA/Au) by dialysis against 200 mL of 0.15 M NaCl (pH 7.5) for 48 h at room temperature.

20.5.2 Electrode pretreatment

Solid electrodes usually need to be cleaned and/or activated to eliminate impurities or adsorbed molecules, as well as to activate the surface generating centers that favor electron transfer. Procedures could be mechanical (e.g., polishing), chemical (treatment with acid or basic solutions or organic solvents), electrochemical (application of highly positive, negative, or both potentials), or a combination. In this case, smoothed glassy carbon surfaces are pretreated before each assay following a chemical/electrochemical procedure as commented below:

1. Pour 20 mL of a 0.1 M HCl solution in the glass electrochemical cell.
2. Immerse the WE (together with the RE and the CE) in the 0.1 M HCl solution and stir while applying a potential of $+1.4$ V for 2 min.
3. Wash the electrode surface with 0.1 M Tris-HCl pH 7.2 buffer solution.

20.5.3 Recording of the analytical signal

After performing the corresponding immunoassay, the analytical signal can be obtained following this procedure:

1. Pour 20 mL of 0.1 M HCl in the glass electrochemical cell.
2. Immerse the three electrodes (WE, RE and CE) in the cell containing 0.1 M HCl and stir. The WE (polished and washed GCE) has HSA immobilized and has been incubated with anti-HSA/Au (in the noncompetitive immunoassay format, see Section

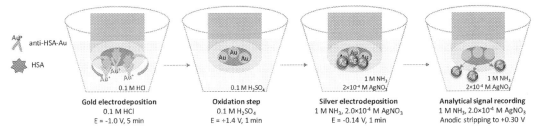

FIGURE 20.1 Schematic diagram of the procedure for recording the analytical signal. *HSA*, human serum albumin.

20.5.4.1) or with a mixture of anti-HSA/Au and HSA (in the competitive immunoassay format, see Section 20.5.4.2).

3. Apply a potential of -1.0 V for 5 min to electrodeposit the gold bound to the anti-HSA/Au on the electrode surface (Au(III) is reduced to Au(0) and is deposited on the electrode).
4. Wash with ultrapure water.
5. Immerse the electrodes in another cell with 20 mL of 0.1 M H_2SO_4.
6. Conduct an oxidation step by applying a potential of $+1.4$ V for 1 min under stirring (in this case, oxide formation favors catalysis).
7. Wash the electrodes with ultrapure water.
8. Immerse the electrodes in a 1.0 M NH_3 solution containing silver nitrate at a fixed concentration (2.0×10^{-4} M).
9. Hold the electrodes at a deposition potential of -0.14 V for 1 min without stirring (Ag(I) is reduced to Ag(0) and deposited over gold).
10. Record cyclic voltammograms from the deposition potential (-0.14 V) to $+0.30$ V at a scan rate of 50 mV/s to obtain the analytical signal (redissolution of Ag(0) that comes back to the solution as Ag(I)).
11. Measure the peak current for each WE (with different concentration of HSA immobilized).

Fig. 20.1 shows a scheme of the above procedure required to obtain the analytical signal (in the case of noncompetitive assay, but the procedure to obtain the analytical signal is the same in both cases).

After each measurement, the WE must be washed by immersing the GCE in another cell (open circuit) containing a stirred solution of 0.1 M KCN for 2 min.

20.5.4 Immunoassays

Assays for HSA can be conducted by following two different formats: noncompetitive and competitive. All the steps are conducted at room temperature. In Fig. 20.2, the schematic diagram of both noncompetitive and competitive assays is represented. In the noncompetitive assay, an excess of anti-HSA/Au is employed. In the competitive format, a mixture of anti-HSA/Au and HSA (in specific concentrations) is added in such a way that both, HSA immobilized on the electrode and free HSA, compete by a limited amount of anti-HSA/Au.

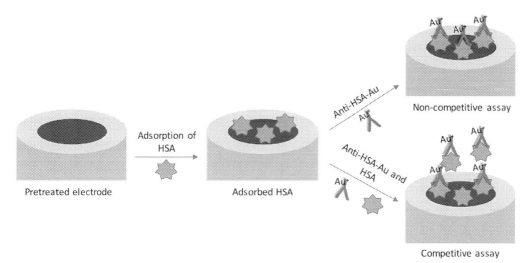

FIGURE 20.2   Schematic diagram of both immunoassay procedures, noncompetitive and competitive. *HSA,* human serum albumin.

The first step is the same for both procedures:

1. Deposit a 50-μL drop of HSA solution on each electrode surface (concentration of HSA varying from 5.0×10^{-10} to 1.0×10^{-8} M in the case of the noncompetitive immunoassay and a fixed concentration of 1.0×10^{-8} M in the case of the competitive immunoassay). Distribute it carefully with the help of the micropipette tip. Leave it there for 15 min.
2. Wash the electrodes with unbuffered 0.15 M NaCl (with pH adjusted to 7.5).

20.5.4.1 Noncompetitive immunoassay

After immobilization of HSA (varying from 5.0×10^{-10} to 1.0×10^{-8} M) on the electrode surface and the washing step, perform the immunological reaction:

1. Place a 50-μL drop of anti-HSA/Au solution (5.0×10^{-8} M in terms of anti-HSA) on the electrode surface and leave it for 60 min.
2. Wash with ultrapure water to remove the excess of antibody that did not react.
3. Perform the electrochemical measurements as indicated in Section 20.5.3.
4. Construct a calibration graph (dose—response curve) for HSA (see Fig. 20.3A as an orientation on the calibration graph, plotting the intensity of the peak current vs. HSA concentration).
5. Study the anti-HSA/Au nonspecific adsorption by substituting HSA by the same volume of buffer. Alternatively, an identical concentration of BSA can be added to evaluate also the specificity (see Fig. 20.3B as guidance about the comparison of the signals obtained when studying nonspecific adsorptions).

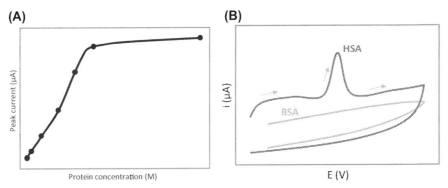

FIGURE 20.3 (A) Example of a calibration plot obtained for human serum albumin in a noncompetitive immunoassay. (B) Example of the cyclic voltammograms obtained for bovine serum albumin (BSA) and human serum albumin (HSA) in the noncompetitive immunoassay. *Reprinted with permission from A. de la Escosura-Muñiz, M.B.González-García, A.Costa-García, Determination of human serum albumin using aurothiomalate as electroactive label, Anal. Bioanal. Chem. 384 (2006) 742–750, with modifications.*

20.5.4.2 Competitive immunoassay

After immobilization of 50 μL of an HSA solution (with a fixed concentration of 1.0×10^{-8} M) on the electrode surface and the washing step, perform the immunological reaction:

1. Place a 50-μL drop of each mixture of anti-HSA/Au (a fixed concentration, 5.0×10^{-8} M in terms of anti-HSA) and HSA (varying from 1.0×10^{-10} M to 1.0×10^{-9} M) on the HSA-modified electrodes. Leave them for 60 min.
2. Wash with ultrapure water to remove the excess of reagents that did not react.
3. Perform the electrochemical measurements as indicated in Section 20.5.3.
4. Construct a calibration graph (dose–response curve) for HSA. The background signals correspond to the ones obtained without HSA (adding just buffer) or adding BSA in the same concentration. Their values are the maximum signals. See Fig. 20.4 as an orientation of the calibration graph, plotting peak current vs. HSA concentration.

20.6 Lab report

Write a lab report following the typical scheme of a scientific article, including a brief introduction, experimental part (materials, equipment, and protocols), results and discussion, and conclusions. The following points should be beard in mind:

1. In the introduction, explain the purpose of the experiment and a short review of the reported methods to tackle this problem. An overview about the advantages and disadvantages of the use of labels in immunosensors (vs. label-free approaches), as well as the different labels that could be employed, should be also included. The label used in this experiment and how it allows obtaining the analytical signal must be explained.

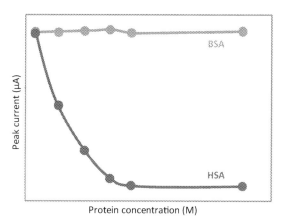

FIGURE 20.4 Example of the effects of human serum albumin (HSA) and bovine serum albumin (BSA) concentrations on the peak current in the competitive immunoassay. *Reprinted with permission from A. de la Escosura-Muñiz, M.B.González-García, A.Costa-García, Determination of human serum albumin using aurothiomalate as electroactive label, Anal. Bioanal. Chem. 384 (2006) 742–750.*

2. Protocols must be suitably detailed including schemes preferentially, where appropriate, as well as the necessary calculations.
3. Discuss the main variables that influence the analytical signals and include figures with representative raw data as well as results presented in tables and graphs for the different studies performed.
4. Include graphs for the calibration curves discussing the values obtained for the figures of merit. Pay attention to the significant figures in each case. Indicate the linear range and the sensitivity of the methodologies. Estimate the limit of detection and quantification of both methods and point out the main characteristics that differentiate them.
5. Indicate and comment, in case it is worth, the results obtained in terms of selectivity and precision.
6. Discuss the incidences happened during the experiment and the main conclusions.

20.7 Additional notes

1. All the solutions should be prepared daily and stored in amber tubes.
2. NaCl, HCl, H_2SO_4, NH_3, and NaOH solutions should be prepared in ultrapure water.
3. KCN reacts with Ag to give complexes and then is used for cleaning the electrodes.
4. The stirring must be homogenous. Place the stir bar just below the WE and procure a homogenous and gentle stirring when adding solutions and in between measurements.
5. A biosensor must include a biological recognition element and a transducer. A capture bioreagent is employed to recognize the analyte from a complex matrix. In this case, and for the sake of simplicity, this step is not considered (the analyte is directly adsorbed on the electrode surface) and then the electrochemical device cannot be considered an immunosensor. With the aim of fabricating a real immunosensor, an anti-HSA antibody (different from the one employed for detection) could be used as

capture reagent. In this case, changing capture/detection antibodies by detection/capture antibodies is an interesting study. Discussion on the use of monoclonal/polyclonal or primary/secondary labeled antibodies could be made.

6. In this specific case where HSA is the analyte, a blocking step is not added. This is usually added after the immobilization of a capture reagent to avoid nonspecific adsorptions and BSA is commonly employed as blocking agent. Other proteins such as casein could be employed. Nonspecific adsorptions should be always checked in all the systems and in case they are relevant, a blocking step should be added.

7. In the noncompetitive immunoassay, studies of the influence of the HSA adsorption time on the analytical signal could be performed by testing an HSA concentration (e.g., 1.0×10^{-8} M) using 5.0×10^{-8} M anti-HSA/Au solution. The peak current can be measured when adsorbing the HSA solution at different times (e.g., 5, 10, 15, 20, 30, or 40 min). Discussion about the optimized adsorption time (and HSA concentration if the study is also performed using diferent concentrations of HSA) can be done.

8. Different studies on precision could be done immobilizing the same HSA concentration on different electrodes, on the same or in different days. The relative standard deviation should be calculated to evaluate the precision.

9. If possible, real urine samples could be used to quantify the HSA. Students should think what dilution is necessary to make to have those samples in the linear range of the calibration plot. They can also add different concentrations of HSA to urine and measure them. By doing so, they will see how the matrix can affect the measurements.

10. This experiment has been performed using GCE electrodes, and therefore, electrode pretreatment is necessary. However, screen-printed electrodes could be used; these electrodes are disposable and so the electrode pretreatment step could be avoided. Moreover, this would allow saving analysis time by parallelizing assays.

11. This experiment is performed to understand immunological reactions based on the HSA/anti-HSA interaction, but any other immunosystem could be checked. In this case, optimization of concentrations and other parameters would have to be done. An example to study the widely used affinity interaction between biotin and streptavidin is through the system: biotinylated goat anti-rabbit IgG and gold-labeled rabbit IgG [20]. Streptavidin could be adsorbed on the electrode and the biotinylated IgG is immobilized. Later, the interaction with rabbit IgG is followed through the gold label. Apart from this simple system, in a current immunoassay, the analyte would be included in between capture and detection immunoreagents. The system could be complicated even more if secondary antibodies are included.

12. It could be interesting, when working with immunoreagents, to revise the commercial sheet indicating the type of antibody, titer, preservation (aliquoting, freezing, etc.), or bioconjugation.

13. It is very convenient to perform a cost analysis in all the cases, especially when bioreagents are included. Consider that longer steps increase analysis time and therefore the cost.

20.8 Assessment and discussion questions

1. Compare the analytical characteristics of this method with others found in the bibliography for the same analytes, using other labels and methodologies.
2. In this experiment, how were the bioreagents immobilized? Search for other methods for immobilization of bioreagents in the literature and discuss the advantages and disadvantages of each one.
3. Look for other labels employed in electrochemical immunosensors and discuss their advantages and disadvantages.
4. How many types of immunoassays exist? Classify them and explain the differences between them.
5. Compare the dose—response curve for the competitive and noncompetitive assays and explain their shape.
6. What is the role of each one of the three electrodes employed in the electrochemical cell?
7. Why is it so important to clean or activate solid electrodes and what type of procedure was employed here?
8. What is the electrochemical technique employed for the detection?

References

[1] D. Kumar, D. Banerjee, Methods of albumin estimation in clinical biochemistry: past, present, and future, Clin. Chim. Acta 469 (2017) 150–160.
[2] W. Guo, J. Song, M. Zhao, J. Wang, Electrochemical immunoassay based on catalytic conversion of substrate by labeled metal ion and polarographic detection of the product generated, Anal. Biochem. 259 (1998) 74–79.
[3] J. Kim, J. Cho, G. Sig, C. Lee, H. Kim, S. Paek, Conductimetric membrane strip immunosensor with polyaniline-bound gold colloids as signal generator, Biosens. Bioelectron. 14 (2000) 907–915.
[4] R. Saber, S. Mutlu, E. Pis, Glow-discharge treated piezoelectric quartz crystals as immunosensors for HSA detection, Biosens. Bioelectron. 17 (2002) 727–734.
[5] F.S. Felix, L. Angnes, Electrochemical immunosensors — a powerful tool for analytical applications, Biosens. Bioelectron. 102 (2018) 470–478.
[6] F. Ricci, G. Adornetto, G. Palleschi, A review of experimental aspects of electrochemical immunosensors, Electrochim. Acta 84 (2012) 74–83.
[7] E.C. Rama, A. Costa-García, Screen-printed electrochemical immunosensors for the detection of cancer and cardiovascular biomarkers, Electroanalysis 28 (2016).
[8] Y. Wan, Y. Su, X. Zhu, G. Liu, C. Fan, Development of electrochemical immunosensors towards point of care diagnostics, Biosens. Bioelectron. 47 (2013) 1–11.
[9] J. Wang, B. Tian, K.R. Rogers, Thick-film electrochemical immunosensor based on stripping potentiometric detection of a metal ion label, Anal. Chem. 70 (1998) 1682–1685.
[10] A. de la Escosura-Muñiz, C. Parolo, F. Maran, A. Mekoçi, Size-dependent direct electrochemical detection of gold nanoparticles: application in magnetoimmunoassays, Nanoscale 3 (2011) 3350.
[11] K.S. Lee, T. Kim, M. Shin, W. Lee, J. Park, Disposable liposome immunosensor for theophylline combining an immunochromatographic membrane and a thick-film electrode, Anal. Chim. Acta 380 (1999) 17–26.
[12] L.X. Tiefenauer, S. Kossek, C. Padeste, P. Thiebaud, Towards amperometric immunosensor devices, Biosens. Bioelectron. 12 (1997) 213–223.

[13] A. Costa-García, M.T. Fernández-Abedul, P. Tuñón-Blanco, Comparative voltammetric study of 2,4-dinitrophenol (DNP), albumin and DNP-Albumin as an analytical approach to the use of DNP as a universal label in immunoelectroanalysis assays, Talanta 41 (1994) 1191–1200.

[14] J.M. Pingarrón, P. Yáñez-Sedeño, A. González-Cortés, Gold nanoparticle-based electrochemical biosensors, Electrochim. Acta 53 (2008) 5848–5866.

[15] M.B. González-García, A.C. García, Adsorptive stripping voltammetric behaviour of colloidal gold and immunogold on carbon paste electrode, Bioelectrochem. Bioenerg. 38 (1995) 389–395.

[16] M.B. González-García, A. Costa-García, Silver electrodeposition catalyzed by colloidal gold on carbon paste electrode: application to biotin − streptavidin interaction monitoring, Biosens. Bioelectron. 15 (2000) 663–670.

[17] A. de la Escosura-Muñiz, M.B. González-García, A. Costa-García, Catalytic effect on silver electrodeposition of gold deposited on carbon electrodes, Electroanalysis 16 (2004) 1561–1568.

[18] D. Hernández-Santos, M.B. González-García, A. Costa-García, Effect of metals on silver electrodeposition. Application to the detection of cisplatin, Electrochim. Acta 50 (2005) 1895–1902.

[19] A. de la Escosura-Muñiz, M.B. González-García, A. Costa-García, Determination of human serum albumin using aurothiomalate as electroactive label, Anal. Bioanal. Chem. 384 (2006) 742–750.

[20] A. de la Escosura-Muñiz, M.B. González-García, A. Costa-García, Electrocatalytic detection of aurothiomalate on carbon electrodes. Application as a non-ezymatic label to the quantification of proteins, Anal. Chim. Acta 524 (2004) 355–366.

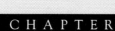

CHAPTER

21

Genosensor on gold films with enzymatic electrochemical detection of a SARS virus sequence

Andrea González-López, M. Teresa Fernández Abedul

Departamento de Química Física y Analítica, Universidad de Oviedo, Oviedo, Spain

21.1 Background

Biosensors, analytical devices incorporating a biological material (e.g., tissue, cell receptors, enzymes, antibodies, nucleic acids, etc.), a biologically derived material (e.g., recombinant antibodies, engineered proteins, aptamers, etc.), or a biomimic (e.g., synthetic receptors, biomimetic catalysts, combinatorial ligands, imprinted polymers, etc.) intimately associated with or integrated within a physicochemical transducer or transducing microsystem, which may be optical, electrochemical, thermometric, piezoelectric, magnetic, or micromechanical [1], have become a very important area of Analytical Chemistry. The majority of current and future analytical requirements will be solved by the use of simple yet sensitive devices, able to be reshaped with new trends such as the integration with, e.g., smartphones [2]. Apart from the well-established enzymatic sensors, the selectivity of affinity interactions, such as those of antigen—antibody or nucleic acid hybridization, is exploited for the respective development of promising immune and nucleic acid assays. Nucleic acid detection is becoming relevant not only in the field of food analysis but also in clinical diagnosis. It is advantageous compared with immunoassay in cases of recent infection or underlying immunodeficiency. Moreover, it is very useful for treatment monitoring because elimination of nucleic acids coming from a pathogen indicates successful handling [3].

Electrochemical biosensors combine the sensitivity of electroanalytical methods with the inherent bioselectivity of the biological component [4]. In DNA sensors, several alternative ways of detecting DNA electrochemically (direct or indirectly) are possible. The intrinsic electroactivity of adenine and guanine can be the basis for direct measurement of nucleic acids

213

[5]. On the other hand, the use of indicator molecules that interact differentially with single- or double-stranded DNA has been widely exploited in indirect approaches. Bioconjugation is commonly not required because electrostatic or hydrophobic interactions take place easily. A different possibility for indirect detection is the use of enzymes, alternative that enhances assay sensitivity enormously because of their inherent amplification. Although enzymes can be directly conjugated to DNA strands, the use of a well-known affinity interaction such as avidin–biotin is an easy alternative. Horseradish peroxidase has been widely used as enzyme label as well as glucose oxidase and alkaline phosphatase (AP). AP (EC 3.1.3.1) is a hydrolase that converts orthophosphoric monoesters into alcohols and thus various phosphate esters can act as substrates. As the name suggests, they are more effective in alkaline media. 3-Indoxyl phosphate (3-IP) has been proposed as an adequate substrate for AP in immunoelectrochemical approaches [6]. Because of its satisfactory electrochemical behavior (see Chapters 5, 30, and 31), it has been used for the development of AP-based enzyme immunoassays. In this experiment, AP is used to detect DNA hybridization events on gold electrodes.

The surface of the genosensor is of paramount importance in the performance of the analytical device because it is where immobilization and transduction take place. Gold has always been an appropriate material for the electrochemical detection of substances, while also allowing different formats: disks, wires, films, etc. Thin-film (thickness in the nm-scale) gold electrodes can be produced by sputtering, a physical vapor deposition methodology in which gold is ejected from a target in a vacuum chamber onto a substrate. Gold also offers the relevant possibility of generating self-assembled DNA monolayers through thiol groups because of the strong binding between sulfur and gold, which may be considered almost covalent.

In this chapter, a DNA hybridization assay with enzymatic electrochemical detection (see Fig. 21.1) is carried out on a 100-nm sputtered gold film that allows working with small volumes. Reducing the cell volume has several advantages, the first being the decrease in the diffusion distances required for analytes to reach their surface-bound receptor partners. A simple, cheap, and easy-to-handle homemade device is presented allowing the performance of the hybridization procedure and sequential detection. Square wave voltammetry (SWV) (see Chapters 3 and 24) is employed for measuring in a fast and sensitive way.

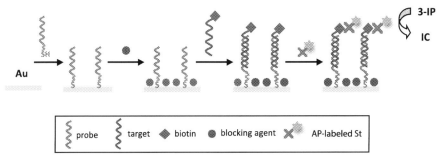

FIGURE 21.1   Schematic illustration of the assay developed on this genosensor. *3-IP*, 3-indoxyl phosphate; *AP*, alkaline phosphatase; *St*, streptavidin; *IC*, indigo carmine.

This experiment is adapted from Ref. [7]. The sequence chosen as target for the development of this genosensor corresponds to a gene that encodes the nucleocapsid protein (422 amino acids) of the SARS (severe acute respiratory syndrome) virus, specifically a short lysine-rich region that appears to be unique to SARS and suggestive of a nuclear localization signal [8]. A 30-mer oligonucleotide with bases comprised between numbers 29,218 and 29,247, both included, was chosen.

A complementary strand to the chosen SARS sequence is labeled with a thiol group and immobilized on the gold surface. The target (30-mer oligonucleotide with a sequence included in the SARS coronavirus) is conjugated to biotin and hybridized with the probe. Addition of AP-labeled streptavidin (St) allows enzymatic detection via the electrochemical signal of the product. The parameters affecting all assay steps are studied, and analytical characteristics, including selectivity, can be discussed.

21.2 Electrochemical cell

The electrochemical cell (see in more detail in Section 21.5.1) consists of 5 cm × 5 cm supports of a 0.125 mm thick polyimide substrate that are sputtered with gold (Emitech sputter coater model K550) and after delimiting areas, act as working electrodes. A miniaturized Ag/AgCl/saturated KCl electrode is employed as reference and a platinum wire as auxiliary electrode. All the three electrodes are connected with alligator clips to the potentiostat.

21.3 Chemicals and supplies

Reagents

- *Oligonucleotide buffer:* Tris-EDTA (TE) buffer pH 8 (0.1 M Tris-HCl buffer solution, 1 mM in EDTA).
- *Enzyme:* AP (conjugated to Streptavidin), AP-labeled St.
- *Substrate and product:* 3-IP and indigo carmine (IC).
- *IC stock solution:* 0.1 M H_2SO_4.
- *3-IP buffer:* 0.1 M Tris-HCl, 10 mM $MgCl_2$ pH 9.8.
- *St-AP buffer:* 0.1 M Tris-HCl, 1 mM $MgCl_2$ pH 7.2.
- *Hybridization buffer (2x SSC):* 30 mM sodium citrate buffer with 300 mM NaCl, pH 7 (Table 21.1).

TABLE 21.1 Commercial DNA sequence, written from 50 to 30, and peptide sequence for the development of the assay.

Description	Sequences (5'-3')
Biotinylated target	ACA-GAG-CCT-AAA-AAG-GAC-AAA-AAG-AAA-AAG-biotin
Biotinylated mismatched target	ACA-GCG-CCT-AAA-AAC-GAC-AAA-AAG-AG-AAG-biotin
Biotinylated and thiolated probe	Biotin-CTT-TTT-CTT-TTT-GTC-CTT-TTT-AGG-CTC-TGT−$(CH_2)_3$−SH

- *Blocking proteins*: Bovine serum albumin (BSA, fraction V), biotin-labeled albumin, casein from sheep milk, and a gelatine derivative called "Perfect Block" in 0.1 M Tris-HCl pH 7.2.
- *Blocking agents*: Methanol, thioctic acid, 1-hexanethiol, 6-mercapto-1-hexanol in ethanol.
- Saturated KCl solution.
- Milli-Q water is employed for preparing solutions.

Materials

- *Electrode fabrication*: Polyimide substrate, conductive wire, epoxy resin (CW2400), self-adhesive washers with 5 mm internal diameter (19.6 mm^2 internal area), crocodile clips, silver wire, platinum wire, syringe rubber piston, micropipette tip, insulating tape.
- *Apparatus*: Weighing scale, pH meter, and incubator.
- *Other materials*: 100-μL and 1-mL micropipettes with corresponding tips, 1.5-mL micro-centrifuge tubes.

21.4 Hazards

Students are required to wear lab coat, appropriate gloves, and safety glasses. Special care should be taken when handling concentrated acids.

21.5 Experimental procedures

21.5.1 Construction of the electrochemical cell

The three-electrode potentiostatic system used for this experiment is schematized in Fig. 21.2. Working electrodes are made on 5 cm × 5 cm supports of a 0.125 mm thick polyimide substrate called Kapton HN. They are covered with gold by a sputtering process explained on Section 21.5.2. A conductor wire is attached to the center of one of the sides by means of an epoxy resin (CW2400) and it is cured at room temperature. The working area is limited by self-adhesive washers of 5 mm internal diameter. The gold film is placed on a support to which an alligator connection is fixed.

Reference and auxiliary electrodes are coupled in a micropipette tip. The reference electrode is an anodized silver wire introduced in the tip through a syringe rubber piston. In turn, the tip is filled with saturated KCl solution and contains a low-resistance liquid junction. The platinum wire acts as auxiliary electrode and it is fixed with insulating tape.

For measurement recording, the tip is fixed on an electrochemical cell Metrohm support, allowing horizontal and vertical movement.

21.5.2 Gold sputtering

The working electrodes made on polyimide supports are covered with gold by a sputtering process. The Kapton slide (5 cm × 5 cm) must be cleaned with ethanol and left drying. Once dried, the slide is covered with gold. Gold atoms are deposited (from the cathode) over the Kapton (placed on the anode) in a vacuum chamber filled with argon. The thickness

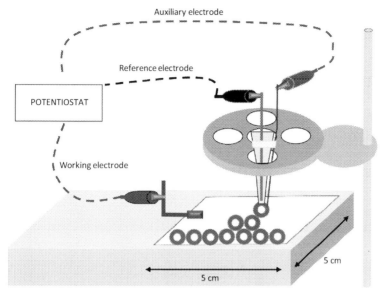

Auxiliary electrode

Reference electrode

POTENTIOSTAT

Working electrode

5 cm

5 cm

FIGURE 21.2   Schematic diagram of the three-electrode potentiostatic system.

of the gold layer is controlled by means of the duration and intensity of the discharge (possible values are a 35 mA discharge applied for 220 s).

21.5.3 Hybridization assay

1. Deposit a volume of 5 μL of the thiolated DNA probe (in TE buffer pH 8) on the gold film (e.g., 1 μM in concentration).
2. Maintain at 37°C for 20 min or at 4°C for 12 h.
3. Wash with 0.1 M Tris-HCl buffer pH 7.2.
4. Add 15 μL of a 2% blocking agent solution (see Section 21.3.1, blocking agent in 0.1 M Tris-HCl pH 7.2 if it is a protein or ethanol if it is a thiol) and wait 10 min.
5. Clean it with 2x SSC buffer.
6. Add the biotinylated complementary strand (target) and wait 60 min at room temperature.
7. Wash the film with 0.1 M Tris-HCl, 10 mM $MgCl_2$ pH 7.2.
8. Add 20 μL of AP-labeled St at a 10^{-9} M concentration (in 0.1 M Tris-HCl, 1 mM $MgCl_2$ pH 7.2) and wait 60 min.
9. Wash the film with 0.1 M Tris-HCl, 10 mM $MgCl_2$ pH 9.8.
10. Add 20 μL of a 3 mM solution of 3-IP (in 0.1 M Tris-HCl, 10 mM $MgCl_2$ pH 9.8) and wait 10 min for its enzymatic hydrolysis.
11. Add, then, 5 μL of concentrated H_2SO_4 to stop the reaction.
12. Finally, add 5 μL of Milli-Q water and perform the electrochemical measurements.

21.5.4 Electrochemical measurement

The tip with the reference and auxiliary electrodes is introduced into the 30-μL drop (20 μL from the 3-IP solution plus 5 μL of the concentrated H_2SO_4 and 5 μL of Milli-Q water) deposited on the gold film. A potential of -0.35 V is applied for 30 s before scanning the potential between -0.15 and $+0.3$ V following a square wave format with frequency of 50 Hz and amplitude of 50 mV. Measure the peak current, which is the analytical signal that can be correlated with the concentration.

21.5.5 Effect of evaporation

Evaporation seems to be a critical condition in the immobilization of SH-DNA. A more rigorous study (step 1−2 of Section 21.5.3) could be done. Using a drop of 5 μL and a strand concentration of 1 μM, different immobilization times can be employed at room temperature (22°C approximately), 37 and 47°C. Note down when evaporation occurred and compare the signals obtained. Select the best immobilization conditions.

21.5.6 Surface blocking

Blocking the surface is one of the most important steps to minimize and control nonspecific adsorptions. Two main types of agents can be considered: proteins and sulfur-containing compounds. The signal/background ratio, and therefore the blocking capacity, should be studied for each compound.

Proteins such as albumin are common blocking agents in bioanalysis and adsorb well onto gold surfaces. Evaluate the nonspecific signal obtained after blocking (step 4 in Section 21.5.3) with aqueous solutions of BSA, casein, and a gelatine derivative called Perfect Block.

The other group of blocking agents are compounds that contain sulfur atoms and therefore present a high affinity for gold. Ethanolic solutions of thioctic acid, 6-mercapto-1-hexanol, and 1-hexanethiol could be tested. It is very interesting to note the effect of these agents on both capacitive and faradaic currents.

21.5.7 Analytical characteristics

A calibration curve for the complementary strand (c-DNA) can be performed under optimized conditions, employing a 1-μM solution of SH-DNA for immobilization. As an example, concentrations in the 0.1−10 nM interval could be tested (step 6 in Section 21.5.3).

SWV measurements for different concentrations are included in Fig. 21.3. Plot the values of the peak current for different concentrations and fit the data to a regression curve. Take into account that it can be nonlinear and fit into logarithmic or a different model. Determine the dynamic range, sensitivity, and limit of detection of the assay.

Precision studies of the platform could be made by obtaining measurements with the genosensor on different working areas, polyimide substrates, days, or groups. Evaluate the results through the value of the relative standard deviation of the peak current.

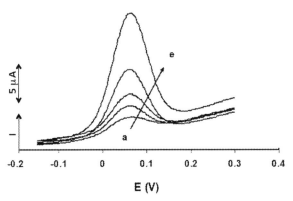

FIGURE 21.3   Response curves for background (nonspecific adsorptions) and several concentrations of bio-tinylated target (from *a* to *e*).

21.6 Lab report

Write a lab report following the typical scheme of a scientific article, including a brief introduction, experimental part (materials, equipment, and protocols), results and discussion, and conclusions. The following points should be beard in mind:

1. In the introduction, explain the purpose of the experiment and do a short review of the methods described to tackle this problem.
2. Protocols must be suitably detailed including schemes preferentially where appropriate and the necessary calculations.
3. Discuss the main variables that influence the analytical signals and include figures with representative raw data and results presented in tables and graphs for optimization studies.
4. Include graphs for the calibration curve, discussing the values obtained for the figures of merit. Pay attention to the significant figures in each case.
5. Indicate and comment in case the results obtained in terms of selectivity (see additional note number 8) and precision.
6. Discuss the incidences during the course of the experiment.

21.7 Additional notes

1. Oligonucleotide solution aliquots must be prepared and maintained at −20°C and working solutions must be conserved at 4°C.
2. IC solutions can be employed to know the electrochemical behavior of the enzymatic product (and the analytical signal). They must be protected from light and kept refrigerated at 4°C. Working solutions must be prepared daily.
3. 3-IP solutions must be prepared daily and kept at 4°C, protected from light.
4. AP-labeled St aliquots must be prepared and maintained at −20°C; working solutions are conserved at 4°C.
5. A drop of 5 μL and a strand concentration of 1 μM are employed in the immobilization steps (step 1 in Section 21.5.1), but both variables could be varied to study their influence.

6. The blocking of the active surface remaining after immobilization is very important in bioassays. Different agents could be checked, as commented in Section 24.5.2. The influence of different concentrations could also be evaluated.

7. SWV is employed for measurement because it is a fast and sensitive electrochemical technique. However, cyclic voltammetry or differential pulse voltammetry could also be evaluated. In particular, cyclic voltammetry should be made initially to know the electrochemical behavior and processes of 3-IP.

8. Selectivity of the genosensor can be studied by evaluating the signal of a, e.g., 3-base mismatch strand: 5′-ACA-GCG-CCT-AAA-AAC-GAC-AAA-AAG-AG-AAG-3′-biotin. Mismatches are located in base numbers 5, 15, and 26. It is also biotinylated at the 3′-end to allow hybridization detection by interaction with AP-labeled St. Adding agents that increase stringency (e.g., 50% of formamide) should be considered.

9. The sensitivity could be improved using different conditions (drop volume, time of the different steps, buffer composition, etc. [7]). Students are encouraged to discuss and evaluate the different variables.

10. In this case, a proof of concept of a biosensor able to detect SARS DNA is presented. The target is labeled with biotin. Then, in a real assay, DNA would have to be amplified using biotinylated primers. Alternatively, a sandwich format (thiolated capture probe—target—biotinylated detection probe) should be employed.

21.8 Assessment and discussion questions

1. Indicate clearly all the steps of the procedure. The use of a scheme is encouraged.
2. Why a DNA strand is functionalized with a thiol group?
3. What is the role of the blocking agent? Enumerate the different possibilities.
4. What is the aim of employing the biotin—avidin interaction here?
5. Explain how the analytical signal is obtained, especially in what concerns to the electro-chemical technique employed.

References

[1] https://www.journals.elsevier.com/biosensors-and-bioelectronics, July 2019.
[2] A. Roda, E. Michelini, M. Zanghery, M. Di Fusco, D. Calabria, P. Simoni, Smartphone-based biosensors: a critical review and perspectives, Trends Anal. Chem. 79 (2016) 317—325.
[3] M. Campas, I. Katakis, DNA biochip arraying, detection and amplification strategies, Trends Anal. Chem. 23 (2004) 49—62.
[4] N.J. Ronkainen, H.B. Halsall, W.R. Heineman, Electrochemical biosensors, Chem. Soc. Rev. 39 (2010) 1747—1763.
[5] K. Wu, J. Fei, W. Bai, S. Hu, Direct electrochemistry of DNA, guanine and adenine at a nanostructured film-modified electrode, Anal. Bioanal. Chem. 376 (2003) 205—209.
[6] C. Fernández-Sánchez, A. Costa-García, 3-Indoxyl phosphate: an alkaline phosphatase substrate for enzyme immunoassays with voltammetric detection, Electroanalysis 10 (1998) 249—255.
[7] P. Abad-Valle, M.T. Fernández-Abedul, A. Costa-García, Genosensor on gold films with enzymatic electrochemical detection of a SARS virus sequence, Biosens. Bioelectron. 20 (2005) 2251—2260.
[8] M.A. Marra, et al., The genome sequence of the SARS-associated coronavirus, Science 300 (2003) 1399—1404.

Aptamer-based magnetoassay for gluten determination

Rebeca Miranda-Castro, Noemí de los Santos Álvarez,
M. Jesús Lobo-Castañón

Departamento de Química Física y Analítica, Universidad de Oviedo, Oviedo, Spain

22.1 Background

The availability of highly sensitive and selective bioreceptors has propelled the progress in the field of biosensors and bioassays. The traditionally used antibodies have, however, shown some weakness, thus opening the door to alternative or complementary bioreceptors, among which aptamers stand out. Aptamers are short single-stranded DNA or RNA sequences evolved in vitro from a chemically synthesized library of different nucleic acid sequences, according to their affinity toward a target molecule, in a process called systematic evolution of ligands by exponential enrichment (SELEX). These synthetic receptors adopt 3D structures defining a ligand-binding site complementary to the desired target [1–3]. Because of the similar functions between aptamers and antibodies, the former ones are generally considered as "chemical antibodies"; although aptamers exhibit prime advantages over classical antibodies, the following stand out: small size, low immunogenicity, as well as fast, reproducible, and inexpensive production. Likewise, aptamers are chemically and thermally stable, and they can be easily modified with all kinds of tracers (e.g., redox active or fluorescent molecules, enzymes) as well as with functional groups (biotin, $-NH_2$, etc.) being easily engineered into biosensors and bioassays, so-called aptasensors and aptamer-based assays, respectively.

One important problem in molecular biorecognition assays is nonspecific adsorption of unwanted interfering biomolecules. It is particularly serious in the case of electrochemical biosensors because the electrode surface acts as both solid support for the sensing phase and as electrochemical transducer and, frequently, the optimal conditions for the binding event and the signal readout come into conflict, forcing the adoption of a trade-off. The use of magnetic microparticles as a solid support for the molecular recognition reaction is one of the most used alternatives because of their high surface area that contributes to

Laboratory Methods in Dynamic Electroanalysis
https://doi.org/10.1016/B978-0-12-815932-3.00022-X

increase assay sensitivity and their magnetic properties that allow easy and efficient separation of the solution by application of an external magnetic field, such as a simple magnet, avoiding nonspecific adsorption [4].

In this experiment, we perform the quantitative determination of the gluten-derived harmful gliadin protein and one of the most immunogenic fragments of this protein, the 33-mer peptide, by using this new kind of receptors (aptamers). In particular, the aptamer to be tested is referred as Gli1 aptamer and it allows detecting the free 33-mer peptide and the entire protein [5].

As antibodies, aptamers can be used in different approaches including direct, indirect, competitive, and sandwich assays [6]. To reliably detect small target molecules lacking multiple binding sites, a competitive assay format becomes highly convenient. This one entails the determination of the native target on the basis of its capacity to compete with its surface-confined counterpart, in this case attached to magnetic beads, for a fixed amount of labeled aptamer (limited binding sites). As a result, the higher the target concentration in the sample, the lower is the electrochemical signal provided by the immobilized labeled receptor we obtain, thus resulting in a signal-off assay. This approach facilitates the detection of the intact gliadin protein and the 33-mer peptide present in hydrolyzed food samples [7]. The principle of a competitive aptamer-based assay is depicted in Fig. 22.1. The affinity interactions are performed on the surface of the magnetic particles in suspension, thus avoiding the contact of the transducer with other sample components.

The immobilized labeled aptamer is then electrochemically quantified by entrapping the modified magnetic particles on the transducer surface by applying a magnetic field. To reveal the binding event, the redox enzyme horseradish peroxidase (HRP) is employed as a reporting molecule because of its large amplifying power (each enzyme molecule is able to catalyze the conversion of thousands of substrate molecules). The enzyme horseradish peroxidase, HRP_{red}, catalyzes the reduction of hydrogen peroxide to generate water, while at the same

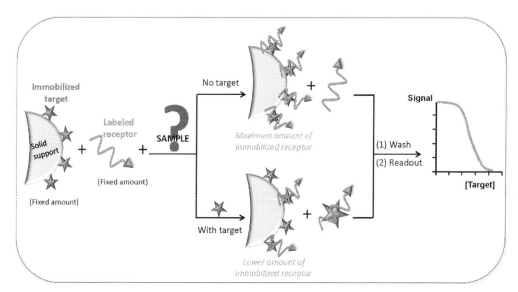

FIGURE 22.1   Schematic illustration of a competitive aptamer-based assay.

time the oxidized form of the enzyme (HRP_{ox}) is obtained. To regenerate the native enzyme form (HRP_{red}), a cosubstrate is incorporated, in this case tetramethylbenzidine (TMB_{red}), which is oxidized to TMB_{ox}, and this species is reduced on the surface of the screen-printed electrode if an adequate potential is applied (Fig. 22.2). The enzyme is specifically attached to the immobilized aptamer by means of the biotin−streptavidin affinity interaction.

The technique employed for the measurement step is chronoamperometry, one of the most widely used for the development of biosensors and bioassays. As explained in Chapter 8, this technique uses as excitation signal an instantaneous change in the potential applied to the working electrode, from an initial value where the species to be measured (in this case TMB_{ox} enzymatically generated) is electroinactive to a final value (E_{app}) where it is electrolyzed (reduced in this example) at a diffusion-limited rate. In this particular situation, because both the enzymatic and the electrochemical oxidation of TMB_{red} give rise to the same oxidation product, the initial potential for the step experiment is fixed at the open-cell potential. The final potential can be selected from a cyclic voltammogram for TMB in solution (Fig. 22.2B). The experiment is performed in the absence of convection, monitoring the current flowing in the cell as a function of time. The typical response curves are also depicted in Fig. 22.2C. The current is sampled at some fixed time window, after a time delay from the application of the potential step where the capacitive current becomes negligible, and the total current is mainly the faradaic current. The mean current during the sampling interval time, at the end of the step potential, is correlated with the concentration of the electroactive species according to the Cottrell equation. In this experiment, the mean current will

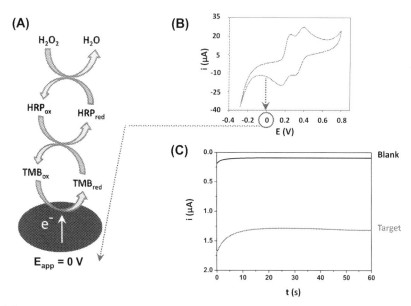

FIGURE 22.2 (A) Reactions involved in the enzymatic and electrochemical processes that take place on the electrode surface. (B) Cyclic voltammogram of TMB_{red} recorded onto screen-printed carbon electrodes (scan rate: 100 mV/s) for selecting the potential to be applied during the chronoamperometric measurement. (C) Typical chronoamperogram obtained in the aptamer-based assay. *HRP*, horseradish peroxidase; *TMB*, 3,3',5,5'-tetramethylbenzidine.

depend on the immobilized enzyme activity and in consequence on the amount of labeled aptamer onto the magnetic particles.

The experiment is directed to Master students. It will give them, during a 4 h laboratory session, an appreciation of the usefulness of the magnetic microparticles for conjugation to biocomponents such as peptides and aptamers and their use in combination with electrochemical transducers to develop assays with high sensitivity and selectivity.

22.2 Chemical and supplies

Reagents

- Biotin-labeled Gli1 aptamer and 33-mer peptide, according to the sequences shown in Table 22.1.
- Streptavidin-modified magnetic beads (1 μm Φ, 10 mg/mL).
- Gliadin standard solution prepared by Working Group on Prolamin Analysis (PWG).
- Streptavidin–peroxidase conjugate (strep-HRP$_2$, enzyme activity: 176 U/mg).
- Biotin.
- Commercial ready-to-use TMB solution including 3,3′,5,5′-tetramethylbenzidine (TMB$_{red}$) and H$_2$O$_2$.
- Tween-20 (70%).
- Phosphate buffer saline (10× PBS).
- 1 M Tris/HCl pH 7.4 solution.
- Sodium chloride.
- Magnesium chloride.
- Milli-Q purified water to prepare all solutions.

Instrumentation and materials

- A potentiostat equipped to perform cyclic voltammetry and chronoamperometry.
- Screen-printed electrochemical cells including a carbon working electrode.
- A specific connector acting as interface between the screen-printed cell and the potentiostat.
- A magnet incorporating a sample rack with 16 tube positions for magnetic separations: the magnetic separator (Fig. 22.3).
- A 12-tube mixing wheel for those steps requiring continuous mixing.
- Vortex.
- Magnets with the same diameter as the working electrode in the screen-printed cell.
- Pipettes, tips, timer, double-sided adhesive tape, 1.5 mL plastic microcentrifuge tubes.

TABLE 22.1   Commercial DNA sequence, written from 5′ to 3′, and peptide sequence for the development of the assay.

Description	Sequences
Biotinylated Gli1 aptamer	Biot-CTAGGCGAAATATAGCTACAACTGTCTGAAGGCACCCAAT
Biotinylated 33-mer peptide	LQLQPFPQPQLPYPQPQLPYPQPQLPYPQPQPF-Lys-Biotin

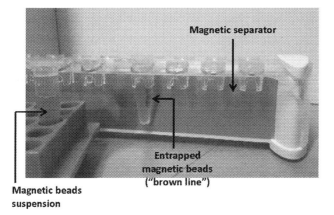

FIGURE 22.3 View of the magnetic beads entrapped on the wall of the tube with the help of the magnetic separator.

Buffer solutions for aptamer-based assay development:

— *Modification buffer* (Bmod): 1× PBS buffer solution pH 7.4 (10-fold diluted from commercial 10× PBS) supplemented with 0.01% Tween-20.
— *Binding buffer* (BB): 50 mM Tris pH 7.4, 0.25 M NaCl, 5 mM $MgCl_2$.
— BBT: Binding buffer supplemented with 0.01% Tween-20.

22.3 Hazards

Reagents handled in this experiment do not pose any hazard to the user, except for sensitive celiac patients.

As required in (bio)chemical laboratories, students are required to wear lab coat, appropriate gloves, and safety glasses.

22.4 Experimental procedure

22.4.1 Modification of streptavidin-coated magnetic microbeads with 33-mer peptide

Microparticles are typically supplied as a liquid suspension in a storage buffer containing preservatives; therefore, a change of medium is required for performing their surface modification. With such a purpose:

1. Mix thoroughly the commercial stock solution in a vortex before taking an aliquot.
2. Transfer 50 μL of the stock solution into an Eppendorf tube and add 950 μL of modification buffer (Bmod) to obtain a 0.5 mg/mL suspension of the magnetic particles.
3. Mix the suspension in the mixing wheel for 2 min at room temperature.
4. Place the vial in the magnetic separator to accumulate the beads in the wall of the tube in contact to the magnet (formation of a brown "line" is observed, see Fig. 22.3).

5. After 2 min, discard the supernatant by aspiration with a micropipette, keeping the vial in the separator.
6. Then, resuspend the microparticles in 1 mL of Bmod.
7. Repeat the steps 4–6, for a total of two washes.

Subsequently, the functionalization of the magnetic beads with the biotinylated peptide is carried out:

8. Remove the buffer carefully using the magnetic separator and the micropipette.
9. Add 1 mL of 2 μM biotinylated 33-mer in Bmod to the beads
10. Incubate the mixture for 30 min at 30 °C under continuous agitation in the mixing wheel.
11. Afterward, wash the modified beads twice in Bmod with the help of the magnetic separator to remove weakly adsorbed peptide fragments.

Then, a treatment with excess biotin is required to block the active streptavidin binding sites that may remain free.

12. Incubate the 33-mer peptide-modified beads with 1 mL of 500 μM biotin in Bmod for 30 min at 30°C under shaking in the mixing wheel.
13. Carry out two washing steps with Bmod as before (steps 4–6 twice).
14. Reconstitute the modified microparticles in 0.5 mL of BB and store at 4 °C until usage. These particles (in a 1 mg/mL suspension) constitute the sensing phase.

22.4.2 Competitive binding assay

A similar protocol is carried out for PWG gliadin protein or 33-mer peptide detection:

1. Take aliquots of 30 μL of the suspension of 1 mg/mL microparticles modified with 33-mer peptide and transfer into Eppendorf vials.
2. Remove the buffer carefully using the magnetic separator and the micropipette.
3. Add to the microbeads 500 μL of BB containing 250 nM biotinylated Gli1 aptamer and different amounts of target: (0.1–10 nM) in the case of the 33-mer peptide and (1–1000 ppb) with PWG gliadin protein acting as calibration standard. Incubate for 30 min in the mixing wheel at 30 °C under stirring conditions.
4. Wash twice the microparticles with 1 mL of BBT to remove nonspecifically adsorbed aptamer.
5. Discard the supernatant and expose the microparticles to 500 μL of BBT containing an excess of enzyme conjugate (2.5 μg/mL strp-HRP$_2$) for 30 min under stirring at 30 °C.
6. Perform two washing steps with BBT and a third one with the same buffer without Tween-20, i.e., BB. (Residues of detergent could adversely affect the subsequent steps.)
7. Resuspend the magnetic beads in 30 μL of BB.

22.4.3 Enzymatic reaction and electrochemical detection

Next steps take place on the screen-printed electrochemical cell:

1. Beforehand, rinse the cell with ethanol and water and dry it using gentle nitrogen stream to get a reproducible starting substrate.

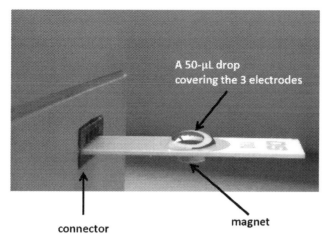

FIGURE 22.4 Experimental setup for chronoamperometric measurement of the enzymatic activity of the enzyme immobilized onto magnetic beads by means of the bioaffinity event. Modified beads are entrapped onto the working electrode of the screen-printed electrochemical cell by using a magnet placed under the working electrode.

2. Fix a magnet under the working electrode of the disposable screen-printed electrochemical cell by using a double-sided adhesive tape (see Fig. 22.4).
3. Deposit 10 μL of modified microparticles on the carbon working electrode and wait for 1 min to ensure their magnetic entrapment.
4. Add 40 μL of commercial TMB solution containing the HRP substrates TMB_{red} and H_2O_2 to cover the three electrodes.
5. Let the enzyme reaction to proceed at room temperature for 30 s (control with a timer) and immediately after, record the chronoamperometric signal due to the reduction of the enzymatically generated product at $E_{app} = 0$ V for 60 s. Detection voltage is selected according to the cyclic voltammogram of TMB_{red} recorded onto the working electrode (Fig. 22.2B).
6. Plot the recorded signal versus the target concentration. Estimate the figures of merit, the dynamic range, and the limit of detection (see next section).

22.4.4 Data collection and analysis

The chronoamperometric signal readout is obtained from the average of the cathodic current recorded for the last 10 s, when a plateau current is reached.

Data can be fitted to a four-parameter logistic model, whose equation is as follows:

$$y = A + \frac{B - A}{\left(1 + \dfrac{x}{EC_{50}}\right)^p}$$

III. Bioelectroanalysis

where the independent variable, x, is the target concentration, the dependent variable, y, is the recorded signal, A the response at an infinite target concentration (the minimum value of y), B the response in the absence of target (the maximum value of y), EC_{50} the target concentration at which the recorded signal is 50% of the maximum signal (half maximal effective concentration, i.e., point on the curve halfway between the horizontal asymptotes A and B), and p the maximum slope (Hill's slope) of the sigmoidal curve at point EC_{50}.

The detection limit of this type of assays can be calculated as the target concentration at which the recorded signal is 95% of the maximum value A.

22.5 Lab report

Write a lab report following the typical scheme of a scientific article, including a brief introduction, experimental part (materials, equipment, and protocols), results and discussion, and conclusions. The following points should be beared in mind:

1. In the introduction, explain the purpose of the experiment and the importance thereof in food safety control field. Do a short review of the described methods to tackle this problem.
2. Protocols must be suitably detailed including schemes where appropriate and the necessary calculations.
3. Include figures with representative raw data (cyclic voltammograms and chronoamperograms) and results presented in tables and graphs (calibration curves), paying special attention to the significant figures in each case.
4. Discuss the values obtained considering the expected results and the incidences during the course of the experiment.

22.6 Additional notes

1. Magnetic microparticles modified with 33-mer peptide should be used within 1 week.
2. Commercial TMB solution is stored in the refrigerator so it should be allowed to reach room temperature protected from light at least 20 min before use.
3. No memory effects are observed when reusing the same screen-printed electrochemical cell for several electrochemical measurements after thorough washing.
4. Special attention should be paid to ensure that the three electrodes in the screen-printed cell are covered with the solution.
5. The low solubility of PWG gliadin in aqueous solutions avoids obtaining a complete sigmoidal curve (gliadin concentrations higher than 1 µg/L cannot be assessed).
6. Different curve fitting programs can be used to obtain the response model (e.g., origin).

22.7 Assessment and discussion questions

1. What are the differences between a biosensor and a magnetobioassay?
2. Identify the type of chemical interactions involved in the aptamer-based assay.

References

[1] C. Tuerk, L. Gold, Systematic evolution of ligands by exponential enrichment: RNA ligands to bacteriophage T4 DNA polymerase, Science 249 (1990) 505–510.

[2] A.D. Ellington, J.W. Szostak, In vitro selection of RNA molecules that bind specific ligands, Nature 346 (1990) 818–822.

[3] M.R. Dunn, R.M. Jimenez, J.C. Chaput, Analysis of aptamer discovery and technology, Nat. Rev. Chem. 1 (2017) 0076.

[4] I.M. Hsing, Y. Xu, W.T. Zhao, Micro and nanomagnetic particles for applications in biosensing, Electroanalysis 19 (2007) 755–768.

[5] S. Amaya-González, N. de-los-Santos-Álvarez, A.J. Miranda-Ordieres, M.J. Lobo-Castañón, Aptamer binding to celiac disease-triggering hydrophobic proteins: a sensitive gluten detection approach, Anal. Chem. 86 (2014) 2733–2739.

[6] S.Y. Toh, M. Citartan, S.C.B. Gopinath, T.-H. Tang, Aptamers as a replacement for antibodies in enzyme-linked immunosorbent assay, Biosens. Bioelectron. 64 (2015) 392–403.

[7] S. Amaya-González, N. de-los-Santos-Álvarez, A.J. Miranda-Ordieres, M.J. Lobo-Castañón, Sensitive gluten determination in gluten-free foods by an electrochemical aptamer-based assay, Anal. Bioanal. Chem. 407 (2015) 6021–6029.

PART IV

Nanomaterials and electroanalysis

Determination of lead with electrodes nanostructured with gold nanoparticles

Graciela Martínez-Paredes[1], María Begoña González-García[2], Agustín Costa-García[3]

[1]OSASEN Sensores S.L, Parque Científico y Tecnológico de Bizkaia, Derio, Spain; [2]Metrohm Dropsens, Parque Tecnológico de Asturias, Edificio CEEI, Asturias, Spain; [3]Departamento de Química Física y Analítica, Universidad de Oviedo, Oviedo, Spain

23.1 Background

Lead is a neurotoxic heavy metal that can be easily absorbed by the human body, especially by ingestion of contaminated food or water; its toxic effects on humans, especially children, have been well documented [1]. Lead is a metal that is purely toxic and plays no role in human metabolism. It has also been identified as an osteotropic or bone seeker element, which can replace by ion exchange the physiological constituents that strengthen the bone such as calcium, especially during remineralization processes. Like calcium, lead remains in the bloodstream and body organs such as the muscle or brain for a few months. What is not excreted is absorbed into the bones, where it can remain for a lifetime. Reports suggest that neurological damage in children may occur at blood lead levels as low as 100 μg/L (ppb). In response to these facts, several governmental agencies have launched campaigns to reduce human lead exposure to a minimum, and Centers for Disease Control and Prevention (CDC) recommended universal screening for children. Thus, there is an increasing demand of accurate, quick, and portable methods for both blood and environmental lead analysis. The determination in blood is aimed to screen large numbers of adults and children at risk of lead poisoning, which are usually measured by graphite furnace atomic absorption spectrometry. Environmental lead analysis tries to quickly find and remediate lead exposure. As lead is persistent and can bioaccumulate in the body over time, its exposure should be minimized, and this is why

Laboratory Methods in Dynamic Electroanalysis
https://doi.org/10.1016/B978-0-12-815932-3.00023-1

regulations affecting lead in drinking water sets the maximum allowable concentration of lead in consumption water at 10 μg/L as the criteria of water quality for human health preservation. Lead can enter into drinking water by corrosion of pipes containing lead, especially where the water has high acidity or low mineral content. The most common problem is with brass or chrome-plated brass faucets and fixtures with lead solder, from which significant amounts of lead can enter into the water, especially hot water [2]. A correct analysis of trace amounts of lead in drinking water by using rapid, simple, portable, and cheap methods seems to be the best approach to monitor lead contamination in water sources.

Electrochemical techniques that introduce a preconcentration step, such as stripping voltammetry (see Chapter 4), have been proved to be powerful electroanalytical techniques for the detection of trace metals in the μg/L range using inexpensive instrumentation [3]. Its high sensitivity can be attributed to the preconcentration step. During preconcentration, the metal analytes are accumulated onto the working electrode (WE). After the preconcentration step, the electrode is scanned linearly toward positive potentials so that the metals are stripped from the electrode and reoxidized at a potential characteristic of each metal. Square wave voltammetry (SWV, see Chapter 3) is often used in this stripping analysis.

Moreover, screen-printed electrodes (SPEs) are nowadays very well-established transducers for the development of electrochemical (bio)sensors because of their disposable character and simple, rapid, and low-cost production. SPEs have a great capability to be modified; these modifications consider not only the metal film formation but also the incorporation of nanomaterials or (bio)molecules. This versatility together with their miniaturized size and the possibility to be connected to portable instrumentation make them very appropriate for on-site determination of target analytes in several fields of application.

With the recent advances in nanotechnology, nanomaterials (materials that range from a few units to hundreds of nanometers in size) have received great interest in the field of (bio)sensors because of their exquisite sensitivity. Many types of nanomaterials have demonstrated their appropriateness for (bio)sensing applications when combined with SPEs. The use of such nanomaterials led to clearly enhanced performance with increased sensitivities and lowered detection limits in several orders of magnitude.

Nanomaterials can be classified based on the number of dimensions that are not confined to the nanoscale range (<100 nm). In this way, 2D nanomaterials, which exhibit plate-like shapes, include nanofilms, nanolayers, and nanocoatings (graphene can be an example); 1D nanomaterials lead to needle-like-shaped nanomaterials, including nanotubes, nanorods, and nanowires; and 0D nanomaterials are usually represented by nanoparticles. Common nanoparticles used in biosensing can be polymeric, ceramic, or metallic. Among the metallic ones, gold nanoparticles in particular have been widely used in many applications. Among them, electrochemical sensors benefit from characteristics such as good biocompatibility, excellent conducting capability, and high surface-to-volume ratio [4].

Using gold nanoparticle—modified screen-printed carbon electrodes (AuNP-SPCEs) has several advantages, which involved catalysis, mass transport, high-effective surface area, and control over electrode microenvironment. The catalytic mechanism of gold nanoparticles for the oxidation of several analytes has been studied extensively [5]. Generally, compared with a bulk gold electrode, the gold nanoparticles on the surface of electrode would enable fast electron transfer kinetics, thus decreasing the overpotential needed for the reduction of metals, which made the voltammograms appear more reversible, and the redox reaction

(A)

(B)

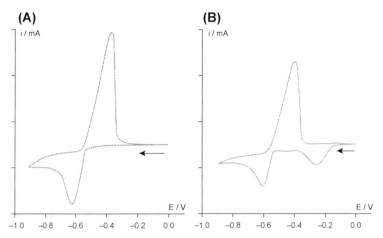

FIGURE 23.1 Cyclic voltammetry recorded in a 200 μg/mL lead solution on: (A) screen-printed carbon electrodes (blue [light gray in print version]) and (B) gold nanoparticle—modified screen-printed carbon electrodes (red
[dark gray in print version]). *From G. Martínez-Paredes, M.B. González-García, A. Costa-García, In situ electrochemical
generation of gold nanostructured screen-printed carbon electrodes. Application to the detection of lead underpotential deposition, Electrochim. Acta 54 (2009) 4801e4808.*

become kinetically viable, as it happens in underpotential deposition (UPD) processes. An
example of the UPD process is shown in Fig. 23.1.

UPD is an electrochemical phenomenon where reduction of a metal cation to a solid metal
can be carried out at a potential less negative than the equilibrium potential for the reduction
of this metal (the potential at which it will deposit onto itself). It can then be understood as
when a metal can deposit onto one material more easily than onto another one because of a
strong interaction between them. This can be observed in the cyclic voltammetry (CV) of
Fig. 23.1, where two additional cathodic peaks and an anodic one appear when using
AuNP-SPCEs instead of SPCEs. The first redox pair, at approximately −0.3 (cathodic peak)
and −0.1 V (anodic peak), corresponds to metal deposition and redissolution processes on
the AuNPs surface, generating a first monolayer of the metal. The formation of a more
compact metal layer onto AuNPs surface is expressed as a second cathodic peak at
around −0.45 V.

In this experiment, gold nanostructured electrodes are used as transducers for the detection of lead based on this lead UPD process. Metal UPD offers advantages over the use of
bulk metal deposition, including better sensitivity and good repeatability of the analytical
response, as UPD processes occur only up to one or two monolayers of the metal, so just a
small quantity of the metal is involved in this process.

The combination of all the techniques stated before provides a methodology that can be
used in relatively cheap, easy-to-use, suitable for in-field use devices and sensitive enough
to perform trace metal analysis. However, the main drawback of these devices is that previous calibration of the underlying electrode is necessary for the analysis of real samples. An
example of a calibration curve for lead detection obtained with this kind of electrodes is
shown in Fig. 23.2.

The analytical characteristics that the methodology developed in this experiment offers allow
carrying out an analysis of lead in water. Considering the linear range and limit of detection of
this methodology, it is possible to perform the analysis with minimum sample dilution.

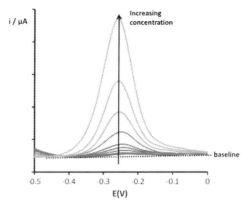

FIGURE 23.2   Example of a calibration curve for lead detection obtained with one gold nanoparticle—modified screen-printed carbon electrode.

In this experiment, adapted from Refs. [6,7], we modify SPCEs with gold to nanostructure them. These electrodes are used to follow the electrochemical UPD of lead over gold nanostructures. The whole experiment can be carried out in a single laboratory session of approximately 3 h. This experiment is appropriate for undergraduate students of Chemistry and Environmental Sciences. After this experiment, students will be able to understand some of the advantages of nanotechnology and get closer to basic electrochemical techniques such as those involving preconcentration steps. Students will also be trained in statistic calculations such as detection limits and regression equations that are valuable for characterizing any analytical method.

23.2 Electrochemical cell

The electrochemical cell consists of SPCEs (ref. DSC110), which are commercially available. An edge connector, from Metrohm DropSens, connects SPEs to the potentiostat.

The electrochemical card incorporates a conventional three-electrode configuration printed on a ceramic substrate (3.4 cm × 1.0 cm). Both WE (disk-shaped, 4 mm in diameter) and counter electrode (CE) are made of carbon ink, whereas the pseudo-reference electrode and electric contacts are made of silver. An insulating layer was printed over the electrode system, leaving uncovered the electric contacts and a working area, which constitutes the reservoir of the electrochemical cell, with a volume of 50 μL.

23.3 Chemicals and supplies

— *Gold nanostructure source:* Standard of gold (III) tetrachloro complex solution in 1 M hydrochloric acid (commercially available from Merck).
— *Components of background electrolyte:* Hydrochloric acid.
— *Analyte:* Lead (II) nitrate.

- *Samples:* Tap water or mineral water.
- *General materials and apparatus:* 1.5-mL centrifuge tubes, micropipettes and corresponding tips, and weighing scale.
- Milli-Q water is employed for preparing solutions and washing.

23.4 Hazards

Hydrochloric acid, used for the preparation of the background electrolyte, is corrosive to metals, causes severe skin burns and eye damage, and may cause respiratory irritation.

Lead nitrate may intensify fire, is harmful if swallowed or inhaled, causes serious eye damage, may damage the unborn child, and is suspected of damaging fertility. Furthermore, it is very toxic to aquatic life with long-lasting effects, so a proper disposal of residues is required, as well as maintaining clean and duty in the working space.

Students are required to wear hand gloves, eye protection, lab coat, and other appropriate protective equipment during this experiment.

23.5 Experimental procedure

23.5.1 Solutions and sample preparation

Prepare the following solutions (estimate the amount of solution required):

- 0.1 M HCl.
- 0.5 mM $AuCl_4^-$ solution prepared in 0.1 M HCl from standard gold (III) tetrachloro complex (1.000 ± 0.002 g of tetrachloroaurate (III) in 500 mL of 1 M HCl).
- Prepare a 200 µg/mL lead stock solution by accurately weighing an exact amount of lead (II) nitrate and dissolving it in the proper amount of 0.1 M HCl.
- Prepare 500 µL of each lead standard solution (e.g., 0, 1, 2.5, 5, 10, 20, 25, 50, 75, 125, and 250 µg/L) in centrifuge tubes by suitable dilution of the lead stock solution in 0.1 M HCl.

Regarding the samples, the only pretreatment required is a proper dilution (e.g., 1:2) in 0.2 M HCl solution to acidify the samples and assure the same background electrolyte as in standards (0.1 M HCl).

23.5.2 Nanostructuration of the electrochemical cell

To nanostructure the WE with gold nanoparticles, proceed as follows:

1. Connect the SPCEs to the potentiostat and place, over the three electrodes, a 50 µL drop of the 0.5 mM $AuCl_4^-$ acidic solution. Apply a constant current intensity of -100 µA for 4 min. After that, and in the same medium, apply a potential of $+0.1$ V during 2 min to desorb the generated hydrogen. Record the chronopotentiogram (E vs. t curve) during the electrodeposition of gold.

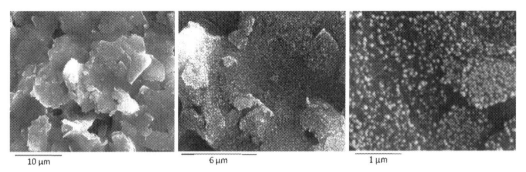

10 μm   6 μm   1 μm

FIGURE 23.3 Scanning electron microscopy image factors of gold nanoparticle—modified screen-printed carbon electrodes (AuNP-SPCEs) formed following the nanostructuration conditions stated before at different amplification factors. The images correspond to different zones of the same AuNP-SPCE. Amplification factor: 3000× (left image), 8000× (center image), and 30,000× (right image).

2. Rinse the AuNP-SPCEs generously with water and let them dry.

Gold nanoparticles are formed over the surface of WE. Scanning electron microscopy images at different amplification factors for Au-SPCEs formed using these nanostructuration conditions can be seen in Fig. 23.3. In these images, AuNPs electrochemically formed and their distribution over the WE carbon surface is shown. These nanostructuration conditions generate AuNPs with a mean size of 70 ± 20 nm in diameter and a density of around 4.4×10^7 AuNPs/mm^2.

23.5.3 Electrochemical measurements

To perform electrochemical measurements employing the SPCEs and AuNP-SPCEs, connect the electrodes to the potentiostat and place an aliquot of 50 μL covering all the three electrodes: WE, reference electrode, and CE.

For CV measurements, scan the potential from 0.0 to -0.9 V at 100 mV/s.

For square wave anodic stripping voltammetry measurements (SWASV), the following parameters were used:

— Preconcentration step at -0.5 V for 90 s
— Potential scan from -0.5 to -0.025 V
— Square wave frequency (f): 50 Hz
— Square wave amplitude (ΔE): 25 mV
— Step potential (E_s): 2 mV

23.5.4 Identification of the underpotential deposition process

To identify the UPD process in which this experiment is based, compare the voltammetric behavior of lead solutions on SPCEs and AuNP-SPCEs. Then,

1. Connect an SPCE to the potentiostat, deposit 50 μL of a 200 μg/mL lead solution, and record the CV.

2. Connect an AuNP-SPCE to the potentiostat and perform a CV measurement in 50 µL of a 200 µg/mL lead solution.
3. Compare both voltammograms and identify the UPD process studied in this experiment.

23.5.5 Calibration curve

1. Pretreat the AuNP-SPCEs performing two or three SWASV measurements in 0.1 M HCl, until the signal obtained is stable and reproducible. Gently rinse the electrode with water between consecutive measurements.
2. Sequentially perform an SWASV measurement for each lead standard solution over the pretreated AuNP-SPCE. Rinse the electrode gently with water between measurements for different standards.
3. Repeat step 2 with two more AuNP-SPCEs so that the analytical signal is obtained by triplicate.
4. Obtain a calibration curve for lead using the peak current as the analytical signal, expressed as an average value with standard deviation, and obtain a regression equation. Estimate also the detection and quantification limits of the method.
5. Evaluate the reproducibility of AuNP-SPCEs as transducers for lead detection.

23.5.6 Determination of lead in water samples

1. Perform measurements in the unknown samples, in triplicate, using the procedure stated before. Use a different pretreated AuNP-SPCE for each measurement.
2. Estimate the concentration of analyte in the unknown sample using the calibration curve by interpolating the peak current into the regression equation.
3. Correct the dilution factor performed to acidify the sample to get the lead content in the sample. Express the results with adequate number of significant figures.

23.6 Lab report

Once the experiment is finished, write a lab report following the style of a scientific article, including an introduction (discussing the role of low-cost mass-produced electrodes as transducers of chemical sensors and the main advantages of the use of nanomaterials on these electrodes), experimental procedures, results obtained and discussion, finishing with main conclusions. In a more detailed manner:

1. Indicate the motivation for doing these experiments and comment the basis of the analytical signal.
2. Include pictures and schematics of the device (dimensions, steps of fabrication, etc.).
3. Include graphical representation of the data obtained along the experiment and remember to include error bars in graphics, whenever possible.

4. The conclusions derived from the experiments should be included. In case a further study of some parameter influences the analytical signal, conclusions for the chosen values should also be included.
5. Represent the calibration curve and include the linear range, sensitivity, limit of detection, and limit of quantification of the sensor.
6. Discuss the reproducibility of the AuNP-SPCEs used as transducers for lead detection.
7. Determine the lead concentration of the samples, indicating the results with adequate number of significant figures.

23.7 Additional notes

1. For rinsing electrodes with water, disconnect them from the potentiostat.
2. During the washing steps, be careful not to spread any liquid over electric contacts of electrodes; in case of spillover, wipe it softly with a tissue paper.
3. The value for different parameters has been already indicated in the text, but it would be interesting to study their influence by varying them in a range (e.g., potential and time in the preconcentration step or step potential, frequency, and amplitude in SWV).
4. Despite the disposable character of SPCEs, the reusability of AuNP-SPCEs has been previously tested and confirmed. Hence, a whole calibration curve of lead can be carried out using the same electrode taking care of rinsing the electrode gently with water after each standard measurement.
5. As tap or mineral water should not contain high quantities of lead according to regulation, water samples can be spiked with a known amount of lead stock solution in such a way that the final concentration fall into the linear range of the method before determination.

23.8 Assessment and discussion questions

1. Explain all the different steps of the methodology with the main parameters that have to be taken into consideration.
2. In the chronopotentiogram recorded during the electrogeneration of gold nanoparticles on SPCEs surfaces, discern, if possible, between nucleation and growth phases of gold nanoparticles.
3. What is the main advantage of using gold nanostructured electrodes instead of gold electrodes in lead water analysis?
4. Discuss the advantages of electrogenerating gold nanoparticles directly on the SPCEs surfaces over physical adsorption of colloidal gold on SPCEs.
5. How could be the detection limit of the method improved in a simple way?

References

[1] M. Ahamed, S. Verma, A. Kumar, M.K. Siddiqui, Environmental exposure to lead and its correlation with biochemical indices in children, Sci. Total Environ. 346 (2005) 48—55.
[2] www.epa.gov/ground-water-and-drinking-water/basic-information-about-lead-drinking-water, July 2018.

[3] X. Niu, M. Lan, H. Zhao, C. Chen Chen, Y. Li, X. Zhu, Electrochemical stripping analysis of trace heavy metals using screen-printed electrodes, Anal. Lett. 46 (2013) 2479–2502.

[4] K. Saha, S.S. Agasti, C. Kim, X. Li, V.M. Rotello, Gold nanoparticles in chemical and biological sensing, Chem. Rev. 112 (5) (2012) 2739–2779.

[5] G.C. Bond, Catalytic properties of gold nanoparticles, Gold Nanoparticles Phys. Chem. Biol. (2012) 171–197.

[6] G. Martínez-Paredes, M.B. González-García, A. Costa-García, Lead sensor using gold nanostructured screen-printed carbon electrodes as transducers, Electroanalysis 21 (2009) 925–930.

[7] G. Martínez-Paredes, M.B. González-García, A. Costa-García, In situ electrochemical generation of gold nanostructured screen-printed carbon electrodes. Application to the detection of lead underpotential deposition, Electrochim. Acta 54 (2009) 4801–4808.

CHAPTER
24

Electrochemical behavior of the dye methylene blue on screen-printed gold electrodes modified with carbon nanotubes

Raquel García-González, M. Teresa Fernández Abedul

Departamento de Química Física y Analítica, Universidad de Oviedo, Oviedo, Spain

24.1 Background

Nanomaterials have been combined, from their discovery and invention, with electrochemistry, and a rich interplay has been established. The literature is replete of many examples, especially in the case of biosensors, where biological/bioderived or bioinspired reagents are included in the recognition area of an electrochemical transducer. Nanomaterials can help in both the immobilization of bioreagents and the transduction of the biological event. Complicate architectures combining countless nanomaterials (single or hybrids) with different elements following various methodologies produce competitive devices, especially in terms of sensitivity and selectivity.

From the classic talk that Nobel Laureate R. Feynman gave on December 29, 1959, at the annual meeting of the American Physical Society at the California Institute of Technology (*There is plenty of room at the bottom* [1]), where he stated: "*I would like to describe a field in which little has been done, but in which an enormous amount can be done in principle ... What I want to talk about is the problem of manipulating and controlling things on a small scale*" to our days, especially since the 1980s, many inventions and discoveries in the fabrication of nanoobjects (materials with the dimensions at the nanoscale) have been made. Nanomaterials have opened a new field, especially because they may have properties different than the same chemical substances with structures at a larger scale. Nanotechnology has been then considered a top science and technology priority. However, it is not entirely a new area, as Nature has

Laboratory Methods in Dynamic Electroanalysis
https://doi.org/10.1016/B978-0-12-815932-3.00024-3

243

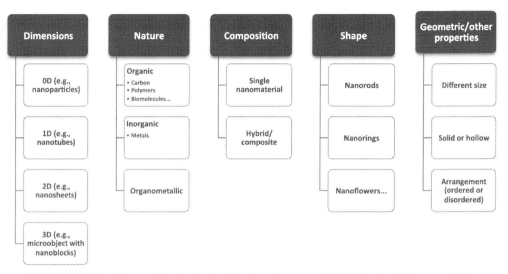

FIGURE 24.1   Scheme with a basic classification of nanomaterials, following different criteria.

many objects and processes that function on a micro- to nanoscale. In this context, combination of "bio" and "nano" has become a priority in many developments and applications.

Nanomaterials can be classified (Fig. 24.1) according to many different criteria. Then, in accordance with the dimensions that are out of the nanoscale, they can be classified as (i) 0D, when all the three dimensions are below 100 nm (nanoparticles, nanospheres, nanoballs, etc.), (ii) 1D, when two dimensions are below 100 nm (nanotubes, nanofibers, nanowires, nanorods, etc.), (iii) 2D, when one dimension is below 100 nm and the object is two-dimensional (graphene, nanosheets, nanofilms, etc.), or finally (iv) 3D materials, when a microobject is made of many nanoblocks joined together (there is no any dimension below 100 nm). Depending on the composition, there are organic and inorganic materials. In the first case, carbon-based materials (carbon nanotubes [CNTs] and nanofibers, graphene) but also polymers (dendrimers) and biomolecules (lipids, proteins) or cellulose nanocrystals (or fibers) are used in different applications. In the second case, metals (gold nanoparticles, quantum dots, metallic films) are mainly employed. There are two major possibilities: (i) combination of organic/inorganic nanomaterials in organometallic structures (e.g., metal organic frameworks) and (ii) generation of hybrid nanomaterials (e.g., Cu@Au nanoparticles) to gather, in both cases, the features of the different materials. Modifications of nanomaterials with different functional groups as well as performing specific treatments are also possible.

Now that many nanomaterials are being synthesized, it is very important to consider, even in the nanoscale, how small they are. In this context the advance in characterization techniques is of paramount importance. Then, depending on the nanomaterial, a different classification can arise. For example, quantum dots can vary from 1 to 50 nm and CNTs from 1 to 10 nm with an aspect ratio (length/diameter) higher than 1000. In any case, the size is continuously decreasing and new interesting materials are arising. These can be solid or hollow, and according to the latter, it is common to find nanopores, nanocavities, nanochannels, etc. According to the shape, objects such as nanoonions, nanoflowers, nano-horns, nanorings, etc., can be found. Taking into account all the possibilities, this field is plenty of opportunities.

Electroanalysis is an area of Analytical Chemistry that has employed nanomaterials to improve the characteristics of the methodologies. Therefore, they have been combined to take advantage of the properties that nanomaterials can provide, mainly in the field of biosensors. This can be due to the high surface/volume ratio that can increase adsorption possibilities and can also improve electron transfer. Specific active places and decrease in diffusion distances can have a part in some examples. In this experiment we are going to observe the effect of the modification of a screen-printed gold electrode with CNTs on the electrochemical behavior of methylene blue (MB). Although in 1978 Wiles and Abrahamson [2] observed a thick mat of fine fibers (4–100 nm) of graphite anodes composed of graphitic layers with a hollow core, the discovery of CNTs was attributed to Iijima [3] who prepared carbon structures consisting of needle-like tubes, each needle comprising coaxial tubes of graphitic sheets. On each tube the carbon-atom hexagons are arranged in a helical fashion about the needle axis, resulting in high-aspect ratio materials. They can be present in different forms and therefore their properties and possibilities are diverse, with electroanalytical applications reported from decades [4–6]. Two main classes can be distinguished: multiwalled carbon nanotubes (MWCNTs), with several concentric tubes, and single-walled CNTs, in which only one graphite sheet is rolled up. Their electronic properties depend on the structure, mainly diameter and chirality. They can be metallic (armchair) or semiconducting (zigzag or chiral). For MWCNTS, hollow tube, herringbone, or bamboo morphological variations can be found. Moreover, closed or open-ended as well as different functionalizations (e.g., −COOH, −OH, −SH, or −NH$_2$) are available. This is an example where a simple chemical composition and atomic bond configuration exhibits extreme diversity and then properties. Apart from these considerations, CNTs could be also generated on a nonconductive surface, able to be used as the working electrode of a potentiostatic system. Depending on the methodology employed for the generation, spaghetti- or forest-like CNTs (Fig. 24.2)

FIGURE 24.2 Field effect scanning electron microscopy images of (A) spaghetti and (B) forest-like carbon nanotubes (CNTs). Reprinted (in part). *With permission from I. Álvarez-Martos, A. Fernández-Gavela, J. Rodríguez-García, N. Campos-Alfaraz, A.B. García-Delgado, D. Gómez-Plaza, A. Costa-García, M.T. Fernández-Abedul, Electrochemical properties of spaghetti and forest like carbon nanotubes grown on glass substrates, Sens. Actautor. B: Chem. 192 (2014) 253–260.*

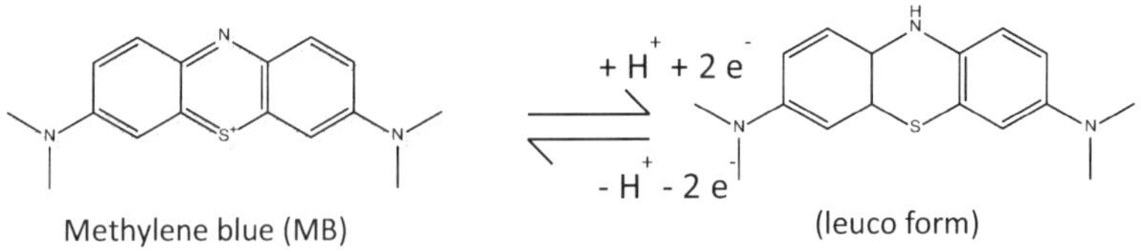

FIGURE 24.3   Structure of methylene blue.

can be obtained with differences in the electrochemical behavior, as demonstrated for dopamine [7].

One of the things that should be considered first is solubilization or dispersion. Although some materials are used as modifiers in a solid form (by rubbing them), most of them are employed as dispersions. The solvent depends on the material and then CNTs could be solved/dispersed in different organic or aqueous media [8,9]. The first group to report the use of CNTs in electroanalysis was Britto and coworkers in 1996 using a carbon paste electrode with bromoform as the binder for studying the behavior of dopamine [10], but the explosion of its use in electroanalysis starts with the work pioneered by Wang with the effect of the modification of a glassy carbon electrode with CNTs for the evaluation of the behavior of nicotinamide adenine dinucleotide [11].

In this experiment, adaptation of reference [12], we will observe the variation in the adsorption and electron transfer phenomena (through the behavior of MB, Fig. 24.3) after modification of the electrode with MWCNTs (amino-functionalized). CNTs have been dispersed in a solution of Nafion/ethanol. The polar chain that has the ionic exchanger favors solubilization [13]. As an example, they will be drop-casted on the surface of a gold screen-printed electrode (SPE).

This assay can be completed in three laboratory sessions of 3–4 h and is appropriate for undergraduate students of advanced Analytical Chemistry or Master students from different fields. With this experiment, the student will learn about nanoanalytical strategies for enhancing the sensitivity of the methodology of MB, a molecule with interesting electrochemical behavior that can be used as label in bioassays. Unlike Chapter 23, where carbon electrodes are nanostructured with gold nanoparticles, in this case gold electrodes are modified with carbon nanomaterials. This approach together with the use of a sensitive electrochemical technique and an accumulation time supposes a triple signal amplification.

24.2 Screen-printed gold electrodes

Screen-printed gold electrodes (AuSPEs) (DropSens, Spain) include a traditional three-electrode configuration on a ceramic card (Fig. 24.4). These SPEs include a gold disk electrode (12.6 mm^2) as working electrode, a gold hooked counter electrode (using for both the same ink), and a silver pseudo-reference electrode. All of them are screen-printed on a ceramic substrate (3.4 × 1.0 × 0.05 cm) and subjected to high-temperature curing (AuSPE-AT). An insulating layer serves to delimit the working area and electric contacts. A specific

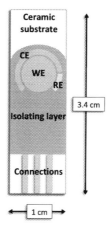

FIGURE 24.4 Scheme of the electrochemical screen-printed gold electrode card. *CE*, counter electrode; *RE*, reference electrode; *WE*, working electrode.

connector allows their easy connection to the potentiostat and alternatively they can be connected using hook clips.

24.3 Chemicals and supplies

- *Analyte*: Methylene blue (MB).
- *Components of the background electrolyte*: Trizma base and sulfuric acid.
- *Electrochemical cell preparation*: Screen-printed gold electrodes, amino-functionalized CNTs, Nafion/ethanol:water (1:1).
- *General materials, apparatus, and instruments*: 10-μL, 100-μL, and 1-mL micropipettes and corresponding tips, 1.5-mL microcentrifuge tubes, ultrasonic bath, centrifuge with interchangeable car for 1.5 and 5 mL tubes, analytical balance, magnetic stirrer, pH meter, incubator, and potentiostat.
- Milli-Q water is employed for preparing solutions and washing.

24.4 Hazards

Sulfuric acid, used for the preparation of the Tris-H_2SO_4 buffer, is corrosive and causes serious burns. Students are required to wear a lab coat, appropriate gloves, and safety glasses. Potential hazardous properties of carbon nanotubes are a matter of ongoing research but there are indications that they can be hazardous if workers or users are exposed through inhalation pathways [6]. Then, precautionary measures should be warranted and an appropriate mask should be used when handling carbon nanostructures.

24.5 Experimental procedure

24.5.1 Solutions and sample preparation

The following solutions are required (estimate the amount of solution required).

- 0.1 M Tris-H$_2$SO$_4$ pH 8.0, buffer solution (all working solutions used along this experiment will be prepared in this buffer solution).
- 10 and 20 μM MB solution.
- 1 mg of MWCNTs in 1 mL of 0.5% Nafion (in EtOH).
- 1 mg of MWCNTs in 1 mL of 0.5% Nafion (in EtOH/H$_2$O, with several ratios 1:1, 1:2, 1:3, 2:1, and 3:1).

24.5.2 Nanostructuration of screen-printed gold electrodes

24.5.2.1 Dispersion of carbon nanotubes

Although prepared dispersions exist in the market, generally CNTs are presented as a black solid powder. Therefore, most of the applications of CNTs, electrode modification among them (especially with SPEs), require a presolubilization for obtaining a homogeneous suspension. Achieving this step is not easy because CNTs are not dispersed in many solvents [1]. The CNTs weight/solvent volume ratio and the dispersion procedure are very important. The following steps are necessary:

1. Prepare a 0.5% Nafion solution in ethanol.
2. Add an amount of 1 mg of MWCNTs to 1 mL of 0.5% Nafion (in ethanol) solution.
3. Perform four cycles of sonication/centrifugation (2 h under sonication and 10 min of centrifugation). After each cycle, discard the precipitate and submit the supernatant to a new cycle. At the end, a completely homogeneous dispersion should be obtained.
4. Use the as-prepared dispersion (medium exchange, scan rate influence, and accumulation studies) or dilute it first with water in appropriate proportion (rest of studies) for electrode nanostructuration.

24.5.2.2 Nanostructuration of screen-printed gold electrodes

1. Deposit a drop (between 1 and 4 μL) of the dispersion of CNTs (this previously prepared or a dilution with water) on the working electrode.
2. Let evaporate the solvent, waiting until dryness (approximately 2 h at room temperature or 1 h at 70°C).

24.5.3 Electrochemical behavior of methylene blue on AuSPEs and MWCNTs-AuSPEs

To ascertain if the process of MB on these electrodes is diffusive or adsorptive in nature, a medium exchange can be performed. Then, cyclic voltammograms (CVs) are recorded as follows:

1. Deposit a 40-μL drop of a 20-μM solution of MB in 0.1 M Tris-H$_2$SO$_4$ pH 8.0 buffer on an AuSPE card, covering all the three electrodes.

2. Record a CV scanning the potential from 0.0 to −0.7 V at a scan rate of 250 mV/s.
3. Wash the card (paying attention not to wet electric connections) with buffer solution and wipe the cell gently with a tissue.
4. Deposit another 40-μL drop of a 0.1 M Tris-H_2SO_4 pH 8.0 buffer solution on the AuSPE, covering all the three electrodes.
5. Record a CV scanning the potential from 0.0 to −0.7 V at a scan rate of 250 mV/s.
6. Compare the signals obtained after steps 2 and 5 and explain if the MB process is controlled by diffusion or adsorption. Discuss if electrodes could be reused or have to be disposed of after measurement.
7. Repeat steps 1−6 but employing AuSPEs that have been nanostructured with MWCNTs (2-μL drop of CNTs dispersion in 0.5% Nafion solution [in ethanol]).
8. Discuss the reversibility of the process of MB in both electrodes.

To confirm the electrochemical behavior (diffusion or adsorption-controlled process), CVs with varying scan rates can be recorded. With this aim:

1. Deposit a 40-μL drop of a 10-μM solution of MB in 0.1 M Tris-H_2SO_4 pH 8.0 buffer on the AuSPEs, covering all the three electrodes.
2. Record CVs scanning the potential from 0.0 to −0.7 V at scan rates varying from 10 to 500 mV/s. Reuse or change the SPE card according to the results of the study.
3. Plot the value of the intensity of the peak current versus the scan rate and versus the square root of the scan rate, to know if the process is governed by diffusion or adsorption.
4. Repeat steps 1−3 but employing AuSPEs that have been nanostructured with MWCNTs (2-μL drop of CNTs dispersion in 0.5% Nafion solution [in ethanol]). Reuse or change the SPE card according to the results of the study.
5. Compare the signals and the graphs and explain the differences in terms of adsorptive or diffusive nature of the process.

24.5.4 Accumulation of methylene blue on MWCNTs-AuSPEs

A way to increase the sensitivity of the methodology for determination of MB species is employing different formats of potential scans (see Chapter 3). Square wave voltammetry (SWV) is a fast and sensitive electrochemical technique, especially adequate for reversible processes.

On the other hand, adsorption of MB on nanostructured electrodes allows also increasing the analytical signal (intensity of the peak current) by accumulating MB previously (Fig. 24.5) before measurement is made.

To know the accumulation behavior and the most appropriate time for this step, the following study is performed:

1. Deposit a 40-μL drop of a 10-μM solution of MB in 0.1 M Tris-H_2SO_4 pH 8.0 buffer on a MWCNTs-AuSPE card (using a 2-μL drop of CNTs dispersion in 0.5% Nafion solution [in ethanol] for the nanostructuration), covering all the three electrodes.

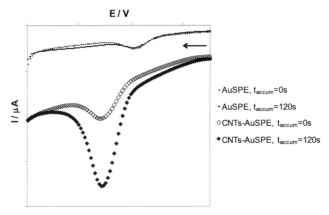

FIGURE 24.5 Enhancement of the signal by using CNTs, a sensitive electrochemical technique (square wave voltammogram) and an accumulation time before measurement (cathodic scan). *AuSPE*, screen-printed gold electrode; *CNT*, carbon nanotube. *From R. García-Goonzález, A. Costa-García, M.T. Fernández-Abedul, Enhanced detection of the potential electroactive label methylene blue by electrode nanostructuration with carbon nanotubes, Sens. Actuator. B: Chem. 202 (2014)129–136.*

2. Record a square wave voltammogram scanning the potential from 0.0 to −0.7 V (frequency, f = 50 Hz; amplitude, A = 0.05 V; step potential, E_s = 0.008 V) with no accumulation time.
3. Compare the intensity of the peak current obtained with CV (see previous section) and SWV.
4. Record SWVs using the same conditions as in step 2 on different MWCNTs-AuSPEs (after depositing the drop of MB solution as in step 1) and different accumulation times without applying any potential (i.e., at the open circuit potential) varying from 0 to 500 s.
5. Represent the accumulation curve by plotting the intensity of the peak current versus the accumulation time.
6. Select the best accumulation time as a compromise between the highest signal and a convenient analysis time.

24.5.5 Optimization of the nanostructuration

Although CNTs are commonly considered as a single reagent, the number of different structures and therefore variation in their properties is enormous. The possibility of employing very different dispersing agents increases this variability. Several parameters mainly referred to the nanostructuration procedure (volume, time, and temperature employed in the drop-casting procedure, concentration of CNTs, etc.) have to be carefully optimized to take advantage of the enhancement in the signal. This, together with the use of a very sensitive electrochemical technique and an accumulation step, implies a triple amplification of the signal, very convenient in most of the methodologies. Below there is a possible procedure for optimizing some of the variables:

24.5.5.1 Ratio of carbon nanotube dispersion/water

1. Prepare dilutions in water of the CNT dispersion with ratios (CNTs dispersion/water (v/v)) of 1:1, 1:2, 1:3, 2:1, and 3:1, homogenizing thoroughly.
2. Nanostructure AuSPEs using a 3-μL drop and let dry until complete solvent evaporation.
3. Deposit a 40-μL drop of a 10-μM solution of MB in 0.1 M Tris-H_2SO_4 pH 8.0 buffer.
4. Record the corresponding CVs scanning the potential from 0.0 to -0.7 V at a scan rate of 100 mV/s after waiting for 120 s (accumulation time).
5. Compare the voltammograms and select the best CNTs dispersion/water ratio.

24.5.5.2 Volume of drop of carbon naotube dispersion

1. Prepare a dilution in water of the CNT dispersion in a 1:1 ratio (CNTs dispersion/water (v/v)). Homogenize.
2. Nanostructure AuSPEs using drop volumes comprised between 1 and 5 μL and let dry until complete solvent evaporation.
3. Deposit a 40-μL drop of a 10-μM solution of MB in 0.1 M Tris-H_2SO_4 pH 8.0 buffer.
4. Record the corresponding CVs scanning the potential from 0.0 to -0.7 V at a scan rate of 100 mV/s after waiting for 120 s (accumulation time).
5. Compare the signals in the voltammograms and select the best drop volume.

24.5.5.3 Time and temperature of the nanostructuration step

1. Prepare a dilution in water of the CNT dispersion in a 1:1 ratio (CNTs dispersion/water (v/v)). Homogenize.
2. Nanostructure AuSPEs using a 3-μL drop and let dry under different conditions; e.g., overnight at room temperature, 1 h at 37°C or 1 h at 70°C.
3. Deposit a 40-μL drop of a 10-μM solution of MB in 0.1 M Tris-H_2SO_4 pH 8.0 buffer.
4. Record the corresponding CVs scanning the potential from 0.0 to -0.7 V at a scan rate of 100 mV/s after waiting for 120 s (accumulation time).
5. Compare the voltammograms and select the best conditions.

24.5.6 Calibration curve

1. Prepare a dilution in water of the CNT dispersion in a 1:1 ratio (CNTs dispersion/water (v/v)). Homogenize.
2. Nanostructure AuSPEs using a 3-μL drop and let dry until complete solvent evaporation.
3. Deposit 40-μL drops of solutions of MB of different concentrations (ranging from 100 nM to 100 μM) in 0.1 M Tris-H_2SO_4 pH 8.0 buffer on bare and nanostructured MWCNTs-AuSPEs.
4. Record the corresponding CVs scanning the potential from 0.0 to -0.7 V at a scan rate of 100 mV/s (in the case of MWCNTs-AuSPEs after waiting for 120 s of accumulation time).

5. Record the corresponding SWVs scanning the potential from 0.0 to -0.7 V (frequency, $f = 50$ Hz; amplitude, $A = 0.05$ V; step potential, $E_s = 0.008$ V, in the case of MWCNTs-AuSPEs after waiting for 120 s of accumulation time).
6. Plot the intensity of the peak current in each case versus the concentration of MB to obtain the calibration curves.
7. Estimate the linear range, sensitivity, limit of detection, and limit of quantitation, comparing the values obtained for bare and nanostructured electrodes with CV and SWV.

24.6 Lab report

Write a lab report following the typical scheme of a scientific article. Thus, the report has to include a short introduction and objective of the experiment, experimental section, results and discussion, and conclusions. To include diagrams, tables and images is highly recommended. The following points should be considered:

1. In the introduction, include the current importance of the nanostructuration of electrode surfaces with advances in (bio)sensor architectures and other nanomaterials that could be employed with this aim. Similarly, include recent reports of nanostructuration of AuSPEs and applications of MB.
2. The experimental part must include representative pictures of the device and procedures employed.
3. In results and discussion, add some interesting raw data such as CVs and SWVs and the representations employed for ascertaining the nature of the electrochemical process on bare and nanostructured electrodes, bar diagrams with the effects of the different variables on the analytical signals, and the calibration plot. Discuss the precision of the methodology (using the RSD) and the figures of merit. They could be compared with others found in the bibliography.
4. Add a pair of important conclusions that have been obtained after performing this experiment.

24.7 Additional notes

1. The buffer solution can be stored at 4°C for 1 week.
2. Because of its low cost and because possibility of adsorption exists, electrodes are considered as single use. All measurements are recorded on three electrodes and error bars must be included in the corresponding graphics. This is very useful also to discuss the precision of the methodology.
3. Although, amino-functionalized MWCNTs are here employed, different carbon nanomaterials, with different functionalization, type, or dimension (carboxylic functionalized, carbon nanofibers, carbon nanoparticles, or graphene).
4. CV and SWV are employed in this experiment, but differential pulse voltammetry ($A = 0.05$ V, $E_s = 0.008$ V) can be also employed for the sake of comparison (see example in Fig. 24.6).

E / V

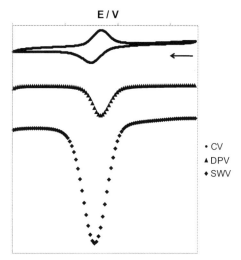

- • CV
- ▲ DPV
- ◆ SWV

FIGURE 24.6 Cyclic (CV), differential pulse (DPV), and square wave voltammograms (SWV) recorded in a solution of the same concentration of methylene blue after performing a 120-s accumulation step. *From R. García-Goonzález, A. Costa-García, M.T. Fernández-Abedul, Enhanced detection of the potential electroactive label methylene blue by electrode nanostructuration with carbon nanotubes, Sens. Actuator. B: Chem. 202 (2014)129–136.*

5. To know the precision of the calibration curve, it can be repeated in different days.
6. These devices can be considered as low-cost platforms (see other in part V). Moreover, it is interesting to discuss the reduction of the costs based on the low volumes used (especially when bioreagents are employed) as well as on the analysis time required.
7. SPEs could be nanostructured time before their use, which would imply a notorious simplification of procedures. Therefore, studies on the storage stability of the modification are relevant. To know the stability of electrode modification, AuSPEs are modified using optimum conditions, stored in a dry and dark place, and electrochemical characterization with SWV is performed for 10 consecutive days.
8. Discussion on the different explanations that have been given to the observed behavior on electrodes modified with CNTs (e.g., thin layer effect on a CNT porous layer generated on a glassy carbon electrode [14]) is very interesting as well as how the use of CNT-modified electrodes has led to a fundamental understanding of the location and nature of electron transfer processes on graphitic electrodes [15].
9. Discussion on different applications of the modification with CNTs in other fields (e.g., in metallic nanomotors [16] or electrofluidic structures [9]) could be also very didactic.

24.8 Assessment and discussion questions

1. A bibliographic search could be made to discuss other types of sensors where MB is involved. Compare main characteristics of each one.
2. CNTs can be employed in very different sensors. Make a table organizing the information.

3. What are the indicators of the reversibility of an electrochemical process?
4. How could you know if an electrochemical process is governed by diffusion or adsorption?
5. Discuss the advantages of the adsorption, a simple physical approach that has been employed to enhance the sensitivity of the sensor.
6. Why is this assay performed in a pH 8 buffer solution?
7. How is the detection potential chosen? What happens for higher or lower potentials?
8. What are the differences between several voltammetry techniques employed in the assay? Draw the excitation signal applied to the working electrode and the response signal in each case.

References

[1] R. Feynman, There is plenty of room at the bottom, Eng. Sci. (1959) 22–36.
[2] P.G. Wiles, J. Abrahamson, Carbon fiber layers on arc electrodes- I: their properties and cool-down behaviour, Carbon 16 (1978) 341–349.
[3] S. Iijima, Helical microtubules of graphitic carbon, Nature 354 (1991) 56.
[4] J. Wang, Carbon-nanotube based electrochemical biosensors: a review, Electroanalysis 17 (2005) 7–14.
[5] M.T. Fernández-Abedul, A. Costa-García, Carbon nanotubes (CNTs) – based electroanalysis, Anal. Bioanal. Chem. 390 (2008) 293–298.
[6] L. Agüí, P. Yáñez-Sedeño, J.M. Pingarrón, Role of carbon nanotubes in electroanalytical chemistry: a review, Anal. Chim. Acta 622 (2008) 11–47.
[7] I. Álvarez-Martos, A. Fernández-Gavela, J. Rodríguez-García, N. Campos-Alfaraz, A.B. García-Delgado, D. Gómez-Plaza, A. Costa-García, M.T. Fernández-Abedul, Electrochemical properties of spaghetti and forest-like carbon nanotubes grown on glass substrates, Sens. Actautor. B: Chem. 192 (2014) 253–260.
[8] R. García-González, A. Fernández-la-Villa, A. Costa-García, M.T. Fernández-Abedul, Dispersion studies of carboxyl, amine and thiol-fucntionalized carbon nanotubes for improving the electrochemical behavior of screen printed electrodes, Sens. Actuator. B: Chem. 181 (2013) 353–360.
[9] M.M. Hamedi, A. Ainla, F. Guder, D.C. Christodouleas, M.T. Fernández-Abedul, G.M. Whitesides, Integrating electronics and microfluidics on paper, Adv. Mat. 28 (25) (2016) 5054–5063.
[10] P.J. Britto, K.S.V. Santhanam, P.M. Ajayan, Carbon nanotube electrode for oxidation of dopamine, Bio-electrochem. Bioenerg. 41 (1996) 121–125.
[11] M. Musameh, J. Wang, A. Merkoci, Y. Lin, Low-potential stable NADH detection at carbon-nanotube-modified glassy carbon electrodes, Electrochem. Comm. 4 (2002) 743–746.
[12] R. García-Goonzález, A. Costa-García, M.T. Fernández-Abedul, Enhanced detection of the potential electroactive label methylene blue by electrode nanostructuration with carbon nanotubes, Sens. Actuator. B: Chem. 202 (2014) 129–136.
[13] J. Wang, M. Musameh, Y. Lin, Solubilization of carbon nanotubes by Nafion toward the preparation of amperometric biosensors, J. Am. Chem. Soc. 125 (2003) 2408–2409.
[14] I. Streeter, G.G. Wildgose, L. Shao, R.G. Compton, Cyclic voltammetry on electrode surfaces covered with porous layers: an analysis of electron transfere kinetics at single-walled carbon nanotube modified electrodes, Sens. Actuator. B: Chem. 133 (2008) 462–466.
[15] C.E. Banks, R.G. Compton, New electrodes for old: from carbon nanotubes to edge plane pyrolytic graphite, Analyst 131 (2006) 15–21.
[16] R. Laocharoensuk, J. Burdick, J. Wang, Carbon-nanotube-induced acceleration of catalytic nanomotors, ACS Nano 2 (2008) 1069–1075.

PART V

Low-cost electroanalysis

CHAPTER

25

Determination of glucose with an enzymatic paper-based sensor

*Olaya Amor-Gutiérrez[1], Estefanía Costa-Rama[1, 2],
M. Teresa Fernández Abedul[1]*

[1]Departamento de Química Física y Analítica, Universidad de Oviedo, Oviedo, Spain;
[2]REQUIMTE/LAQV, Instituto Superior de Engenharia do Porto, Instituto Politécnico do
Porto, Porto, Portugal

25.1 Background

Paper has been used as substrate for analytical devices for centuries, mainly for lateral flow immunoassays (e.g., pregnancy test). However, the use of paper as the basis of microfluidic assays dates from 2007, when Whitesides and coworkers reported the first microfluidic paper-based analytical device (μPAD) [1]. As substrate material, paper shows unique advantages over other traditional materials (e.g., polymers, glass or ceramics): (i) it is a low-cost material available everywhere; (ii) it is light and flexible, making easy its transportation and storage; (iii) it has the ability of moving fluids passively by capillarity; (iv) it has a very high surface area because of its microfiber composition, which favors adsorption; (v) it is porous and biocompatible, allowing the immobilization and storage of reagents in active form; (vi) it has the ability of filtrating and then separating microscopic components; (vii) it can be found in a broad variety of thicknesses, porosities and compositions (e.g., modified with metallic or carbon materials); (viii) it is compatible with a wide range of printing technologies; (ix) it can be easily modified to become hydrophobic, conductive, etc.; (x) it can be stacked and folded to create 3D structures; and (xi) it is easy to dispose by incineration [2–4]. All these advantages make the paper very attractive as substrate for designing analytical platforms as demonstrated in the large amount of articles related to applications of μPADs in clinical, food or environmental fields [5].

In what concerns the material, although cellulosic flexible sheets are generally understood as paper, other noncellulosic porous materials can be employed (e.g., glass fiber or polyester). Among cellulosic papers, the most widely used are manufactured from shorter

Laboratory Methods in Dynamic Electroanalysis
https://doi.org/10.1016/B978-0-12-815932-3.00025-5

cotton fibers [6] or from those coming from wood. Depending on the type of process to generate the paper pulp, mechanical or chemical, the paper will vary its composition. Cellulose fibers are hollow tubes 1.5-mm long, 20-μm wide, with a wall thickness of 2 μm. Two macroscopic properties describe paper: the thickness or caliper, τ (m), and the basis weight, bw (g/m^2), which is the mass of dry paper *per* square meter. Whatman No. 1 filter paper (the most common in analytical devices) has values of 180 μm and 87 g/m^2 for τ and bw, respectively, which supposes a density of 483 kg/m^3. Porosity arises from spaces between the fibers, uncollapsed fiber lumens, and the intrinsic porosity of the fiber walls. If the fibers are dried after pulping, some of the pores collapse irreversibly. It has to be taken into account that, as paper is formed in a filtration process, the fibers are approximately layered in the x,y plane and the mass distribution is usually not constant in the z (thickness) dimension, with the maximum density in the center. Thus, fluid transport along a strip of paper may depend on the angle at which the paper was cut. On the other hand, immobilization of bioreagents can be influenced by the porosity and specific surface area of the paper structure that is accessible to the bioreagent.

Regarding the detection, colorimetry has been the dominant detection principle in μPADs because of its simplicity and potential for developing instrument-free devices. Moreover, the white color of the paper makes it ideal for colorimetric tests because it provides high contrast with a colored substrate. Nevertheless, it has limited dynamic ranges, low sensitivity and variability with environmental illumination. Therefore, it has been mainly employed with qualitative/semiquantitative purposes. In this context, electrochemical techniques offer an alternative detection system for μPADs that can improve sensitivity, selectivity and detection limits. Electrochemical detection is very suitable for μPADs because (i) it does not require complex/expensive materials or instrumentation; (ii) electrodes can be easily miniaturized and fabricated onto paper; and (iii) portable potentiostats are already available for on-site analysis [7]. Hence, electrochemical detection coupled to μPADs (μPEDs or EμPADs) offers a good match for developing inexpensive, portable and sensitive devices. The first example of EμPAD was reported by Henry and coworkers in 2009 and was based on the use of screen-printed carbon electrodes [8].

On the other hand, glucose is one of the most important biological compounds as it is involved in a multitude of reactions. Glucose determination is of paramount importance not only in the clinical field, in blood analysis, and in other biological fluids (e.g., tears, sweat, etc.) because of the incidence of diabetes mellitus, but also in food and beverages to know their composition. In the last years, the sensor market for food analysis has experienced an important expansion because of the increasing customers' interest in knowing what they are eating. Hence, glucose sensors are also interesting for customers who want to follow a healthy diet. Because of that, scientific articles related to the development of sensors/devices for glucose analysis is enormous [9]: for example, when the term "glucose sensor" is searched in the Google Scholar data base (July 27, 2019), close to 23,200 documents are found from year 2017. The most popular device for glucose analysis is the electrochemical glucometer employed for diabetes patients to determine their glucose concentration in blood. Most of the commercial glucometers are based on electrochemical readout employing screen-printed enzyme electrode test strips [10]. Taking this into account and considering the amount of glucose analysis that are daily made, many people will benefit from low-cost simple platforms.

The cost of the analysis depends on many factors, mainly materials and reagents, instrumentation, personnel and analysis time. The use of low-cost materials, the low volume of reagents required, the use of inexpensive instrumentation, the simplicity of procedures that favor the use by nonqualified personnel, and the considerable decrease in the analysis time (favored by the miniaturization of devices, integration of steps, reduction of complexity, etc.) decrease enormously the cost. Moreover, additional expenses such as those related to disposal management can be weaved and mass production of devices decreases the price per unit. Then, in general, analysis with low-cost devices is being democratized, equaling the opportunities for developed and developing countries as well as remote or central locations.

In this chapter, a paper-based enzymatic sensor for glucose determination in beverages, adapted from reference [11], is described. The enzymes glucose oxidase (GOx) and horseradish peroxidase (HRP) are employed together with ferrocyanide as electron transfer mediator (see Fig. 25.1). The analytical signal is the intensity of the current due to the reduction of the ferricyanide enzymatically produced. GOx catalyzes the oxidation of glucose in the presence of oxygen, producing hydrogen peroxide, substrate of HRP. Then, in the subsequent enzymatic indicator reaction, H_2O_2 is reduced to H_2O and ferrocyanide is converted to ferricyanide. Therefore, the concentration of ferricyanide, which is measured chronoamperometrically (see Chapter 8) at a potential where it is reduced, is proportional to the initial concentration of glucose. The paper-based electrochemical platform is fabricated employing a piece of chromatographic paper and an inexpensive commercial connector header. Hydrophobic barriers are created on the paper in an easy way using a wax printer.

This experiment can be completed in two laboratory sessions of 3—4 h and is appropriate for undergraduate students of advanced Analytical Chemistry or Master students from different fields where analysis is required. With this experiment, the student will learn not only about the development of a glucose biosensor (optimizing parameters and studying the analytical characteristics), but also about the fabrication of paper-based electrodes and their integration in miniaturized and easy-to-handle platforms.

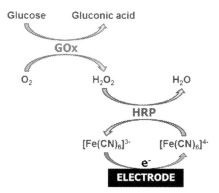

FIGURE 25.1   Schematic of the enzymatic reactions and the electrochemical reduction of ferricyanide produced on the surface of the biosensor. *GOx*, glucose oxidase; *HRP*, horseradish peroxidase.

25.2 Electrochemical cell design

The electrochemical cell consists of a paper-based working electrode (WE) combined with two-wire counter (CE) and reference electrodes (RE) (Fig. 25.2). The WE consists of a circular paper area modified with carbon ink (see Section 25.5.2). This area is defined by a wax hydrophobic barrier. Wire electrodes (RE and CE) are part of a gold-plated commercial connector header. This connector header also provides the necessary connection for the paper-based WE as shown in Fig. 25.2A. The solution will be deposited on the opposite side of the ink and, therefore, the wire acting as connection of the WE does not contact RE and CE. Similarly, RE and CE do not contact the carbon ink. Ionic contact is established when the solution wets all the three electrodes (carbon ink at the bottom of the paper and wire electrodes at the surface). Moreover, the connector header allows the use of commercial connectors as interface between the electrochemical cell designed and the potentiostat (see Fig. 25.2B).

25.3 Chemical and supplies

- *Enzymes:* Glucose oxidase (GOx) and horseradish peroxidase (HRP).
- *Enzyme mediator:* Potassium ferrocyanide.
- *Components of background electrolyte:* Trizma base and nitric acid.
- *Analyte:* Glucose.
- *Samples:* Orange juice and cola beverage.
- *Electrochemical cell preparation:* Carbon paste, dimethylformamide (DMF), Whatman chromatographic paper grade 1, gold-plated connector headers.
- *General materials, apparatus and instruments:* 1-mL Eppendorf tubes, 100-μL and 1-mL micropipettes and corresponding tips, wax printer, ultrasound bath, hot plate, weighing scale, pH meter and potentiostat.
- Milli-Q water is employed for preparing solutions and washing.

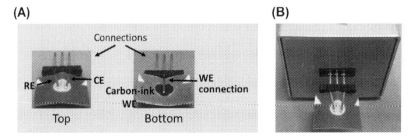

FIGURE 25.2 Pictures of the (A) paper-based electrochemical cell containing the three electrodes and (B) electrochemical cell inserted in a commercial interface that connects electrodes to the potentiostat. *CE,* counter electrode; *RE,* reference electrode; *WE,* working electrode.

25.4 Hazards

Nitric acid, used for the preparation of Tris-HNO$_3$ buffer, is corrosive and causes serious burns. Potassium ferrocyanide in contact with acids releases a very toxic gas. DMF is highly toxic by inhalation and contact; therefore, it has to be handled in a fume hood. Students are required to wear a lab coat, appropriate gloves and safety glasses.

25.5 Experimental procedure

25.5.1 Solutions and sample preparation

The following solutions are needed (estimate the amount of solution required):

– 0.1 M Tris-HNO$_3$ pH 7.0 buffer solution (all working solutions used along this experiment will be prepared in this buffer solution).
– 1 mM potassium ferrocyanide solution.
– Mixtures of enzymes and ferrocyanide with concentrations:
    – 0.8 U/µL GOx, 2.5 U/µL HRP and 0.1 M ferrocyanide
    – 1.6 U/µL GOx, 2.5 U/µL HRP and 0.1 M ferrocyanide
    – 0.8 U/µL GOx, 5.0 U/µL HRP and 0.1 M ferrocyanide
    – 1.6 U/µL GOx, 5.0 U/µL HRP and 0.1 M ferrocyanide
– 0.1 M glucose stock solution. From this one, different glucose solutions of 0.3, 0.5, 1.0, 3.0, 5.0, 7.5, 10.0 and 15.0 mM concentration should be prepared by dilution.

Regarding the samples, the only pretreatment required is degasification of the cola beverage with the help of a stirrer. Then, both samples (orange juice and cola beverage) are diluted 1:20 in the Tris-HNO$_3$ buffer solution with the aim of achieving a glucose concentration within the linear range of the biosensor constructed.

25.5.2 Fabrication of the electrochemical cell

To fabricate the electrochemical cell, the following steps are required:

1. Print on the chromatographic paper the desired pattern (a circular area of a 6-mm diameter) employing a wax printer. Then, diffuse the wax by heating in a hot plate for 1 min at 100 °C (Fig. 25.3A).
2. Deposit 2 µL of carbon ink (23% (w/w) of carbon paste diluted in DMF and sonicate for 1 h) on the circular area (Fig. 25.3B) and let dry for 12 h.
3. Separate three wires from the connector header (Fig. 25.3C) and form with them a clip to support the paper. The one in the middle is the connection of the WE. Bend the wires that will act as RE and CE (those at the ends) one against the other and separate them from the plane where WE is (Fig. 25.3C). Thus, the paper WE can be inserted in the clip that wires acting as RE and CE form with the wire for WE connection (Fig. 25.2A).

(A)

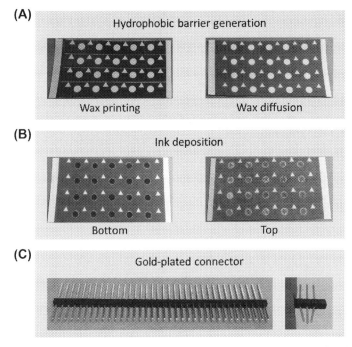

FIGURE 25.3   (A) Wax pattern printed on the Whatman Grade 1 paper before and after the diffusion step. (B) Printed paper after deposition of the carbon ink. (C) Gold-plated connector header as delivered (left) and ready to be integrated with the paper-based working electrode (right).

The paper-based WE is single use, while only one connector header is required all along this experiment if it is washed with Milli-Q water after each use.

25.5.3 Electrochemical measurements

To perform electrochemical measurements employing the paper-based electrochemical cell, an aliquot of 10 μL of the working solution (sample or standard) is deposited on it on the opposite side where the carbon ink was added.

25.5.4 Study of the electrochemical behavior of ferrocyanide

Chronoamperometry is performed measuring, with time, the intensity of the current produced after the application of a potential, negative enough to produce ferricyanide reduction. To choose this suitable potential, cyclic voltammograms (CVs) are recorded on a paper-based device (unmodified) by scanning the potential between −0.2 and +0.5 V at a scan rate of 25 mV/s in 10 μL of 1 mM ferrocyanide solution (in the buffer solution, which acts as background electrolyte). In addition, a CV is recorded in this buffer solution to assure that there are no interferences (see Fig. 25.4A). Then, −0.1 V can be selected as the appropriate potential for recording chronoamperograms (i-t curves), as the reduction of the ferricyanide enzymatically produced is warranted.

(A)

(B)

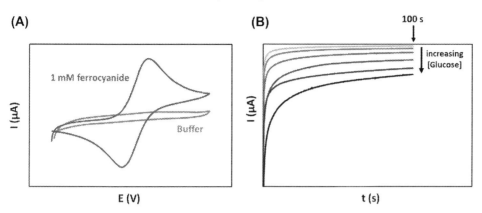

FIGURE 25.4 Schematic representation of the (A) cyclic voltammograms obtained for 1 mM ferrocyanide and buffer solutions between −0.2 and +0.5 V at 25 mV/s and (B) chronoamperograms obtained for different glucose concentrations by applying −0.1 V for 100 s.

25.5.5 Construction of the biosensor and signal readout procedure

The construction of the biosensor is based on the immobilization of the enzymes (GOx and HRP) as well as the enzyme mediator, potassium ferrocyanide, by simple adsorption on the paper-based WE. The procedure is as follows: deposit 10 μL of a mixture of GOx, HRP and ferrocyanide on the WE (on the opposite side of the carbon ink) and then let dry for 30 min.

To obtain the analytical signal that correlates to glucose concentration, chronoampero-grams are recorded by applying −0.1 V for 100 s (see Fig. 25.4B). The intensity is due to the reduction of the ferricyanide enzymatically generated and it is related to the concentration of glucose in the sample. The higher the concentration of glucose, the higher the concentration of hydrogen peroxide and also of ferricyanide. The analytical signal is the current measured at a fixed time (e.g., 20, 50 or 100 s).

25.5.6 Optimization of the biosensor

To optimize the concentration of the enzymes, the mixtures of enzymes and potassium ferrocyanide previously prepared are employed for constructing different biosensors. These are tested recording chronoamperograms in both the background electrolyte and a 2.5 mM glucose solution. Thus, the concentration of enzymes that provides the best signal to background ratio is chosen.

25.5.7 Calibration and determination of glucose in beverages

Once the concentration of enzymes and mediator is optimized, a calibration plot is performed between 0.3 and 15.0 mM glucose solutions (using those previously prepared and a different paper-based WE for each measurement). To know the precision of the calibration curve, it can be repeated in different days.

To know the glucose concentration in samples of beverages (orange juice and cola), record chronoamperograms in the dilutions previously prepared. Several replicates have to be measured to obtain the average and standard deviation of the result.

25.6 Lab report

Write a lab report following the typical scheme of a scientific article. Thus, the report should include a short introduction, experimental part, results and discussion, and conclusions. It is highly recommended to include diagrams, tables and images. The following points should be considered:

1. In the introduction, include the current importance of the paper as substrate material for analytical platforms. Remark the main advantages and search recent publications on (i) electrochemical detection based on paper focusing on the design of the electrochemical cell, (ii) other interesting analytes that can be determined electrochemically with paper devices, and (iii) other low-cost approaches employed in electroanalytical devices.
2. Include representative pictures of the device and graphs representing the data obtained along the experiment: (i) CV recorded in the background electrolyte and in a potassium ferrocyanide solution, (ii) bar diagrams with the intensities of the currents obtained for different concentrations of enzymes and mediator, (iii) chronoamperograms for different glucose concentration and calibration curve obtained representing the intensity of the current at a fixed time for different concentrations of glucose.
3. Discuss the values obtained for the linear range, sensitivity, limit of detection and limit of quantification of the biosensor constructed (they can be compared to other found in the literature).
4. Determine the glucose concentration in the samples and report the result including the average and standard deviation. Perform a statistical t-student test to compare the values obtained with those indicated in the package. Compare also the results obtained by different (groups of) students with the aim of calculating the accuracy and precision of the results.

25.7 Additional notes

1. The buffer solution and the enzyme stock solutions can be stored at 4°C for 1 week.
2. Because of the low cost of the platform, all the measurements can be performed in triplicate to allow the students to observe the precision of the biosensor. An interesting study is to prepare several devices to study their repeatability/reproducibility.
3. This experiment can be considered as low-cost analysis. Then, the cost of the device could be calculated considering the price of the paper, headers, carbon ink, enzymes, etc. Moreover, it is interesting to discuss the importance of the small volumes used (especially when bioreagents are employed) as well as the analysis time required.

25.8 Assessment and discussion questions

1. A bibliographic search on paper-based glucose sensors could be made to discuss other types (e.g., colorimetric, nonenzymatic, etc.).
2. This is a bienzymatic approach, using ferrocyanide as mediator. Classify the enzymes (following the Enzyme Commission (EC) numbers) and indicate other possible mediators.
3. In this case, a simple physical approach (adsorption) has been employed for enzyme immobilization. Discuss other possible approaches.
4. The experiment is performed in a buffer solution of pH 7. Why?
4. How is the detection potential chosen? What happens for higher or lower potentials?
5. Cyclic voltammetry and chronoamperometry are the electrochemical techniques employed here. What are the differences between the two? Draw the excitation signal applied to the WE and the response signal in each case.

References

[1] A.W. Martinez, S.T. Phillips, M.J. Butte, G.M. Whitesides, Patterned paper as a platform for inexpensive, low-volume, portable bioassays, Angew. Chem. Int. Ed. 46 (2007) 1318–1320.
[2] E.J. Maxwell, A.D. Mazzeo, G.M. Whitesides, Paper-based electroanalytical devices for accessible diagnostic testing, MRS Bull. 38 (2013) 309–314.
[3] A.W. Martinez, S.T. Phillips, G.M. Whitesides, E. Carrilho, Diagnostics for the developing world: microfluidic paper-based analytical devices, Anal. Chem. 82 (2010) 3–10.
[4] J.P. Rolland, D.A. Mourey, Paper as a novel material platform for devices, MRS Bull. 38 (2013) 299–305.
[5] Y. Yang, E. Noviana, M.P. Nguyen, B.J. Geiss, D.S. Dandy, C.S. Henry, Paper-based microfluidic devices: emerging themes and applications, Anal. Chem. 89 (2017) 71–91.
[6] R. Pelton, Bioactive paper provides a low-cost platform for diagnostics, TrAC Trends Anal. Chem. 28 (2009) 925–942.
[7] A. Ainla, M.P.S. Mousavi, M.-N. Tsaloglou, J. Redston, J.G. Bell, M.T. Fernández-Abedul, G.M. Whitesides, Open-source potentiostat for wireless electrochemical detection with smartphones, Anal. Chem. 90 (2018) 6240–6246.
[8] W. Dungchai, O. Chailapakul, C.S. Henry, Electrochemical detection for paper-based microfluidics, Anal. Chem. 81 (2009) 5821–5826.
[9] C. Chen, Q. Xie, D. Yang, H. Xiao, Y. Fu, Y. Tan, S. Yao, Recent advances in electrochemical glucose biosensors: a review, RSC Adv. 3 (2013) 4473.
[10] J. Wang, Electrochemical glucose biosensors electrochemical glucose biosensors, Chem. Rev. 108 (2008) 814–825.
[11] O. Amor-Gutiérrez, E. Costa Rama, A. Costa-García, M.T. Fernández-Abedul, Paper-based maskless enzymatic sensor for glucose determination combining ink and wire electrodes, Biosens. Bioelectron. 93 (2017) 40–45.

Determination of arsenic (III) in wines with nanostructured paper-based electrodes

Estefanía Núñez Bajo[1], M. Teresa Fernández Abedul[2]

[1]Department of Bioengineering, Royal School of Mines, Imperial College London, London, United Kingdom; [2]Departamento de Química Física y Analítica, Universidad de Oviedo, Oviedo, Spain

26.1 Background

Electroanalysis has moved from conventional cells that include mercury and solid electrodes (metallic or carbon electrodes) to more advantageous formats. Glass or polymeric cylinders that were commonly the body of main conductive elements (e.g., carbon paste) with an inner connection to the potentiostat turned into flat surfaces where all the three electrodes are included in a small area. Thick- and thin-film technologies can be used to pattern the working or all the three electrodes of a complete electrochemical cell on polymers, ceramics, glass, hydrophobic/hydrophilic paper, or other substrates. A container is not required and drops of the electrolyte can be deposited for the measurement directly on the surface, although an adhesive, a dielectric layer, or a resin can be employed for delimiting the cell area. On the other hand, hydrophilic materials can transport passively solutions in an easy way, and barriers to define the working area can be made by wax diffusion [1] or drawn with permanent markers [2]. Apart from other advantages, miniaturization and low cost of electroanalytical devices allow approaching analysis to zones (e.g., remote, dangerous, or not accessible) and situations (e.g., testing at the point of care, monitoring environmentally polluted areas, or tracing a toxin through the food chain) that were unthinkable years ago.

The first electrochemical paper-based analytical device [3], based on carbon screen-printed electrodes (SPEs), was rapidly followed by different electrochemical transducers. In this work we have drop-casted a carbon ink suspension on paper, eliminating the use of masks and special equipments. The paper-based working electrode (PWE) was then coupled to SPEs. In this

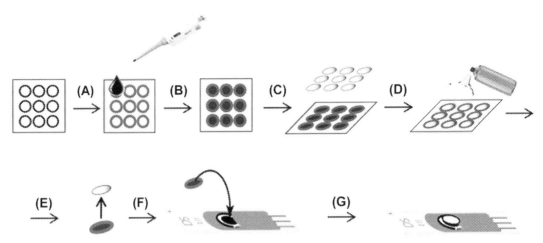

FIGURE 26.1 Schematic representation of the fabrication and coupling of paper-based carbon working electrodes (PCWEs) with SPCEs: (A) wax melting after printing, (B) carbon ink deposition, (C) protection of the conductive layer with plastic covers, (D) addition of spray adhesive, (E) removal of the plastic cover after cutting the PCWE, (F) placement of the PCWE over the screen-printed carbon electrode (SPCE) to get (G), the final platform composed by the PCWE with external reference and counter electrodes from the SPCE. *Reprinted with permission from E. Núñez-Bajo, M.C. Blanco-López, A. Costa-García and M.T. Fernández-Abedul, Electrogeneration of gold nanoparticles on porous-carbon paper-based electrodes and application to inorganic arsenic analysis in white wines by chronoamperometric stripping, Anal. Chem. 89, 2017, 6415–6423.*

way, apart from support, the SPE card provided the other two electrodes to complete the cell (reference and auxiliary electrodes) [4]. The design employed (Fig. 26.1) allows (i) adequate connection between the two carbon surfaces (that of the SPE and that of the paper substrate), to generate gold nanostructures on the carbonaceous inner surface of the paper and (ii) reusability of the SPE (if wanted) because the carbon area of the paper covers all the surface of the SPE working electrode (WE) and its thickness impedes the wetting.

As seen in the section devoted to nanomaterials, these are very convenient in Electroanalysis and can be also combined with paper-based platforms. Among others, gold nanoparticles (AuNPs) are the most common because of their high surface-to-volume ratio and surface energy that enhance the conductivity and effective area of the electrochemical transducer. Nanostructuration of paper-based electrodes with gold is usually made through the deposition of previously prepared AuNPs by chemical reduction of tetrachloroauric (III) acid [5,6]. They could be used either as a seed for the further growth of interconnected particles [5] or added to the paper substrate together with the carbon component [6]. However, nanoparticles could be generated in situ on the WE in a simple, fast, reproducible, controlled, and localized way. This electrochemical generation of AuNPs does not require ligands such as citrate to stabilize the resulting nanoparticles [7]. In this ligand-free approach, the surface of the nanoparticles is not covered by molecules that decrease their specific surface activity. In this work, AuNPs electrogeneration is made on paper and was applied to the determination of arsenic.

Electrochemical methods based on the use of nanostructured transducers have experienced enormous growth. These have been applied to the determination of heavy metals (HMs) in different samples, such as industrial or drinking water, soils, food, etc.

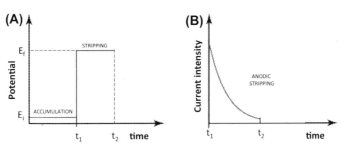

FIGURE 26.2   Anodic stripping chronoamperometry: (A) potential versus time waveform for the accumulation and stripping steps, where E_i is the potential at which the species is reduced to the elementary state, E_f the stripping potential, t_1 the time at which the potential is switched, and t_2 the time of the anodic stripping step (measurement time), and (B) current intensity versus time (chronoamperogram) for the stripping step.

They commonly make use of anodic stripping voltammetric techniques where the HM is electrochemically (cathodically) deposited on the transducer and, later, anodically redissolved achieving very low detection limits because of the preconcentration on the electrode (see Chapters 4 and 35). Current advances in detection systems based on low-cost electroanalytical devices have favored a large use of screen-printed carbon electrodes (SPCEs) nanostructured with AuNPs for the determination of HMs because of the adsorption of elemental metals on AuNPs (see Chapter 23). Anodic stripping techniques combined with paper-based devices are promising tools for fast and low-cost analysis of arsenic in wine samples, where the maximum total concentration permissible is 200 ppb (μg/L) according to the European regulation [8]. Therefore, in this experiment, a low-cost yet sensitive sensor is fabricated by electrogeneration of AuNPs on paper-based carbon electrodes and applied to the analysis of inorganic arsenic in wines. For this, As(III) is electrochemically reduced to elemental As and then it is stripped anodically. In this case, and for the sake of simplicity, the oxidation step is performed by chronoamperometry (see Chapter 8 and 25). Fig. 26.2 shows the excitation signal (E-t curve) corresponding to the accumulation (preconcentration or deposition) step and to the stripping (or redissolution) step in the anodic stripping chronoamperometry of the metal. In the accumulation step, a potential negative enough where the reduction happens is applied. After this, a potential is applied to produce the oxidation for a fixed time, and the corresponding chronoamperogram (i-t curve) is recorded.

The method, based on reference [4], (i) requires very simple instrumentation, (ii) is sensitive enough to evaluate if the current normative is complied with, (iii) is based on the use of AuNPs that allow reproducible oxidation during the stripping, (iv) requires very small volume of gold solution for nanostructuration, which increases the sustainability, (v) avoids possible errors due to the color and turbidity of the samples, and (vi) allows to carry out speciation analysis by differential chronoamperometric measurements.

The experiment is directed to Graduate or Master students of courses related to the development of analytical methodologies (Chemistry, Biotechnology, etc.). It will give them, during two 4-h laboratory sessions, an appreciation of the basis of the fabrication of paper-based electrodes, the modification with electrogenerated AuNPs, and the application to anodic stripping methodologies.

26.2 Chemicals and supplies

Reagents

— *PWE*: Carbon ink (e.g., C10903P14 from Gwent), dimethylformamide (DMF), transparency sheets, adhesive spray (e.g., mounting spray from 3 MM).
— *Plating solution*: Titrisol Gold Standard 10.15 mM $HAuCl_4$ in 2.0 M HCl solution.
— *Analyte standard*: 1000 mg/L As(III) standard solution in 2% HCl. Dilutions are prepared in 3 M HCl.
— Hydrochloric acid (37% HCl).
— Milli-Q purified water is employed to prepare all the solutions.

Samples

Arsenic is determined in white wines. They could be commercial or certified.

Instrumentation and materials

— Potentiostat and SPCEs such as STAT3000 and DRP-110 SPCEs from Metrohm DropSens.
— A Xerox wax printer is employed for the fabrication of the paper-based electrode.
— Ultrasonic water bath and weighing scale.
— Micropipettes (1−10, 10−100, and 100−1000 µL) with tips.
— Microcentrifuge tubes (0.5−1.5 mL), volumetric flasks, beakers, and other volumetric materials.

26.3 Hazards

As(III) solution in HCl is corrosive to metals (category 1) and carcinogenic (category 1A). In addition, solutions have to be disposed in an appropriate container because of the chronic aquatic toxicity (category 3). Use appropriate skin and eye protection.

DMF is flammable in liquid and vapor phase, harmful in contact with skin or if inhaled, and causes serious eye irritation. Manipulation must be performed in a fume hood.

The 37% HCl solution is corrosive to metals; it causes severe skin burns and eye damage and causes respiratory irritation. Manipulation must be performed in a fume hood.

Students are required to wear gloves, eye protection, and lab coat during this experiment.

26.4 Experimental procedure

26.4.1 Fabrication of the electrochemical cell

1. Perform wax printing on filter paper with a pattern of 6-mm diameter circles with 0.7 mm of drawing stroke.

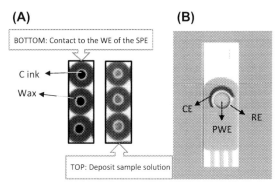

FIGURE 26.3   Pictures of (A) conductive layers on filter paper fabricated by drop-casting of carbon ink. The paper-based working electrode (PWE) is coupled to the screen-printed carbon electrode (SPCE) by direct contact of carbon films. (B) Final device with the PWE coupled to the SPCE. *CE*, carbon counter electrode; *PWE*, paper working electrode; *RE*, silver pseudo-reference electrode.

2. Melt the wax by putting the printed paper on a hot plate or oven at 110 °C for 2 min to create hydrophobic barriers. Hydrophilic circles with 4 mm of diameter are obtained after wax melting.
3. Prepare a dispersion of 24% carbon ink in DMF (w/w) and sonicate it for at least 15 min.
4. Drop a volume of 2 μL of this suspension over each hydrophilic circle leaving the solvent to evaporate for 1 h at 70 °C. A picture is shown in Fig. 26.2.
5. Protect the conductive layers with plastic covers (4-mm circles of transparency sheet) and apply adhesive spray covering the hydrophobic (wax) zones.
6. Remove the plastic covers and cut the conductive layers into 4.5-mm round discs (using a paper punch or scissors). As a result, adhesive conductive discs with 4-mm hydrophilic area and 0.5-mm hydrophobic adhesive circular crowns are obtained. These paper-based electrodes will act as WEs.
7. Wash SPCEs with Milli-Q water, dry them, and place the PWEs just on top of the WEs of the SPCEs. The adhesive side is located over the ceramic surrounding the WE in such a way that the conductive layer is in direct contact with the SPE carbon spot. The silver pseudo-reference and carbon counter electrodes of the SPCE card act as external electrodes. A picture is shown in Fig. 26.3.

26.4.2 Gold nanostructuration of the carbon paper-based electrode

1. Connect the electrochemical cell (with the PCWE coupled to the SPCE) to the potentiostat.
2. Deposit a volume of 50 μL of a 10.15 mM $HAuCl_4$ solution in 2.0 M HCl over the whole cell (ensure full coverage of the PWE, silver pseudo-reference electrode, and carbon counter electrode).
3. Generate AuNPs by multicyclic voltammetry, sweeping the potential 45 times from +0.5 to +0.1 V at a scan rate of 100 mV/s.

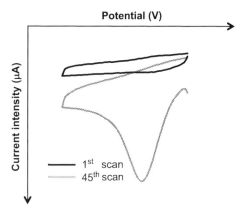

FIGURE 26.4 First and last cyclic voltammograms recorded in the plating solution using a paper-based carbon working electrode coupled to a screen-printed carbon electrode.

4. Rinse with Milli-Q water and remove the adsorbed hydrogen by applying +0.12 V for 240 s.
5. Peel off the gold-modified paper-based carbon working electrode (AuNPs-PCWE) from the SPCE and keep it in a dry environment at room temperature.
6. Represent one of the voltammograms recorded during the nanostructuration and explain the redox processes involved. An example is shown in Fig. 26.4.

26.4.3 Electrochemical characterization of As(III) by cyclic voltammetry using nanostructured paper-based devices

1. Prepare a 200 mg/L As(III) solution in 3.0 M HCl.
2. Place one nanostructured PWE (AuNPs-PWE) as prepared in Section 26.4.2 just on top of a new SPCE in such a way that the adhesive side of the paper is located over the ceramic surrounding the carbon spot. Then, both carbon conductive films are in direct contact.
3. Connect the electrochemical cell (with the PCWE coupled to the SPCE) to the potentiostat.
4. Deposit a volume of 50 μL of the As(III) solution over the whole cell (PWE, silver pseudo-reference electrode, and carbon counter electrode).
5. Record a cyclic voltammogram by sweeping the potential from −0.6 to +1.3 V at 100 mV/s.
6. Peel off the nanostructured PWE and dispose it in an appropriate container.
7. Repeat the procedure using a PWE on a new SPCE, without generating previously AuNPs.
8. Compare both cyclic voltammograms indicating the redox processes involved. Determine the peak potentials and currents.
9. Considering the processes involved, discuss the potentials that have to be applied for anodic stripping voltammetric determination (in both steps, deposition and redissolution).

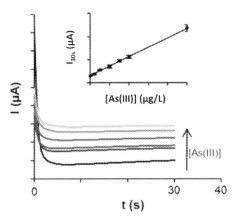

FIGURE 26.5 Stripping chronoamperograms obtained by using AuNPs paper-based carbon working electrodes for As(III) determination. The inset shows the calibration curve.

26.4.4 Calibration curve by chronoamperometric stripping and analytical features of the sensor

1. Prepare several As(III) solutions in 3.0 M HCl with concentrations ranging from 2 to 200 µg/L.
2. Follow steps 2–4 of Section 26.4.3.
3. After adding the drop of solution of the lowest concentration of As(III), apply the pre-concentration potential chosen in Section 26.4.3 (e.g., −0.3 V) during 180 s.
4. Apply the potential chosen in Section 26.4.3 (e.g., +0.4 V) during 30 s for the anodic stripping of As(0) and record the i-t curve (Fig. 26.5).
5. Once the chronoamperogram (i-t curve) is recorded, peel off the nanostructured PWE and dispose it in an appropriate container.
6. Repeat the necessary steps for all the concentrations of As(III) and also for the blank (3.0 M HCl).
7. Obtain the calibration curve by representing the current intensity at 30 s (I_{30s}) versus As(III) concentration (Fig. 26.5, inset).
8. Estimate the dynamic linear range, limit of detection, limit of quantitation, and sensitivity.

26.4.5 As(III) determination in white wines

1. Mix a volume of 500 µL of unfiltered sample with 250 µL of 9.0 M HCl solution (sample solution).
2. Mix a volume of 500 µL of unfiltered sample with 7.5 µL of 1 mg/L As(III) standard solution (to ensure an As(III) content higher than 10 µg/L) and add a volume of 9.0 M HCl solution until a final volume of 750 µL (spiked sample solution).
3. Repeat the same procedure explained in Section 26.4.4 for both sample solutions.
4. Determine the concentration of As(III) in both solutions and discuss if the sample fulfills the legislation.

26.5 Lab report

Write a lab report that includes an abstract, a brief introduction explaining the purpose of the experiment, a detailed experimental section, results and discussion, and conclusions. Include tables, graphics, or figures wherever necessary. The following points should be considered:

1. In the introduction, include a revision of reported works with some examples of (i) the use of AuNPs on paper-based electrochemical devices, (ii) other designs of paper-based electroanalytical platforms that employ different nanomaterials, and (iii) other methodologies that would allow decentralized arsenic determination.
2. In the experimental section, explain the protocols to fabricate the electrochemical cell and also to generate AuNPs and to perform the electrochemical measurements for arsenic determination.
3. In the results and discussion section, show the most representative curves and include the incidences during the course of the experiment. Include a figure with the calibration curve. Discuss the possibility of performing a speciation analysis (for As(III) and As(V) determination) and explain the procedure.
4. In the conclusions, highlight the most representative results and the main advantages.

26.6 Additional notes

1. The use of a wax printer is not crucial. However, hydrophobic barriers are completely necessary because the adhesive can diffuse through the paper contaminating the analytical zone. Barriers could be also generated using a permanent marker. However, care should be taken while drawing by hand to ensure precision.
2. The same SPCE can be used for several measurements (e.g., five different AuNPs-PWEs). Washing with water is made in between. Precision is adequate but the acidic medium could corrode the silver pseudo-reference electrode.
3. White wines are supposed to not have arsenic and then a spiked sample is prepared but, in case there is some, spiking is not required. The standard addition methodology should be performed in case there are matrix effects. If the slopes of the external calibration curve and the standard addition methodology are similar, no matrix interferences exist and the external calibration curve could be employed.
4. This chapter is thought for two sessions of 4 h, but if the lab time wants to be extended and the practice completed, the optimization of the gold nanostructuration can be performed (see Chapter 35).
5. An estimation of the cost of the determination can be done by students.

26.7 Assessment and discussion questions

1. Comment some examples of paper-based devices employed for the determination of HMs as those included in the introduction of the report.

2. Discuss if the sensor can be applied for the speciation analysis of As(III)/As(V). Explain a possible procedure.
3. Why AuNPs are required for As(III) determination?
4. Indicate the different steps of the electrochemical measurement (AuNP generation, As(III) deposition as As(0), and As(0) redissolution to As(III)), as well as the potentials required.
5. Indicate different possible techniques for the two steps of the determination and represent the excitation and response signals.
6. Indicate how the determination is made if external calibration or standard addition methodologies are employed.

References

[1] E. Carrilho, A.W. Martinez, G.M. Whitesides, Understanding wax printing: a simple micropatterning process for paper-based microfluidics, Anal. Chem. 81 (2009) 7091−7095.
[2] E. Núnez-Bajo, M.C. Blanco-López, A. Costa-García, M.T. Fernández-Abedul, Integration of gold-sputtered electrofluidic paper on wire-included analytical platforms for glucose biosensing, Biosens. Bioelectron. 91 (2017) 824−832.
[3] W. Dungchai, O. Chailapakul, C.S. Henry, Electrochemical detection for paper-based microfluidics, Anal. Chem. 81 (2009) 5821−5826.
[4] E. Núnez-Bajo, M.C. Blanco-López, A. Costa-García, M.T. Fernández-Abedul, Electrogeneration of gold nanoparticles on porous-carbon paper-based electrodes and application to inorganic arsenic analysis in white wines by chronoamperometric stripping, Anal. Chem. 89 (2017) 6415−6423.
[5] S. Ge, W. Liu, L. Ge, M. Yan, J. Yan, J. Huang, J. Yu, In situ assembly of porous Au-paper electrode and functionalization of magnetic silica nanoparticles with HRP via click chemistry for Microcystin-LR immunoassay, Biosens. Bioelectron. 49 (2013) 111−117.
[6] B. Guntupalli, P. Liang, J.-H. Lee, Y. Yang, H. Yu, J. Canoura, J. He, W. Li, Y. Weizmann, Y. Xiao, Ambient filtration method to rapidly prepare highly conductive, paper-based porous gold films for electrochemical biosensing, ACS Appl. Mater. Interfaces 7 (2015) 27049−27058.
[7] I. Sultana, A. Razaq, M. Idrees, M.H. Asif, H. Ali, A. Arshad, S. Iqbal, S.M. Ramay, S.Q. Hussain, Electrodeposition of gold on lignocelluloses and graphite-based composite paper electrodes for superior electrical properties, J. Electron. Mater. 45 (2016) 5140−5145.
[8] OIV Code Sheet − Issue 2015/01. Maximum acceptable limits. http://www.oiv.int/public/medias/3741/e-code-annex-maximum-acceptable-limits.pdf.

Pin-based electrochemical sensor

Estefanía Costa-Rama[1,2], *M. Teresa Fernández Abedul*[2]

[1]REQUIMTE/LAQV, Instituto Superior de Engenharia do Porto, Instituto Politécnico do Porto, Porto, Portugal; [2]Departamento de Química Física y Analítica, Universidad de Oviedo, Oviedo, Spain

27.1 Background

As commented in many pages along this book, current trends in Analytical Chemistry lead to the development of new low-cost devices for decentralized analysis. If we want to perform analyses at the point of need, no matter if clinical, environmental, or food samples are involved, small, fast, and simple platforms should be used. It would be excellent if they are also autonomous and have low energy requirements. There are not only many possible strategies that include commercial electrochemical cells (e.g., Chapters 6 or 30) and miniaturized instrumentation, but also innovative approaches using, e.g., paper (Chapter 25, 26 or 29) or transparency films (Chapter 4).

In our daily life, there are many common supplies that we encounter everywhere and use almost everyday in the office, in the kitchen, etc., for keeping things in order, for sewing, for do-it-yourself projects, etc. Interestingly, while they were designed for specific applications, one can find innovative out-of-box applications. This is an important exercise of creativity that leads to higher research productivity, cost savings, and convenience [1]. In the case of electroanalysis, conductive materials are especially useful because they can transport and transfer electrons. Then, as electroanalysis (electrodics) is an interfacial phenomenon, they only have to offer a small surface where electron transfer could happen. This is the case of commercial stainless steel pins commonly employed for sewing tasks. In this chapter, these low-cost (<0.01 €/unit) materials are used for the construction of an electrochemical biosensor.

Pins are small and simple sewing elements. However, they have several interesting parts (Fig. 27.1): (i) a sharp tip that can be used to drill films or fabrics and also as a connection point, (ii) a shaft that provides connection points or can act as an electrode (a thread wrapped around it can lead liquid samples toward it [2]), and (iii) a head that can act also as electrode or can work as a stop to maintain the pin attached to a surface. When dealing with electroanalytical applications, the material becomes very important. Then, stainless steel (alloy of iron with a minimum of 10.5% chromium) is preferable to other alloys to avoid undesirable

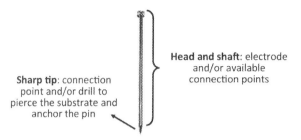

FIGURE 27.1   Scheme of the parts of a stainless steel pin with possible uses.

redox processes [3]. Chromium produces a thin layer of oxide ("passive layer") on the surface that prevents any further corrosion. It contains also varying amounts of carbon, silicon, manganese and other elements that can be added to bring useful properties [4]. However, it is said that it is "stainless" but not "stain impossible" and then, although in normal atmospheric or water-based environments it will not corrode, in more aggressive conditions, the basic stainless steel should be changed by other types.

On the other hand, apart from highly conductive, pins are easily modifiable with conductive inks and thus, stainless steel pins allow the construction of low-cost electroanalytical platforms with highly versatile designs [4–6], which could even mimic ELISA plates [2]. As they are made entirely of steel, they offer readily accessible connection points allowing to be connected to the potentiostat by common alligator or hook clips. Moreover, pins match easily with different materials such as paper or flexible polymers to fabricate electrochemical platforms. In this experiment, the pinhead is modified with carbon ink and is used as working electrode (WE). The sharp tip serves to drill the substrate and the shaft allows connecting to the potentiostat through a three-pin Dupont female cable. Although alligator clips can be also employed, the Dupont cable is preferred because the connection of the pins becomes easier. Hydrophobic paper was proposed as a foldable substrate for constructing pin-based immunoelectrochemical cells [2] but, in this experiment, a piece of transparency film, in another out-of-box application (it is commonly used in projections after drawing or photocopying), is employed as substrate to insert the pin-based electrodes and to deposit the working solution.

The experiment here commented is adapted from Ref. [4] where an electrochemical device combining pin-based electrodes and transparency film is constructed. With the aim of demonstrating the usefulness of the device, an enzymatic sensor for glucose determination is developed. The procedure to generate the recognition layer is the same as in Chapters 17 and 25 but, in this case, a very low-cost homemade cell is employed. A bienzymatic mixture (glucose oxidase [GOx] and horseradish peroxidase [HRP]) is used together with potassium ferrocyanide as electron transfer mediator. The reagents are immobilized by adsorption on the surface of the pin-based WE depositing the enzyme/mediator cocktail.

This experiment can be completed in two laboratory sessions of 3–4 h and is appropriate for undergraduate students of Analytical Chemistry or Master students from different fields where analysis is required. With this experiment, the student will learn about electroanalytical techniques and the development of a glucose biosensor using very low cost and mass-produced elements that are of common use.

27.2 Electrochemical cell design

The pin-based electrochemical cell consists of two unmodified pins as reference (RE) and counter (CE) electrodes and a carbon-coated pin as WE (Fig. 27.2A). The pins are made of stainless steel and their head has a 1.5-mm diameter. Transparency film rectangles of approximately 3×2 cm are utilized as substrates where the pins are inserted and the working solution is deposited (Fig. 27.2B). A gold-plated connector header (as the one employed in Chapter 25), but after removing its pins (Fig. 27.2C), is used for aligning the stainless steel pins at the same distance in between.

As commented in the introduction, to connect the stainless steel pins to the potentiostat, alligator clips or other common interfaces can be employed. However, the employment of a three-pin Dupont female cable allows an easier and quicker connection.

27.3 Chemicals and supplies

— *Enzymes:* Glucose oxidase (GOx) and horseradish peroxidase (HRP).
— *Redox probe and electron transfer mediator:* Potassium ferrocyanide.
— *Components of background electrolyte:* Monobasic and dibasic sodium phosphate and sodium hydroxide.
— *Analyte and possible interferents:* Glucose, fructose, and ascorbic acid.
— *Samples:* Honey and orange juice.
— *Electrochemical cell preparation:* Carbon paste, dimethylformamide (DMF), isopropyl alcohol, stainless steel pins, transparency film for photocopier, three-pin Dupont female cables, and gold-plated connector headers.

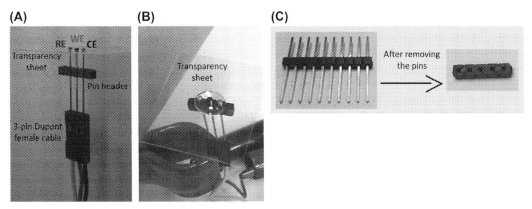

(A)   **(B)**   **(C)**

FIGURE 27.2   Pictures of the (A) pin-based electrochemical cell, (B) electrochemical cell after depositing the working solution, and (C) the gold-plated pin connector that is used as alignment tool, before and after removing the pins.

- *General materials, apparatus, and instruments:* 10-µL, 100-µL, and 1-mL micropipettes with corresponding tips, 1-mL microcentrifuge tubes, weighing scale, pH meter, oven, ultrasound bath, and potentiostat.
- Milli-Q water is employed for preparing the solutions and washing.

27.4 Hazards

DMF is highly toxic by inhalation and contact; hence a fume hood has to be employed. Potassium ferrocyanide releases a very toxic gas in contact with acids. Students are required to wear lab coat, appropriate gloves, and safety glasses.

27.5 Experimental procedure

27.5.1 Solutions and sample preparation

The following solutions are needed (the amount of solution required has to be previously estimated):

- A 0.1 M phosphate buffer of pH 7.0 (all working solutions used along this experiment will be prepared in this buffer solution).
- A 0.5 mM potassium ferrocyanide solution.
- A mixture of enzymes and ferrocyanide with concentrations: 3 U/µL GOx, 5 U/µL HRP, and 20 mM ferrocyanide.
- A 0.1 M glucose stock solution. From this, glucose solutions of concentrations 50, 75, 100, 250, 500, 750, 1000, 2500, and 5000 µM will be prepared.
- A 500-µM fructose solution.
- A 500-µM ascorbic acid solution.
- A solution 500 µM in glucose and 500 µM in fructose.
- A solution 500 µM in glucose and 100 µM in ascorbic acid.

Regarding the samples, honey and orange juice, the only pretreatment required is their dilution in phosphate buffer to achieve a glucose concentration comprised in between the linear calibration range. For honey, 1 g (exactly weighed) is dissolved in 50 mL and then diluted 150 times. For orange juice, a 500-time dilution is needed.

27.5.2 Fabrication of the electrochemical cell

The procedure to fabricate the electrochemical cell is as follows:

1. Clean the stainless steel pins by sonication in isopropyl alcohol for 20 min.
2. Prepare the carbon ink by making a 50% (w/w) suspension of the carbon paste in DMF. To achieve a homogeneous suspension, it is sonicated for 1 h at 37 kHz of frequency and 320 W of power.

3. Coat the pins that will be used as WE with carbon ink. With this aim, immerse them in the ink and then let to dry in an oven at 70°C for 15 min. Repeat this process three times. The effect of the drying time after the third immersion will be studied.
4. Place the pins as in Fig. 27.2A and connect them to the potentiostat using the three-pin Dupont female cable. The transparency is drilled with the pins using a pin header without the pins (Fig. 27.2A) for aligning the pins (they are at the same distance, 1.5 mm approximately).

In this configuration, the WE is located in the middle, the CE has an area that is similar to this of the WE (to not limit the current), and the RE is located to the other side of the WE (Fig. 27.2B).

To perform electrochemical measurements with the pin-based electrochemical cell, an aliquot of 70 μL of the working solution is deposited on the transparency film, covering all the three pins.

Because of the low cost and robustness of the pin-based platform, the three electrodes can be changed after each measurement or they can be employed for several measurements. When the pin-based WE is modified with the mixture of enzymes and mediator (ferrocyanide), it is only used for one glucose measurement. However, CE and RE can remain.

27.5.3 Evaluation of the effect of the drying time

To study the effect of the drying time, pin-based WEs are modified with the carbon ink and left to dry (after the third immersion in the carbon ink) for 15 min, 1 h, and 12 h. Those pin-based WEs are employed to construct cells for recording cyclic voltammograms (CVs) in a 0.5 mM ferrocyanide solution sweeping the potential between -0.2 and $+0.7$ V at a scan rate of 50 mV/s.

The best drying time is chosen considering the values of faradaic (i_f) and capacitive (i_c) currents, trying to obtain always the highest i_f/i_c ratio.

27.5.4 Assessment of the detection potential for glucose analysis

As ferrocyanide is used as mediator in the pin-based glucose sensor, it is necessary to know its redox potentials. Thus, considering the CVs previously obtained (Section 27.5.3, Fig. 27.3A) for the best drying time, a potential that assures electrochemical reduction of the enzymatically generated ferricyanide should be chosen.

27.5.5 Construction of the biosensor and calibration curve of glucose

To fabricate the glucose biosensor, the mixture of enzymes (GOx and HRP) and ferrocyanide is immobilized onto the pin-based WE surface by adsorption (Fig. 27.3B). The procedure consists of depositing 3 μL of the mixture onto the WE surface (carbon ink—coated pinhead) and then it is left until dryness (approximately 40 min).

The analytical signal is the intensity of the current measured at 50 s in the chronoamperogram (i-t curve, Fig. 27.3C) recorded at -0.2 V (potential previously chosen). Thus, this

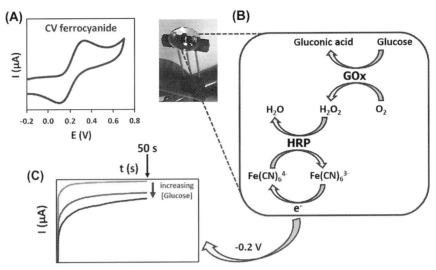

FIGURE 27.3   Schematic representation of the (A) cyclic voltammogram recorded in a 1 mM ferrocyanide solution between −0.2 and +0.7 V at 50 mV/s, (B) enzymatic reactions and electrochemical reduction of the ferricyanide enzymatically generated on the surface of the pin-based working electrode, and (C) chronoamperograms obtained for different glucose concentrations by applying −0.2 V for 50 s.

current is measured for glucose solutions of concentrations comprised between 50 and 5000 μM. Each concentration is measured in triplicate to estimate the precision of the methodology. A calibration curve is obtained by plotting the current measured (at 50 s) for each concentration against the concentration of glucose. Also, the linear range, the sensitivity (as the slope of the calibration curve), the limit of detection (as the concentration corresponding to a signal that is three times the standard deviation of the intercept [or of the estimate] in the lower range of concentrations), and the limit of quantification (as the concentration corresponding to a signal that is 10 times the standard deviation of the intercept or estimate) are obtained.

27.5.6 Study of the precision

When analytical methodologies are developed, the precision of the devices has to be studied. Disposability is based on precision; different devices can be employed if they provide statistically similar analytical signals. This is very important in the case of miniaturized electrochemical cells that employ pseudo-reference electrodes and/or handmade WEs. Because of the low cost of the platform and the simplicity of the design, here it is very easy to perform several measurements allowing estimating the precision. Then, successive voltammograms could be recorded (e.g., seven times):

1. In 1 mM potassium ferrocyanide (in 0.1 M PB pH 7.0) with the same trio of pins. In this case, although measurements can be performed in the same drop, as stirring is not

possible, it is more convenient to change the drop between measurements to warranty the solution contacting the electrode surface is renewed.
2. In 1 mM potassium ferrocyanide (in 0.1 M PB pH 7.0) but changing in this case the pin acting as RE and then changing all the three electrodes. In this way, the suitability of using pins as REs can be studied. It would be convenient to check the potential of several pins (that will act as RE) against a saturated calomel electrode (as made in Ref. [5]) (see Chapter 1).

In all the cases, the parameter that allows evaluating the precision is the relative standard deviation.

27.5.7 Interference evaluation

To study the selectivity of the sensor, two potential interferences such as fructose and ascorbic acid are evaluated. They are chosen as interferences because of its usual presence in food samples that contain glucose (e.g., honey, juices, wines, etc.). Thus, several biosensors are prepared to record chronoamperograms (at -0.2 V for 50 s) in the following solutions: 500 µM glucose, 500 µM fructose, 500 µM ascorbic acid and mixtures 500 µM in glucose and 500 µM in fructose, as well as 500 µM in glucose and 100 µM in ascorbic acid.
In view of the results achieved, the effect of the presence of these interferences is discussed.

27.5.8 Determination of glucose in real food samples

Biosensors are prepared to record chronoamperograms (at -0.2 V for 50 s) in the different sample solutions prepared to obtain the analytical signal that correlates to glucose concentration. Calculate the glucose concentration in honey and orange juice by using the external calibration curve. Several replicates are measured to obtain the average and standard deviation of the results.

27.6 Lab report

At the end of the experiment, write a lab report that includes an introduction (discussing the role of low-cost mass-produced items as substrates of analytical platforms), experimental procedures, results obtained and discussion, finishing with main conclusions. In a more detailed manner, the following points should be considered in the report:

1. Include pictures and schematics of the device (dimensions, steps of fabrication, etc.).
2. Indicate the basis of the analytical signal.
3. Include graphical representation of the data obtained along the experiment. The motivation for doing these studies should be commented first. The conclusions derived from the studies and the values chosen regarding the WE coating and the detection potential for performing the chronoamperograms should be also included.
4. Represent the calibration curve and include the linear range, sensitivity, limit of detection, and limit of quantification of the biosensor.
5. Determine the glucose concentration of the samples, indicating the result with adequate number of significant figures.

6. Discuss the selectivity of the methodology explaining why fructose and ascorbic acid are chosen as important interferences for this biosensor.
7. Discuss the precision and accuracy of the electrochemical platform.

27.7 Additional notes

1. Buffer solution and stock enzyme solutions can be stored at $4°C$ for at least 1 week.
2. Regarding the selectivity, the biosensor does not respond to the presence of fructose at the same concentration than glucose (signal change is lower than 10% with respect to the background). Ascorbic acid shows an important analytical signal when is at the same concentration than glucose. However, in food samples, the ascorbic acid is present at much lower concentration (in the order of mg/mL) than glucose (in the order of g/mL). Because of this, a mixture of 500 μM glucose and 100 μM of ascorbic acid is tested and a signal change lower than 10% with respect to the background is obtained, indicating that ascorbic acid has no influence to determinate glucose concentration in real food samples. This discussion can be interesting and other possible interferents can be considered.
3. Depending on the budget, methodology could be validated analyzing samples with a methodology based on a different principle of detection. There are enzymatic kits in the market that allow optical determination of glucose by measuring the absorbance of an enzymatic colored product. A statistical t-student test could be performed to evaluate the absence (or presence) of systematic errors.
4. As this is a low-cost system, it could be very interesting to detail in the lab report the cost of analysis per device and discuss the importance of using low volumes of samples and reagents, especially when bioreagents are involved.
5. Different designs of electrochemical cells could be employed [2] and they can be discussed/checked. Moreover, in biosensing, it is very important to develop multi-plexed devices [4]. In this case, the design could be varied to include one more (or even two or three) WE that could be modified with a different oxidase. Discussion on the different possibilities is encouraged.
6. The modification with nanomaterials to improve the analytical characteristics of the device could be evaluated. In this way, the influence of, e.g., the modification with carbon nanotubes (by including them on the ink [2] or depositing them on the surface [4]) on the analytical signal could be tested.
7. Discussion on the possible use of other conductive elements, commonly employed in the office, kitchen, or do-it-yourself tasks is encouraged to stimulate creativity.

27.8 Assessment and discussion questions

1. Estimate the apparent Michaelis–Menten constant (K_M) and compare it with K_M of other glucose biosensors previously reported.

2. What are the main issues regarding area and position of electrodes that have to be considered when designing an electrochemical cell?
3. Why can a simple pin be considered a pseudo-reference electrode?
4. What are the advantages/disadvantages of pin-based electrodes when compared with conventional electrodes and screen-printed electrodes?
5. What are the two parts of the biosensor? In this work, how bioreagents were immobilized? Discuss the advantages and disadvantages of the method for immobilization employed in this work versus other methods for immobilization of enzymes found in the bibliography.
6. What are the enzymatic reactions involved and why a mediator is employed? How is the detection potential chosen? Which is the analytical signal?
7. What are the principles of the electrochemical techniques employed in this experiment? What are the excitation and response signals in this case?
8. Design a multiplex pin-based electrochemical cell.

References

[1] A. Tay, Out of the box, finding creative new uses for common laboratory supplies, Lab. Manag. 13 (2018) 28–32.
[2] A.C. Glavan, A. Ainla, M.M. Hamedi, M.T. Fernández-Abedul, G.M. Whitesides, Electroanalytical devices with pins and thread, Lab. Chip 16 (2016) 112–119.
[3] British Stainless Steel Association (BSSA). https://www.bssa.org.uk/, 2018.
[4] E.C. Rama, A. Costa-García, M.T. Fernández-Abedul, Pin-based electrochemical glucose sensor with multiplexing possibilities, Biosens. Bioelectron. 88 (2017) 34–40.
[5] E.C. Rama, A. Costa-García, M.T. Fernández-Abedul, Pin-based flow injection electroanalysis, Anal. Chem. 88 (2016) 9958–9963.
[6] A. García-Miranda Ferrari, O. Amor-Gutiérrez, E. Costa-Rama, M.T. Fernández-Abedul, Batch injection electroanalysis with stainless-steel pins as electrodes in single and multiplexed configurations, Sens. Actuator. B Chem. 253 (2017) 1207–1213.

CHAPTER

28

Flow injection electroanalysis with pins

Estefanía Costa-Rama[1,2], *M. Teresa Fernández Abedul*[2]

[1]REQUIMTE/LAQV, Instituto Superior de Engenharia do Porto, Instituto Politécnico do Porto, Porto, Portugal; [2]Departamento de Química Física y Analítica, Universidad de Oviedo, Oviedo, Spain

28.1 Background

As mentioned before in this part of the book, currently there is a widespread interest in employing mass-produced, inexpensive, and disposable materials for developing electroanalytical devices. Reducing the size and cost of the elements employed in the construction of an electrochemical cell is one of the main trends in electroanalysis. On the other hand, automation, which intends to decrease human participation, is also an important tendency in current Analytical Chemistry because of the constant increasing demand of information. Development of high-throughput analytical methodologies (with high number of samples analyzed per unit of time) is desirable when the number of samples is elevated. When continuous monitoring is required, an automated system is also very advantageous. Samples can be introduced in the system and measured at fixed times. Continuous monitoring of a variable in a stream is another possibility.

Flow injection analysis (FIA), as was commented in Chapters 5 and 9, was introduced by Ruzicka and Hansen in 1975 [1,2]. It is based on the injection of a reproducible volume of sample into a nonsegmented continuous laminar flow of a carrier solution that delivers it to any detector without intervention of the operator. A valve is employed to inject the sample in such a way that the flow of the stream is not disturbed. FIA arose as evolution of the continuous flow injection, which is based on the continuous introduction and analysis of the sample [3,4]. FIA has become a mature and important analytical technique widely employed in different fields such as environmental, food, and clinical [5,6]. It allows the automation of analyses increasing the accuracy of the results and decreasing the analysis time.

On the other hand, nowadays, there is a marked trend toward decentralization of analysis as well as to perform "self-analytics," involving in many cases devices fabricated following "do-it-yourself" approaches [7–9]. In this context, FIA results very attractive because of its great versatility that allows customizing analytical methodologies by incorporating different elements, such as valves, auxiliary channels, connectors, heaters, dialyzers, etc., to perform different steps and procedures (reactions, washings, etc.) required in the analytical process. Among the detectors, many different principles have been combined with FIA [5]. The simplicity of the electrochemical detection fits perfectly with the purposes of FIA and electrochemical cells have been carefully designed with different configurations: wall jet, thin layer, etc. [10,11] (see also Chapters 5,9 and 16). The combination has produced systems very adequate for many interesting applications [5].

Electroanalytical measurements, in general, can be performed in stationary solutions (quiescent solutions, where diffusion is the only means of mass transport of electroactive species to the electrode surface) or under forced convection (stirred solutions, rotating electrodes, or, as in this case, a flow system). Then, here we are performing measurements under hydrodynamic conditions. When a potential, high enough to produce the electrolysis of the analyte, is applied and the concentration gradient ($C-C^s$, C being the concentration in solution and C^s the concentration at the electrode surface in an approximated linear gradient) is maximum (C^S is zero because the analyte is rapidly electrolyzed), then the current (limiting current, $i_L = nFADC/\delta$, where n is the number of electrons, F the Faraday's constant, A the electrode area, D the diffusion coefficient, and δ the thickness of the diffusion layer) is proportional to the concentration in solution (see also Chapters 9 and 16).

In this experiment, adapted from reference [12], an electrochemical FIA system is developed. As in Chapter 27, stainless steel pins are employed as electrodes to construct an innovative electrochemical detector for the FIA system. The system has conventional components: (i) a propulsion system, (ii) an injector, (iii) tubing, (iv) a detector consisting of a flow cell coupled to an instrumental system (in this case, a potentiostat connected with the three pin-based electrode potentiostatic cell), and (v) a control system (a computer acting as an active/passive interface that can control flow, injection, and detection parameters and record, store, and process data). Thus, the electrochemical (amperometric) cell is here fabricated by inserting directly the pins in a piece of tubing through which solutions flow. Therefore, this electrochemical detector is very simple and inexpensive and moreover, pin-based electrodes can be easily and cheaply replaced. In this case, and because no reaction is required, the flow system is employed as a means to transport the sample to the detector in a low-dispersion system. When the potential is applied and the sample goes through the detector, a current due to the redox process appears. As monitoring is continuous, the current decreases back to the baseline when the sample leaves the detector. Then, the record (fiagram, i-t curve) has a peaked shape with the current at the maximum directly proportional to the concentration of the electroactive species.

The feasibility of this system is evaluated employing ferrocene carboxylic acid as it is a common redox probe with a well-known electrochemical behavior. Although the low cost of the pin-based electrodes makes them disposable, in this case, these electrodes can be employed for many measurements maintaining the electroanalytical signal and its accuracy. Finally, to demonstrate the usefulness of this FIA system, it is applied for glucose determination in beverages by an enzymatic assay combining glucose oxidase (GOx), horseradish peroxidase (HRP), and ferrocyanide as electron transfer mediator (see other possibilities in Chapters 17, 25 and 27).

This experiment can be completed in two laboratory sessions of 3—4 h and is appropriate for undergraduate students of advanced Analytical Chemistry or Master students from different fields where analysis is required. With this experiment, the student will learn about the components of an FIA system, the main variables that influence the analytical signal, and the principles of quantitative analysis employing FIA. Because of the versatility of FIA, students will improve their creativity, especially when simple pins are employed in an out-of-the-box application (see also Chapter 29 for a staple-based system).

28.2 Flow injection analysis and electrochemical cell design

This basic FIA system requires the following elements (Fig. 28.1A):

— A peristaltic pump to provide a constant flow of carrier solution.
— A six-port rotary valve to inject a constant and reproducible volume of sample in a short time without disturbing the flow carrier.
— Tubing to connect all the components.
— A detector to obtain the analytical signal that can be correlated to analyte concentration. It consists of a flow cell connected to a potentiostat.

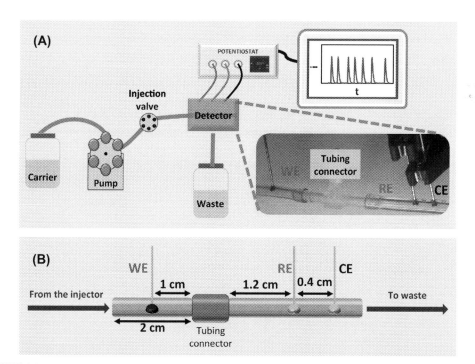

FIGURE 28.1 (A) Schematic representation of a flow injection system (photograph of the pin-based electrochemical detector in the inset). (B) Scheme with the distances to fabricate the pin-based electrochemical detector employing two pieces of tubing. CE, counter electrode; RE, reference electrode; WE, working electrode.

All these components are connected using tubing with a specific diameter. The material is also important and depends on the type of solutions that will flow through the system. The peristaltic pump uses also specific tubing with fixed diameter. Depending on the diameter and on the rotation speed of the pump drums, the flow rate (i.e., µL/min or mL/min) is fixed.

The electrochemical detector is handmade and consists of three pin-based electrodes: two unmodified pins as reference (RE) and counter (CE) electrodes and a carbon-coated pin as working electrode (WE). The pins, as discussed in Chapter 27, are made of stainless steel and their head has a 1.5-mm diameter.

The three pin-based electrodes are directly inserted in two pieces of PVC pumping tube (Fig. 28.1A, inset): one for the WE and the other for CE and RE, which are located downstream. Then, the WE could be exchanged, if required, meanwhile RE and CE can be reused. One of the considerations that has to be taken into account when designing an electrochemical cell is that the substances produced in the redox process that take place on the CE surface do not diffuse to the WE generating interferences. In this case, as CE is located downstream the WE, products are taken away with the flow to the waste. A piece of PVC tubing is used to connect the pieces of the specific pumping tubing where the electrodes are inserted (in between them and with the tube connections coming from the injector). To connect the pins to the potentiostat, alligator clips or other common interfaces can be employed.

28.3 Chemical and supplies

— *Enzymes*: Glucose oxidase (GOx) and horseradish peroxidase (HRP).
— *Redox probe*: Ferrocene carboxylic acid ($FcCO_2H$).
— *Electron transfer mediator*: Potassium ferrocyanide.
— *Components of background electrolyte:* Monobasic and dibasic sodium phosphate and sodium hydroxide.
— *Analyte*: Glucose.
— *Samples*: Orange juice and cola beverages.
— *Electrochemical cell preparation*: Carbon paste, dimethylformamide (DMF), isopropyl alcohol, pumping tubes, and stainless steel pins.
— *General materials, apparatus, and instruments*: 1-mL microcentrifuge tubes, 10-µL, 100-µL and 1-mL micropipettes and corresponding tips, 1-mL microcentrifuge tubes tubing, peristaltic pump with 1.65 mm i.d. tubing, injector (equipped with a 100-µL loop), potentiostat, weighing scale, pH meter, oven, and ultrasound bath.
— Milli-Q water is employed for preparing solutions.

28.4 Hazards

DMF is highly toxic by inhalation and contact, so it has to be handled in a fume hood. Potassium ferrocyanide releases a very toxic gas in contact with acids. Students are required to wear a lab coat, appropriate gloves, and safety glasses.

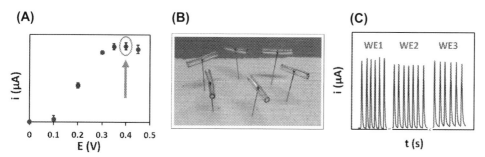

FIGURE 28.2 (A) Schematic representation of a hydrodynamic curve performed injecting three times a 0.25 mM FcCO$_2$H solution (at 1.5 mL/min) applying different potentials between 0.0 and +0.45 V. (B) Photograph of several pieces of pump tubing with carbon-coated pins (WEs) inserted, ready to be used in the flow injection analysis system. (C) Schematic representation of a precision study performing successive injections employing three different pin-based WEs (with the same RE and CE). CE, counter electrode; RE, reference electrode; WE, working electrode. *Figure B reprinted with modifications from E.C. Rama, A. Costa-García, M.T. Fernández-Abedul, Pin-based flow injection electroanalysis, Anal. Chem. 88 (2016) 9958–9963.*

28.5 Experimental procedure

28.5.1 Fabrication of the electrochemical cell

The procedure to construct the electrochemical cell is as follows:

1. Clean the stainless steel pins by sonication in isopropyl alcohol for 20 min.
2. Prepare the carbon ink mixing carbon paste and DMF at 50% (w/w). Sonicate for 1 h at 37 kHz of frequency and 320 W of power to homogenize it.
3. Coat the pins that will be used as WEs by immersion in the carbon ink and leave it to dry at 70°C for 15 min. Repeat this process three times but letting to dry for 12 h after the last immersion.
4. Cut pieces of pumping tube and insert the pins inside as in Fig. 28.1B. The pins are easily inserted in the tubes because of their thin sharp tip that allows drilling the tubing: first introduce the sharp tip into the inner part of the tubing to drill it and then pull the pin until only the head of the pin is inside the tubing. The WEs go in one piece of tubing (see several in Fig. 28.2B), while RE and CE go in a separate piece.
5. Place and connect the pieces of tubing with the intermediate connector as in Fig. 28.1B.
6. Connect the pin-based electrodes (by the shaft of the pins) to the potentiostat using alligator clips. Pins are placed upside down to favor their connection through the alligator clips and also to hold the tubing system more easily.

28.5.2 Solutions and sample preparation

The following solutions are needed (estimate the amount of solution required):

− A 0.1 M phosphate buffer of pH 7.0 (this buffer is used as carrier solution and to prepare all working solutions used along this experiment).
− A 10 mM FcCO$_2$H stock solution to prepare diluted solutions with concentrations comprised between 0.01 and 5 mM.

- A 40 mM potassium ferrocyanide solution.
- A mixture of enzymes and ferrocyanide with concentrations: 0.12 U/μL GOx, 0.1 U/μL HRP, and 20 mM ferrocyanide.
- A 0.1 M glucose stock solution to prepare diluted solutions with concentrations ranging from 25 to 500 μM.

Regarding the samples, orange juice and cola beverage, the only pretreatment required is their dilution in phosphate buffer to achieve glucose concentrations comprised in between the linear calibration range.

28.5.3 Hydrodynamic curve

The hydrodynamic curve (or voltammogram, HDV) is very useful for selecting the appropriate detection potential. It is obtained by repetitive injections of the sample with the electrode potential held at a different value for each injection. In this case, a hydrodynamic curve (i vs. E curve) is recorded injecting a 0.25 mM FcCO$_2$H solution in a continuous flow of 1.5 mL/min applying potentials comprised between 0.0 V and +0.45 V.

According to the hydrodynamic curve obtained (peak current vs. potential), the suitable potential to perform the measurements with FcCO$_2$H solution is chosen. The selection of potential influences the sensitivity. Optimum sensitivity is achieved by holding the potential in the limiting current region of the HDV for the analyte in question (see Fig. 28.2A). Because a plateau is obtained, this potential usually corresponds with the most precise measurements.

28.5.4 Evaluation of the pin-based flow injection analysis system

To test the response of the pin-based FIA system, solutions of different concentrations of FcCO$_2$H (e.g., nine values from 0.01 to 5 mM) are injected in a continuous flow of 1.5 mL/min. Represent the analytical signal (current at the maximum of the fiagram) versus the FcCO$_2$H concentration with the aim of confirming if the FIA system responds linearly to the concentration of the redox probe.The reproducibility of the pin-based FIA system can be studied injecting a 0.25 mM FcCO$_2$H solution in a continuous flow of 1.5 mL/min and monitoring the current at +0.4 V. With this aim:

1. Perform seven successive injections employing the same pin-based electrodes.
2. Perform successive injections with three different pin-based WE and the same RE and CE (e.g., seven for each different WE, see Fig. 28.2C).
3. Perform injections changing all the three electrodes: employ three different trios of electrodes and perform seven injections for each trio.

Calculate the relative standard deviation (RSD) in each case.

28.5.5 Calibration curve and glucose determination in real beverage samples

To perform the calibration for glucose concentration, the mixture of GOx, HRP, and ferrocyanide (0.12 U/μL, 0.1 U/μL, and 20 mM, respectively) previously prepared is added to the glucose solution (or the sample) in a glucose solution/mixture ratio (v/v in %) of 95:5. After leaving to react for 1 min, the mix is injected in the FIA system. The measurements are performed in a continuous flow of 1.5 mL/min applying -0.1 V (Fig. 28.3).

Obtain the analytical signal for glucose solutions of concentrations ranging from 25 to 500 µM. Calculate the equation for the calibration curve and estimate the sensitivity, the limit of detection, and the limit of quantification.

Employing the calibration curve obtained, calculate the glucose concentration in the orange juice and the cola beverage, performing adequate dilutions.

28.6 Lab report

At the end of the experiment, the students should write a lab report including an introduction, experimental, results and discussion, and conclusions sections. The following points should be considered in the report:

1. Include pictures of the pin-based detector fabricated for the FIA system.
2. Include the HDV for $FcCO_2H$ justifying the potential that was chosen. Explain how the detection potential of ferrocyanide can be selected.
3. Represent the calibration curve for $FcCO_2H$ and determine the linear range, sensitivity, limit of detection, and limit of quantitation. Discuss the precision of the system.
4. Represent the calibration curve for glucose determination with the pin-based FIA system indicating the linear range, sensitivity, limit of detection, and limit of quantitation.
5. Determine the glucose concentration of the samples.

28.7 Additional notes

1. Buffer solution and stock enzyme solutions can be stored at 4°C for at least 1 week.
2. In FIA systems, all the measurements are performed in triplicate to obtain precise results, with adequate RSD values.
3. In case an injector is not available, a two-channel pump can be employed with one of the streams for pumping the sample. In case the sample is aspirated, the volume introduced in the system depends on the sampling time.
4. In this work two different pieces of tubing were employed, one for the WE and another for RE and CE. As the flow cleans the surface of WE, replacement is not required in many cases (it will depend on the analyte/system). Actually, the flow cell containing the WE in this experiment was employed for more than 300 measurements without requiring any change. Then, the system could be simplified using one piece of tubing for all the three electrodes.
5. Studies of the influence of the flow rate (performing successive injections varying the flow rate) are interesting to study sample dispersion. Similarly, the influence of the volume of the sample can be evaluated varying the volume of the sample loop.

(A)   **(B)**

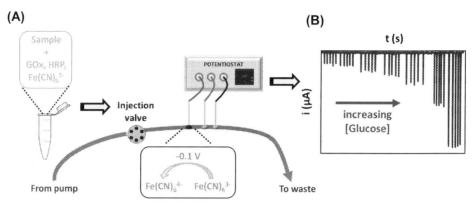

FIGURE 28.3 Schematic representation of (A) the enzymatic assay for glucose determination. (B) Fiagram (i vs. t curve) including signals for injections of solutions with glucose concentration ranging from 25 to 500 μM glucose concentrations.

6. According to our work, the best potential to perform the measurement for the injections $FcCO_2H$ solution is +0.4 V because, although a lower potential could be applied, this potential is chosen to assure the precise oxidation of the $FcCO_2H$.

7. Positive currents are obtained in the characterization studies with $FcCO_2H$ as an oxidation (anodic) process is involved (Fig. 28.2C). However, in glucose determination, negative currents are obtained due to the reduction of ferricyanide (to ferrocyanide) that was enzymatically produced (Fig. 28.3B).

8. The time of the enzymatic reaction is controlled (1 min) but it could be also varied and its influence studied. The time inside the FIA system, from injection to detection, is always the same as long as the flow rate remains constant and the same setup is employed.

9. Sample throughput can be determined (h^{-1}) measuring the time between injection and the return to the baseline.

10. Results obtained for beverage samples can be compared (applying the t-student's test) with the results obtained by other (group of) students.

28.8 Assessment and discussion questions

1. Indicate the main components of this FIA system.
2. How can sample throughput be increased?
3. What happens if the flow rate is increased? And what if it is the sample loop?
4. Explain the shape of the IIDV. What happens if measurements were made at a different potential?
5. In this case, an offline enzymatic assay is consideree. How could this be made with an online configuration? Discuss the possibilities considering advantages and disadvantages.

References

[1] J. Ruzicka, E.H. Hansen, Flow injection analyses. Part I. A new concept of fast continuous flow analysis, Anal. Chim. Acta 78 (1975) 145–157.

[2] J. Ruzicka, J.W.B. Stewart, Flow injection analysis part II. Ultrafast determination os phosphorus in plant material by continuous flow spectrophotometry, Anal. Chim. Acta 79 (1975) 79–91.

[3] C. Vakh, M. Falkova, I. Timofeeva, A. Moskvin, L. Moskvin, A. Bulatov, Flow analysis: a novel approach for classification, Crit. Rev. Anal. Chem. 46 (2016) 374–388.

[4] A.A. Kulkarni, I.S. Vaidya, Flow injection analysis: an overview, J. Crit. Rev. 2 (2015) 19–24.

[5] M. Trojanowicz, K. Kołacińska, Recent advances in flow injection analysis, Analyst 141 (2016) 2085–2139.

[6] J. Ruzicka, E.H. Hansen, Retro-review of flow-injection analysis, TrAC — Trends Anal. Chem. 27 (2008) 390–393.

[7] A.A. Rowe, A.J. Bonham, R.J. White, M.P. Zimmer, R.J. Yadgar, T.M. Hobza, J.W. Honea, I. Ben-Yaacov, K.W. Plaxco, Cheapstat: an open-source, "do-it-yourself" potentiostat for analytical and educational applications, PLoS One 6 (2011) e23783.

[8] A. Ainla, M.P.S. Mousavi, M.N. Tsaloglou, J. Redston, J.G. Bell, M.T. Fernández-Abedul, G.M. Whitesides, Open-source potentiostat for wireless electrochemical detection with smartphones, Anal. Chem. 90 (2018) 6240–6246.

[9] Habitatmap. http://habitatmap.org/habitatmap_docs/HowToBuildAnAirCasting AirMonitor.pdf, 2018.

[10] MicruXfluidic. http://www.micruxfluidic.com/, 2018.

[11] DropSens. http://www.dropsens.com/, 2018.

[12] E.C. Rama, A. Costa-García, M.T. Fernández-Abedul, Pin-based flow injection electroanalysis, Anal. Chem. 88 (2016) 9958–9963.

CHAPTER

29

Staple-based paper electrochemical platform for quantitative analysis

Andrea González-López[1], Paula Inés Nanni[2],
M. Teresa Fernández Abedul[1]

[1]Departamento de Química Física y Analítica, Universidad de Oviedo, Oviedo, Spain;
[2]Laboratorio de Medios e Interfases, Universidad Nacional de Tucumán, Tucumán, Argentina

29.1 Background

It is already known that miniaturization, automation, rapidity, simplicity, and low-cost analysis are main goals in Analytical Chemistry. In this context, the use of low-cost substrates, such as paper (see also Chapters 25 and 26), and the out-of-box applications of materials commonly employed for different purposes, such as stainless steel pins [1,2] (see Chapters 27 and 28), are appropriate alternatives to conventional electrochemical cells. Regarding the use of paper, after the pioneer work of Whitesides' group in microfluidic paper-based analytical devices [3], Henry's group integrated an electrochemical detection based on carbon ink electrodes [4]. Since then, the combination of paper with electrochemistry has attracted significant interest. Wax printing [5] allowed the generation of hydrophilic regions delimited by hydrophobic barriers, although in a homemade lower cost version, a permanent marker could also be employed [6].

In this chapter, we will combine a low-cost paper-based detection with portable and simple instrumentation allowing analysis decentralization. It will be possible by integrating an electrochemical detection, which fits perfectly with the concept of point-of-care tests because of its simplicity and low cost while maintaining sensitivity and selectivity.

The design of the electrochemical cell is one of the main steps in the development of the electroanalytical platform. In this case, a three-electrode system is employed: one of them with an appropriate conductive area (carbon ink) acting as working electrode (WE), an adequate pseudo reference electrode (RE) for applying potentials precisely, and an auxiliary or counter electrode (AE) to close the circuit. The most innovative breakthrough in this case is the use of stainless steel staples as electrodes, which is very advantageous, as they are

Laboratory Methods in Dynamic Electroanalysis
https://doi.org/10.1016/B978-0-12-815932-3.00029-2

(i) low-priced, (ii) reduced in size, (iii) easy to store, (iv) able to be modified with conductive ink, and (v) appropriate as RE and AE without any modification. In addition, previous works proposed stainless steel as a suitable material for electrochemical purposes [7].

This experiment, adapted from [8], has two objectives: on one hand, the fabrication of a very low-cost electrochemical platform, using common materials, and on the other hand, the demonstration of its usability through the studies on the electrochemical behavior of ferrocene carboxylic acid (FCA), a common redox probe. It can be completed in two laboratory sessions of 3-4 hours and is appropriate for undergraduate students of Analytical Chemistry or Master students from different fields where analysis is required, especially when low cost approaches for decentralization want to be shown.

29.2 Electrochemical setup

The staple-based electrochemical cell consists of two unmodified staples as reference (RE) and counter (CE) electrodes and a carbon-coated staple as WE (Fig. 29.1). To connect the stainless steel staples to the potentiostat, alligator clips or other common interfaces can be employed. However, the employment of three-pin Dupont female cables, embedded in

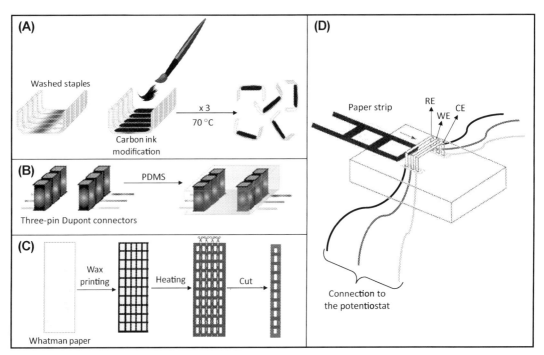

FIGURE 29.1 Schematic representation of the fabrication process of the stapled paper-based platform: (A) Modification of staples with carbon ink (WEs), (B) design of the PDMS holder, (C) wax-printed paper platform procedure, and (D) PDMS holder with the connectors, the three stainless steel staples inserted, and the paper strip placed on the platform ready to slide. *CE*, counter electrode; *RE*, reference electrode; *WE*, working electrode. *Adapted from P.I. Nanni, A. González-López, E. Nunez-Bajo, R.E. Madrid, M.T. Fernández-Abedul, Staple-based paper electrochemical platform for celiac disease diagnosis, ChemElectroChem 5 (2018) 4036−4045.*

a PDMS block allows an easier and quicker connection. A paper-based platform located over the PDMS and below the staples is used for electrochemical measurement.

29.3 Chemicals and supplies

- *Reagents*: FCA (solutions prepared in 0.1 M phosphate buffer [PB] pH 7.0), acetone.
- *Materials*: Stainless steel staples (*STANLEY Stainless steel 1/4" 6 mm* were the staples used for this work), two three-pin Dupont female connectors, polydimethylsiloxane (PDMS): Silicone Elastomer kit Sylgard 184, Whatman chromatographic grade 1 paper (100×300 mm^2, 180 μm thick), laminated cards 60×301 mm^2, double-sided tape, and a small brush.
- *Carbon ink*: Carbon paste and N,N-dimethylformamide (DMF).
- *Equipment*: Vortex, oven, hot plate, sonicator, solid ink printer (ColorQube 8570DN-42PS Xerox), and potentiostat.
- Ultrapure Milli-Q water is employed throughout the work.

29.4 Hazards

Concentrated acids are corrosive and cause serious burns. Students are required to wear a lab coat, appropriate gloves, and safety glasses.

29.5 Experimental procedure

29.5.1 Fabrication of the electrochemical platform

29.5.1.1 Staples pretreatment

The staples that are going to be used as electrodes in the final device must be cleaned with acetone until the polymeric layer that covers their surface is gone. The staples will be ready when they achieve a polish and glossy finish.

29.5.1.2 Modification of staples with carbon ink

The staple used as WE must be modified with carbon ink in the inner part of its crown. With this aim:

1. Dilute the carbon paste in DMF in a 1:1 ratio (w/w) (i.e., 50%) to generate the carbon ink for the modification.
2. Vortex the mixture for a few seconds and then sonicate for 1 h.
3. Adhere all the staples together on a double-sided tape with their inner part up to be mass-modified.
4. Paint them with the carbon ink using a small brush (see Fig. 29.1A).
5. Place in an oven at 70°C for 15 min to evaporate the solvent.
6. Repeat steps 4 and 5 two more times but the last time, after the last coating, place in the oven for 1 h. They must be painted three times to ensure the full coverage of the stainless steel surface.

The staples acting as RE and counter electrode do not need to be covered with the carbon ink, so they will be ready to use after the pretreatment step.

29.5.1.3 Design of the electrochemical cell (PDMS holder)

1. Mix the PDMS base and the cross-linker agent (Sylgard 184) in a 10:1 ratio (w/w).
2. Before curing, place the two three-pin Dupont female connectors inside the mixture with a 1.5 mm of distance between them (to ensure the staple fitted in between).
3. Cure at room temperature for 24 h (see Fig. 29.1B).
4. Connect the Dupont connectors to the potentiostat through alligator clips (as can be seen in Fig. 29.1D).
5. Place the staples on the corresponding connectors of the PDMS holder in such a way that the WE is in the middle.

29.5.1.4 Wax-printed paper platform

1. Design an adequate pattern using the corresponding software. The aim is achieving well-defined separated working areas where the sample will be disposed and not dispersed out.
2. Print the pattern on a Whatman chromatographic grade 1 paper (100×300 mm^2, 180 μm thick) with a solid ink printer.
3. Heat at 110°C for 5 min on a hot plate to ensure wax diffusion through the paper. This has to be considered to obtain final working areas of approximately 3×7 mm^2. The procedure is schematized in Fig. 29.1C.
4. Adhere a rigid thin polymer backing (laminated cards) to the opposite side of these wax-printed areas to avoid liquid dispersion through the back or cross contamination between samples. It will also allow handling and sliding easily the paper strip on the holder.
5. Place the paper strip over the PDMS and below the staples (already inserted in the Dupont connectors) in such a way that it can be easily slid until the staples are over a hydrophilic working area.

29.5.2 Study of the analytical characteristics

29.5.2.1 Evaluation of the staple modification with carbon ink

1. Mix the carbon paste and DMF in different ratios (26, 50, and 60%) to compare their performance.
2. Modify the staples according to Section 29.4.2.
3. Prepare a 10^{-3} M FCA solution in 0.1 M PB pH 7.0.
4. Prepare the holder with the staples and be sure it is connected to the potentiostat. Place the paper strip conveniently, as described in the previous section.
5. Add 18 μL of the 0.1 M PB pH 7.0 solution on the top, making sure all the solutions are in contact with the three staples.
6. Record a cyclic voltammogram (CV) between -0.6 V and $+0.8$ V with a scan rate of 100 mV/s (see Chapter 2 on CV).
7. Slide the paper strip to place a new hydrophilic working area below the staples.
8. Add 18 μL of the FCA solution on the top, again making sure all the solutions are in contact with the three staples.
9. Record a CV between -0.6 V and $+0.8$ V with a scan rate of 100 mV/s.

10. Slide the paper strip and change the staple acting as WE by another one with a different carbon paste:DMF ratio.
11. Repeat steps 5–9.
12. Measure the intensities of anodic and cathodic peak currents. Compare the results for the different ratios.

29.5.2.2 Reproducibility

To check the precision of the low-cost platform (using the same staple as WE and different paper areas and also using different staples and different paper areas), several steps could be followed:

1. Prepare a 10^{-3} M FCA solution in 0.1 M PB pH 7.0 and place the staples (this acting as WE, modified with carbon ink [50% in carbon paste]) in the PDMS block that already holds a paper strip with delimited hydrophilic working areas.
2. Add 18 μL of the FCA solution on the top of the device.
3. Record a CV between −0.6 V and +0.8 V with a scan rate of 100 mV/s.
4. Slide the paper strip to situate a new working area under the staples.
5. Wash the three staples with 0.1 M PB pH 7.0 solution.
6. Slide again the paper strip.
7. Record a new CV after adding 18 μL of the FCA solution.
8. Repeat steps 4–7 until seven different paper strips have been evaluated with FCA.
9. Measure the intensity of the peak currents and calculate the relative standard deviation (RSD).

To obtain the precision for different staples, change the staple that acts as WE by a new one before step 7 and calculate the RSD for measurements performed with seven different staples.

29.5.2.3 Calibration curve

1. Prepare solutions of FCA in 0.1 M PB pH 7.0 (e.g., 0.1, 0.25, 0.5, 0.75, and 1 mM).
2. Place the staples (this acting as WE modified with carbon ink [50%]) in the PDMS block that already holds a paper strip with delimited hydrophilic working areas.
3. Add 18 μL of 0.1 M PB pH 7.0.
4. Record a CV between −0.6 V and +0.8 V with a scan rate of 100 mV/s.
5. Slide the paper strip to a new area.
6. Add 18 μL of the most diluted FCA solution on the top of the device.
7. Record a new CV.
8. Slide the paper strip and wash by adding 18 μL of 0.1 M PB pH 7.0.
9. Slide the paper strip again and add 18 μL of the FCA solution that follows in concentration and record the corresponding voltammogram.
10. Repeat steps 8 and 9 until having recorded CVs for all the FCA solutions.
11. Measure the intensity of the peak current (e.g., anodic process) and represent again the concentration. Indicate the linear range, the sensitivity, and the limit of detection of the methodology.

29.6 Lab report

Write a lab report following the typical scheme of a scientific article, including a brief introduction, experimental part (materials, equipment, and protocols), results and discussion, and conclusions. The following points should be beard in mind:

1. In the introduction, explain the purpose of the experiment and do a short review of other low-cost analytical devices.
2. Protocols must be suitably detailed including schemes preferentially where appropriate and drawings of the electrochemical platform.
3. Discuss the main variables that influence the analytical signals and include figures with representative raw data and the graph with the calibration curve.
4. Discuss the precision of the device.
5. Calculate the cost of the device, considering the components for the fabrication.
6. Discuss the incidences during the course of the experiment.

29.7 Additional notes

1. The pressure of the staples on the paper is an important parameter to be considered. Staples stop when inserted in the connections just to leave appropriate room for the paper strip and a liquid film of solution. The rigid backing has an adequate thickness to approximate paper and staples but without contacting.
2. The area of the WE is very important. If the same backing, connectors, and sample volume are employed, similar electrode area is obtained.
3. The effect of the sample volume and its evaporation influences the analytical signal. Several optimizations confirmed that 18 μL was the best sample volume to be deposited over the device to obtain the higher intensity of the analytical signal.
4. Cyclic voltammetry has been employed as electrochemical technique. Other techniques such as chronoamperometry or square wave voltammetry could be evaluated.

29.8 Assessment and discussion questions

1. One of the staples is modified with carbon ink to act as WE. Explain how the staple is prepared and how they are arranged in the electrochemical platform.
2. Staples are disposable elements. A debate on the disposability/reusability of devices can be very interesting.
3. In this experiment, staples are used in an out-of-box application. This can be an appropriate experiment to stimulate the creativity. Then, discussion about possible designs including nonconventional elements can also be interesting.

References

[1] A.C. Glavan, A. Ainla A, M.M. Hamedi, M.T. Fernández-Abedul, G.M. Whitesides, Electroanalytical devices with pins and threads, Lab. Chip 16 (2016) 112—119.

[2] E.C. Rama, A. Costa-García, M.T. Fernández-Abedul, Pin-based electrochemical glucose sensor with multiplexing possibilities, Biosens. Bioelectron. 88 (2017) 34—40.

[3] A.W. Martinez, S.T. Phillips, M.J. Butte, G.M. Whitesides, Patterned paper as a platform for inexpensive, low-volume, protable bioassays, Angew. Chem. Int. Ed. 119 (2007) 1340—1342.

[4] W. Dungchai, O. Chailapakul, C.S. Henry, Electrochemical detection for paper-based microfluidics, Anal. Chem. 81 (2009) 5821—5826.

[5] S. Altundemir, A.K. Uguz, K. Ulgen, A review on wax printed microfluidic paper-based devices for international health, Biomicrofluidics 11 (2017) 041501.

[6] E. Nunez-Bajo, M.C. Blanco-López, A. Costa-García, M.T. Fernández-Abedul, Integration of gold-sputtered electrofluidic paper on wire-included analytical platforms for glucose biosensing, Biosens. Bioelectron. 91 (2017) 824—832.

[7] H. Ayoub, V. Lair, S. Griveau, P. Brunswick, F. Bedioui, M. Cassir, Electrochemical characterization of stainless steel as a new electrode material in a medical device for the diagnosis of sudomotor dysfunction, Electroanalysis 24 (2012) 1324—1333.

[8] P.I. Nanni, A. González-López, E. Nunez-Bajo, R.E. Madrid, M.T. Fernández-Abedul, Staple-based paper electrochemical platform for celiac disease diagnosis, ChemElectroChem 5 (2018) 4036—4045.

PART VI

Multiplexed electroanalysis

Simultaneous measurements with a multiplexed platform containing eight electrochemical cells

Estefanía Costa-Rama[1,2], *M. Teresa Fernández Abedul*[2]

[1]REQUIMTE/LAQV, Instituto Superior de Engenharia do Porto, Instituto Politécnico do Porto, Porto, Portugal; [2]Departamento de Química Física y Analítica, Universidad de Oviedo, Oviedo, Spain

30.1 Background

The continuously increasing demand of analytical information involves performing more analysis in less time. Multiplexed devices are an attractive way to achieve this goal. However, to analyze different species with the same device, or alternatively, perform simultaneous analysis of the same analyte in different samples, is still a challenge. This is especially important when bioassays are performed because several steps are usually involved, which can increase enormously the analysis time. Several multiplexing strategies are possible. In the case of two analytes, A and B, the following approaches could be followed [1]:

1. *Determination of the analytes as a whole*, without differentiation among them. A positive response may indicate that A, B, or both are present, but without differentiation between them. This is very important in the case of determining, e.g., the presence of pathogens in food, where the presence of only one of them implies a health hazard. It does not matter if A or B is present; the food has to be disposed or treated in a different manner.
2. *Spatially resolved assays or measurements*, with capture/specific reagents for A and B immobilized in different areas. The same label (indicator species) could be employed (e.g., conjugated to secondary anti-A and anti-B antibodies) because the signal will be produced in different areas.
3. *Sequential detection*, i.e., separation in time of the measurements. In this case, the same surface (multispecific surface) is employed for capturing A and B. The same label is

used for both A and B, but measurement for A is taken first. After finishing this measurement, the one that correlates to B is performed. If the signal had not returned to the initial value after measuring A, then the difference between the signals for B and A is proportional to the concentration of B.

4. *Multiple labels* can be employed for multiplexing purposes (in this example, one for A and a different one for B). Possibilities are enormous and new attempts are being continuously reported. When enzymes are involved, care should be taken to maintain their activities because optimum conditions usually vary for different enzymes.

5. *Separation-based* methodologies are a different possibility to perform multiplexed analysis.

In this experiment (as in Chapter 31), spatial resolution is employed. Then, parallel measurements or assays can be performed simultaneously. As mentioned in several previous chapters, miniaturization is an important trend in Analytical Chemistry. Electrochemical techniques fit perfectly with this trend because they allow the development of multiplexed devices of small size based on miniaturized, cheap, and mass-fabricated electrodes. These multiplexed devices can be based on electrochemical platforms: (i) where two or more working electrodes (WEs) share reference (RE) and counter (CE) electrodes (see Chapter 31, same measurement solution) [2—6]; (ii) where there are many electrochemical-independent cells (examples with 8 or 96 cells can be found) [7—9]. The development of these kind of platforms has been possible because of the great progress in electrochemical instrumentation: potentiostats that are able to perform several measurement simultaneously in the two modes: different WEs sharing CE and RE, or several electrochemical cells working independently [10]. When this instrumentation is not available, sequential measurements should be performed. In Fig. 30.1, a multiplexed platform based on screen-printed electrodes is depicted.

This platform is employed in this experiment to perform different multiplexed alternatives:

— *Simultaneous measurements for the same analyte.* This is very useful when several samples have to be analyzed. Moreover, it is also very important in which regards the precision,

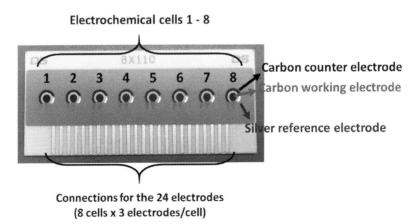

FIGURE 30.1   Picture of electrochemical arrays consisting of eight three-electrode cells based on screen-printed electrodes.

especially for complex sensor architectures (e.g., with bio- and/or nanomodified surfaces) or unstable reagents, as measurements are performed at the same conditions (temperature, humidity, instrumentation, etc.). In this case, several cyclic voltammograms will be recorded in solutions of potassium ferrocyanide, a well-known redox probe.

— *Multianalyte determination.* Obviously, the interest in this kind of multiplexed analysis is continuously growing. This is of extreme importance especially in clinical applications where obtaining simultaneous information about several parameters/biomarkers can be crucial for early disease diagnosis. In this chapter, this kind of multiplexed measurements will be performed in two ways:

  * Employing different detection parameters. As the platform has eight different electrochemical cells, four different concentrations of potassium ferrocyanide and dopamine (an important neurotransmitter) will be evaluated. These electroactive species show different redox processes; therefore, different parameters have to be applied.
  * Employing the same detection parameters. In this case, two different bioassays will be performed, one for biotin (vitamin H) detection and the other for IgG (immunoglobulin G) detection. Thus, the bioassay for each analyte will be different. However, the label employed in this case is the same, and therefore, the same electrochemical parameters could be employed for obtaining the analytical signal.

In this experiment, two affinity bioassays will be performed. For biotin detection, the bioassay is based on the biotin—streptavidin interaction employing streptavidin labeled with the enzyme alkaline phosphatase (AP). For IgG detection, an anti-IgG antibody, also labeled with AP, is employed. With the AP as enzymatic label, a mixture of 3-indoxyl phosphate disodium salt (3-IP) and silver ions (Ag^+) is used as substrate (see Chapter 31) [11]. AP hydrolyzes 3-IP generating an indoxyl intermediate that reduces the silver ions resulting in metallic silver (Ag^0) and indigo blue. Then, the metallic silver enzymatically generated is determined through the redissolution peak by anodic stripping voltammetry (Fig. 30.2).

This experiment, which can be completed in two laboratory sessions of 3—4 h, is useful for undergraduate students of advanced analytical (bio)chemistry or electroanalysis. This experiment shows the advantages of multiplexed devices and also different possibilities. Moreover, as the multiplexed platform used is based on screen-printed electrodes, students also learn how to use it and their advantages such as low cost, small size, portability and easy handling.

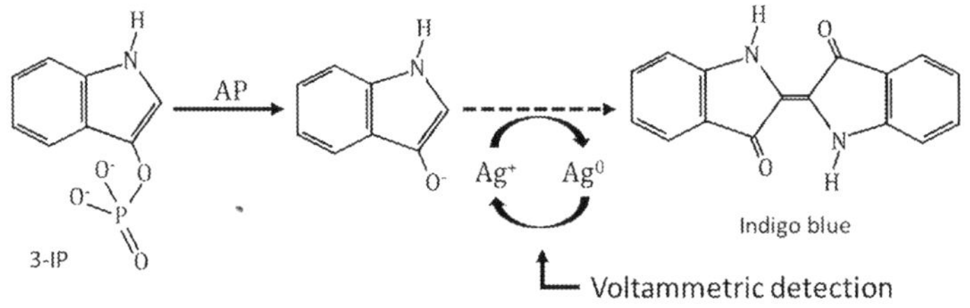

FIGURE 30.2 Scheme of the metallic silver generation by the enzymatic reaction catalyzed by alkaline phosphatase (AP) [11].

30.2 Electrochemical platform

The platform employed in this work is a commercial electrochemical array consisted of eight screen-printed three-electrode electrochemical cells (Fig. 30.1). The dimensions of the platform are $3.4 \times 7.9 \times 0.1$ cm. Each one of the electrochemical cells consists of a silver-based RE and a CE and a WE made of carbon ink.

30.3 Chemical and supplies

— *Redox probes:* Potassium ferrocyanide and dopamine.
— *Biological reagents:* Biotin, streptavidin conjugated to alkaline phosphatase (streptavidin–AP), mouse IgG antibody (IgG), anti-mouse IgG conjugated to AP antibody (anti-IgG-AP) and bovine serum albumin (BSA).
— *Enzymatic substrates:* Silver nitrate ($AgNO_3$) and 3-indoxyl phosphate disodium salt (3-IP).
— *Components of background electrolytes:* Potassium chloride (KCl), Trizma base and nitric acid.
— *General materials, apparatus and instruments:* 1-mL microcentrifuge tubes, simple and eight-channel micropipettes of 10-µL, 100-µL and 1-mL volumes and corresponding tips, multipotentiostat and an adequate edge connector, analytical balance, pH meter and fridge.
— Milli-Q water is employed for preparing solutions (and washing when indicated).

30.4 Hazards

Nitric acid, used for the preparation of Tris-HNO_3 buffer, is corrosive and causes serious burns. Potassium ferrocyanide in contact with acids releases a very toxic gas. Students are required to wear lab coat, appropriate gloves and safety glasses.

30.5 Experimental procedure

30.5.1 Preparation of solutions

The following solutions are needed (estimate the amount of solution required).

— 0.1 M KCl solution (working solutions of redox probes will be prepared in this solution).
— 0.1 M Tris-HNO_3 pH 7.0 buffer solution (buffer 1). Working solutions of biological reagents will be prepared in this buffer solution.
— 0.1 M Tris-HNO_3 pH 9.8 buffer solution containing 20 mM $Mg(NO_3)_2$ (buffer 2). Solutions of silver nitrate and 3-IP will be prepared in this buffer solution.
— 0.1 M potassium ferrocyanide and dopamine stock solutions. From these, solutions of potassium ferrocyanide and dopamine of 0.05, 0.1, 0.5, and 1.0 mM concentration will be prepared.

- Biotin solutions of the following concentrations: 0.1, 1.0, and 10 nM.
- 0.2 nM streptavidin–AP solution.
- 1:25,000 diluted anti-IgG-AP antibody solution.
- IgG antibody solutions of following concentrations: 0.1, 0.5, and 1.0 μg/mL.
- 2% BSA solution.
- 0.1 mM 3-IP, 0.4 mM silver nitrate (mixed) solution.

30.5.2 Electrochemical measurements

To perform electrochemical measurements employing the eight-array electrochemical platform, an aliquot of 25 μL of the working solution is deposited on each one of the electrochemical cells covering the three electrodes (WE, RE, and CE).

30.5.3 Study of the electrochemical behavior of ferrocyanide and dopamine

To explore this possibility, and as a proof of concept, measurements will be performed under different conditions for two different species, ferrocyanide and dopamine. To know their redox processes, cyclic voltammograms are recorded for each species separately (each in a different cell) as follows:

1. Deposit a 25-μL drop of a 0.5-mM solution of potassium ferrocyanide (in 0.1 M KCl) in one of the cells of the platform (cell A).
2. Deposit a 25-μL drop of a 0.5-mM solution of dopamine (in 0.1 M KCl) in other cell of the platform (cell B).
3. Record cyclic voltammograms at 100 mV/s, scanning the potential from −0.2 V to +0.5 V for cell A (ferrocyanide) and from −0.2 V to +0.9 V to for cell B (dopamine) (Fig. 30.3).
4. Measure the peak potential and current intensity for both cathodic and anodic peaks.

30.5.4 Study of the inter- and intra-array reproducibility

The precision of the analytical devices is essential, especially for those disposable, as in this case. Thus, once the redox process for potassium ferrocyanide is known, this species is

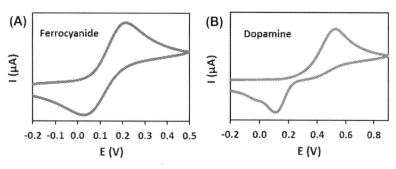

FIGURE 30.3   Schematic representation of the cyclic voltammograms recorded, at 100 mV/s, (A) in a ferrocyanide solution (in 0.1 M KCl) between −0.2 V and +0.5 V and (B) in a dopamine solution (in 0.1 M KCl) between −0.2 V and +0.9 V.

employed to study the reproducibility of the electrochemical platform. Moreover, different platforms are evaluated to determine if they provide statistically similar analytical signals.

Then, successive cyclic voltammograms are recorded as follows:

1. Deposit a 25-μL drop of a 0.5 mM potassium ferrocyanide solution (in 0.1 M KCl) in each one of the eight cells of the array.
2. Record a cyclic voltammogram for each cell between −0.2 and 0.5 V at 100 mV/s.
3. Repeat the steps 1 and 2 several times (e.g., 7 times) employing the same platform. Although measurements could be performed in the same drop, as stirring is not possible, it is more convenient to wash the platform with water and change the drop between measurements to ensure the solution is renewed.
4. Repeat the steps 1 and 2 employing different array platforms (e.g., 3).

In all the cases, the intensity of current for both anodic and cathodic peaks, as well as the peak potential, is measured. Thus, the relative standard deviation for those parameters could be estimated (i) for different drops (measurements) using the same electrochemical cell, (ii) for eight different electrochemical cells of the same platform (intra-array) and (iii) for the three different platforms (inter-array).

30.5.5 Simultaneous calibration of ferrocyanide and dopamine

Once the redox process for potassium ferrocyanide and dopamine has been studied and also the reproducibility has been evaluated, a calibration plot for potassium ferrocyanide and dopamine is performed employing the same platform.

With this aim, linear sweep voltammograms (LSVs) are recorded as follows:

1. Drop 25 μL of potassium ferrocyanide solutions (in 0.1 M KCl) in the concentration range comprised from 0.05 to 1.0 mM in four of the electrochemical cells of the array (cells 1−4).
2. Drop 25 μL of dopamine solutions (in 0.1 M KCl) in the concentration range comprised from 0.05 to 1.0 mM in the other four electrochemical cells of the array (cells 5−8).
3. Record LSVs, at 100 mV/s, from −0.2 V to +0.5 V for cells 1−4 (ferrocyanide) and from −0.2 V to +0.9 V to for cells 5−8 (dopamine).
4. Repeat this procedure employing three different platforms to estimate the reproducibility of the measurements.

The analytical signal is the anodic peak current intensity; representing this intensity of current versus the species (ferrocyanide or dopamine, respectively) concentration, a calibration curve for each is obtained. Determine the equation of the calibration curve (linear regression), as well as the sensitivity, the limit of detection, and the limit of quantification for both species.

30.5.6 Simultaneous bioassays

Two different bioassays, but employing the same label, are performed simultaneously in the same array platform: one based on the interaction of streptavidin−biotin and other based on the interaction of mouse IgG and anti-mouse IgG (Fig. 30.4A).

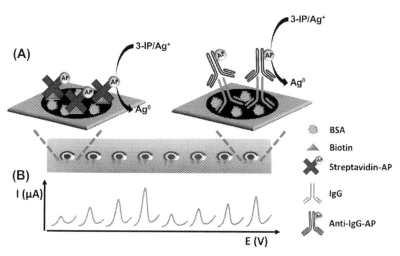

FIGURE 30.4 Schematic representation of (A) simultaneous bioassays performed on the electrochemical platform and (B) linear sweep voltammograms obtained for each electrochemical cell. *3-IP*, 3-indoxyl phosphate; *AP*, alkaline phosphatase; *BSA*, bovine serum albumin.

The procedure to carry out both bioassays in the same platform is as follows:

1. Deposit 4 μL of 0.0, 0.1, 1.0 and 10 nM biotin solutions in four cells of the platform (cells 1—4) and 4 μL of 0.0, 0.1, 0.5 and 1.0 μg/mL IgG antibody solutions in the remainder four cells (cells 5—8). Incubate overnight at 4°C.
2. After washing the platform with buffer 1, block the surface of the electrodes depositing 25 μL of 2% BSA solution for 30 min.
3. Carry out another washing step with buffer 1 and drop 25 μL of 0.2 nM of streptavidin—AP solution in the cells 1—4 and 25 μL of 1:25,000 diluted anti-IgG-AP solution in the cells 5—8. Let it react for 30 min.
3. Wash the platform with buffer 2 and drop 25 μL of the mix 3-IP/silver nitrate (0.1 mM/0.4 mM, respectively). Let it react for 20 min.
4. Record LSVs from 0.0 V to +0.4 V at 50 mV/s (Fig. 30.4B).

The analytical signal is the intensity of current of the anodic peak obtained at ≈ +0.25 V. Representing this peak current intensity versus the concentration of biotin and IgG antibody, respectively, both calibration curves for biotin and IgG antibody are obtained.

30.6 Lab report

A lab report following the typical scheme of a scientific article should be written by the students. Thus, the report should contain the following sections:

1. An Introduction with the advantages of miniaturized electrodes (and comparison with conventional ones) and the importance of developing devices for multiplexed analysis. The use of schemes and diagrams is highly recommended.

2. An Experimental section that lists all the reagents, materials, and apparatus employed along the work, and explains how the (bio)assays are performed.
3. A section containing results and discussion. In this part, the use of selected raw data, graphs or tables is encouraged. The values for precision attained for each assay with particular emphasis on this point because of its importance for multiplexed and disposable devices should be included and discussed.
4. A final section that includes the most important conclusions of the work and the student feelings about it.

30.7 Additional notes

1. The buffer solutions can be stored at 4°C for 1 week.
2. Keep ferrocyanide and dopamine solutions away from light.
3. Suitable concentrations for antibodies may vary depending on the chosen antibodies. In this case, a simple IgG and anti-IgG (labeled with AP) was including, but a different pair of antibodies or immunological system could be evaluated.
4. Cyclic voltammograms could be recorded at different scan rates for ferrocyanide and dopamine solutions with the aim of ascertaining if the processes are diffusive or adsorptive in nature (see Chapters 3 or 24). The reversibility of the redox process for both species can also be studied.
5. For ferrocyanide and dopamine, if the precision studies give good results when the same platform is employed, it can be reused for several measurements. The adsorption of reagents (in the case of bioassays where capture reagents are immobilized) or products of the electrochemical reaction (the silver enzymatically generated that is deposited onto the electrode) changes the surface of the electrode and then the analytical signal (intensity of current). Consequently, in that case, the platform must be discarded after measuring. Although these platforms are considered disposable, discussion on the reusability/disposability is encouraged.
6. It is convenient to repeat measurements in triplicate to evaluate the precision of the methodology. In the case of the calibration curve, this could be performed in three different platforms (even in different days and/or by different students) to estimate the robustness.
7. In Section 30.5.5, two different analytes are considered as a proof of concept, but the eight electrochemical cells could be employed for performing a calibration curve with different concentrations of the same analyte. Standards and samples could be measured in the same card. Similarly, the number of analytes could be increased up to eight. Students are encouraged to find in the bibliography different interesting species (antioxidants, pigments, pharmaceuticals, etc.) to be directly determined with multiplexed platforms like the one employed in this work.
8. One of the advantages of using eight different cells is the decrease in analysis time. The use of an eight-channel micropipette (that fits with this platform) is then very convenient.
9. In this case, the same enzyme is employed as label in the different cells, but because the electrochemical cells are independent (see the difference with Chapter 31), different enzymes could even be employed. Thus, the versatility of this platform is very interesting: for example, for determination of analytes such as glucose, lactate or

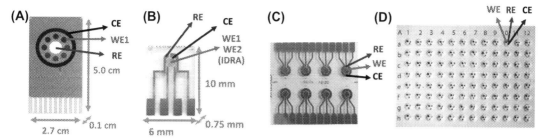

FIGURE 30.5   Picture of commercial (A) thick-film electrodes with 8 working electrodes (WEs) sharing reference electrode (RE) and counter electrode (CE) [10], (B) thin-film electrodes with 2 interdigitated WEs, sharing RE and CE [12] *IDRA*, interdigitated ring array, (C) 8 thin-film separate electrochemical cells [12], and (D) 96 pin-based separate electrochemical cells [7].

alcohol in the same sample. Then different enzymatic assays with, e.g., glucose, lactate and alcohol oxidase/HRP and mediator could be performed in each electrochemical cell. Each enzyme is deposited in two screens (for the shake of precision), and measurements for the three analytes (same sample) could be performed in the same platform.

10. In this experiment, the product of the enzymatic reaction is solid (indigo blue and metallic silver) and deposits over electrode. If two WEs sharing the same RE and CE are employed (as in Chapter 31), care should be taken if products are not deposited but diffuse to the solution. The detection time has to be carefully selected to avoid crosstalk between electrodes. In this case, since independent cells are used, both situations are valid: product can adsorb on the electrode or diffuse to the solution.

11. The use of a multipotentiostat is encouraged. In case a common potentiostat is employed, sequential measurements have to be taken. If the drops are added at the same time, rapid measurements are aimed to avoid solvent evaporation (a cover can be prepared). Times for bioreactions have to be also considered.

13. In this work, a commercial array with eight independent electrochemical cells with carbon-based WEs is employed. However, nowadays, different multiplex electrochemical platforms can be found: for example, screen-printed electrodes (thick-film) with two (see Chapter 31), four or eight WEs (Fig. 30.5A [10]) sharing the same RE and CE, as well as thin-film electrodes with two WEs (that could be even interdigitated electrodes (Fig. 30.5B, see Chapter 6)), separated thin-film (Fig. 30.5C [12]) or pin-based (Fig. 30.5D, [7]) electrochemical cells.

30.8 Assessment and discussion questions

1. Discuss the advantages of multiplexed analysis/devices over single-measurement devices.
2. Comment the different strategies for approaching multiplexed analysis with examples of each. A bibliographic search could be made to discuss the different types, as well as their advantages and inconveniences.

3. What is the difference between cyclic and linear sweep voltammetry? Draw the excitation and response signal for each of the main potential scans.
4. Discuss how the study of the precision of the platforms can be made.

References

[1] A. Brecht, R. Abuknesha, Multi-analyte immunoassays application to environmental analysis, Trends Anal. Chem. 14 (1995) 361–371.

[2] M.M.P.S. Neves, M.B. González-García, C. Delerue-Matos, A. Costa-García, Multiplexed electrochemical immunosensor for detection of celiac disease serological markers, Sens. Actuator. B Chem. 187 (2013) 33–39.

[3] R. García-González, A. Costa-García, M.T. Fernández-Abedul, Dual screen-printed electrodes with elliptic working electrodes arranged in parallel or perpendicular to the strip, Sens. Actuator. B Chem. 198 (2014) 302–308.

[4] E.C. Rama, A. Costa-García, M.T. Fernández-Abedul, Pin-based electrochemical glucose sensor with multiplexing possibilities, Biosens. Bioelectron. 88 (2017) 34–40.

[5] A. García-Miranda Ferrari, O. Amor-Gutiérrez, E. Costa-Rama, M.T. Fernández-Abedul, Batch injection electroanalysis with stainless-steel pins as electrodes in single and multiplexed configurations, Sens. Actuator. B Chem. 253 (2017) 1207–1213.

[6] W. Tedjo, J.E. Nejad, R. Feeny, L. Yang, C.S. Henry, S. Tobet, T. Chen, Electrochemical biosensor system using a CMOS microelectrode array provides high spatially and temporally resolved images, Biosens. Bioelectron. 114 (2018) 78–88.

[7] A.C. Glavan, A. Ainla, M.M. Hamedi, M.T. Fernández-Abedul, G.M. Whitesides, Electroanalytical devices with pins and thread, Lab. Chip 16 (2016) 112–119.

[8] D. Martín-Yerga, M.B. González-García, A. Costa-García, Electrochemical immunosensor for anti-tissue transglutaminase antibodies based on the in situ detection of quantum dots, Talanta 130 (2014) 598–602.

[9] J. Biscay, M.B. González-García, A. Costa-García, Electrochemical biotin determination based on a screen printed carbon electrode array and magnetic beads, Sens. Actuator. B Chem. 205 (2014) 426–432.

[10] Metrohm DropSens, http://www.dropsens.com/ (July 2018).

[11] P. Fanjul-Bolado, D. Hernández-Santos, M.B. González-García, A. Costa-García, Alkaline phosphatase-catalyzed silver deposition for electrochemical detection, Anal. Chem. 79 (2007) 5272–5277.

[12] Micrux Technologies, http://www.micruxfluidic.com/ (July 2018).

Simultaneous detection of bacteria causing community-acquired pneumonia by genosensing

Graciela Martínez-Paredes[1], María Begoña González-García[2], Agustín Costa-García[3]

[1]OSASEN Sensores S.L., Parque Científico y Tecnológico de Bizkaia, Derio, Spain; [2]Metrohm Dropsens, Parque Tecnológico de Asturias, Edificio CEEI, Asturias, Spain; [3]Departamento de Química Física y Analítica, Universidad de Oviedo, Oviedo, Spain

31.1 Background

Mycoplasma pneumoniae (MP) has been associated with a variety of clinical manifestations, including those involving the respiratory tract. It is a significant cause of respiratory tract infections in humans and an important pathogen in acute respiratory illnesses in children and adults, accounting for as many as 20–40% of all cases of community-acquired pneumonia [1]. However, these types of infections are not only associated to this bacterium but also to several other bacteria: *Streptococcus pneumoniae* (SP), *Legionella pneumophila*, or *Chlamydophila pneumoniae* [2] to name but a few, and the treatment differs depending on the causing bacteria. A rapid and precise diagnosis of the infection cause helps to select the treatment best suited to the situation, thus fighting against the increasing problem of antibiotic resistance, which leads to higher medical costs, prolonged hospital stays, and increasing mortality [3].

In clinical analysis, the determination of single biomarkers (protein, RNA/DNA sequences, etc.) has limited diagnostic value and most of the diseases are identified by the simultaneous presence of various biomarkers. The information provided by evaluating single biomarkers is limited, and multiple markers may be more useful for disease screening and assessing multiple physiological pathways that contribute to disease activity and prognosis. Furthermore, there are considerable technical benefits in enabling simultaneous quantification of multiple analytes. Single analyte measurement can be laborious, time-consuming, and costly, whereas the simultaneous screening of various analytes permits reduced sample consumption, technician time, and reagent volumes and increases sample throughput while enabling a rapid, low-cost, and reliable quantification. Moreover, multiplexed diagnostics is a strong request from international public health organizations and can be used in several ways: to simultaneously detect several pathogens that are responsible for similar symptoms, to improve the sensitivity toward one disease by targeting several biomarkers or RNA/DNA strands, to get more information about a pathogen such as the identification of mutants, etc. Therefore, multiplexed testing, which is simultaneous detection of different analytes from a single specimen, has become more important for clinical diagnostics in the last decades [4].

Nucleic acid detection is becoming relevant in clinical diagnosis because a DNA/RNA test not only addresses the question of whether a patient is infected with a particular pathogen or not but also that is very useful for treatment monitoring because elimination of pathogen nucleic acids indicates successful handling. End-point polymerase chain reaction (PCR) allows the production of multiple copies of DNA from the amplification of a single copy or a few copies of a DNA template. Owing to its high sensitivity, nucleic acid amplification has been widely used for the identification and detection of pathogens in clinical samples, being considered as an alternative to conventional microbiological culture techniques. Additionally, a careful design of the primers used as starters for the DNA amplification process can lead to the amplification of similar DNA sequences belonging to different specimens, thus making possible the replication, at the same time, of DNA from different pathogens [5].

The development of electrochemical nucleic acid hybridization biosensors has attracted considerable research efforts in the last years. Nucleic acid sensing applications require high sensitivity through amplified transduction of the hybridization reaction. Electrochemical devices offer elegant routes for interfacing at molecular level, nucleic acid recognition, and signal transduction elements and are uniquely qualified for meeting the low-cost, low-volume, and power requirements of decentralized nucleic acid diagnostics. In addition, electrochemical biosensors have the capability of multiplexing needed for the simultaneous detection of various analytes at the same time [6]. Screen-printed electrodes (SPEs), which are miniaturized, tailor-made, and portable transducers, fulfill the requirements needed for the electrochemical transduction of nucleic acid hybridization biosensors.

In this experiment, adapted from Ref. [7], a nucleic acid hybridization assay with enzymatic electrochemical detection is carried out on a gold-nanostructured dual screen-printed carbon electrode (AuNPs-d-SPCE). With this aim we modify dual screen-printed carbon electrodes (d-SPCEs) [8] generating gold nanoparticles (AuNPs) over the surface of working electrodes, taking advantage of the unique properties that nanoscale materials offer for interfacing biological recognition events with electronic signal transduction (see Chapter 23). The electrodes used in this experiment are commercially available d-SPCEs, aimed at detecting two signals simultaneously, allowing (differential) measurement of up to two analytes in the solution.

Before starting the hybridization procedure, d-SPCEs are gold-nanostructured and modified with streptavidin to allow a proper orientation of the biotinylated oligonucleotide probes. It is well-known that the incorporation of AuNPs to the transducer electrodes improves the biocompatibility between the electrode surface and the proteins as well as facilitates electron transfer between them. Additionally, it has demonstrated a considerable improvement of the analytical signal for similar assays while minimizing the use of biological reagents. After a blocking step, the electrode surface is ready for the hybridization reaction

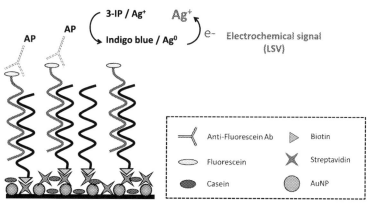

FIGURE 31.1   Scheme of the electrochemical genosensor format used in this experiment. *AP*, alkaline phospha-tase; *3-IP*, 3 indoxylphosphate; *LSV*, linear sweep voltammetry; *AuNP*, gold nanoparticles.

with labeled oligonucleotide probes, which have been selected to complement the target sequences (Fig. 31.1). An enzymatic amplification reaction is used to follow the hybridization reaction. In this experiment an enzymatic substrate system which enzymatic reaction is described in Fig. 31.2 is used (see also Chapters 5, 20, and 21). The substrate 3-indoxyl phosphate (3-IP) produces an intermediate indoxyl compound able to reduce silver ions in solution into a metallic deposit meanwhile the indoxyl intermediate generates the insoluble compound indigo blue. The current generated during the anodic stripping of the silver constitutes the analytical signal. This metallic silver deposit is located where the enzymatic label alkaline phosphatase (AP) is attached, avoiding cross-talk between electrodes and therefore making possible to use the same label in both electrodes. Then, the approach employed here for multiplexing purposes is the spatial separation of assays (two working electrodes are used, one for each assay) using only one label. Simultaneous detection simplifies enormously the procedure, and measurements from both working electrodes can be recorded at the same time (Fig. 31.3B).

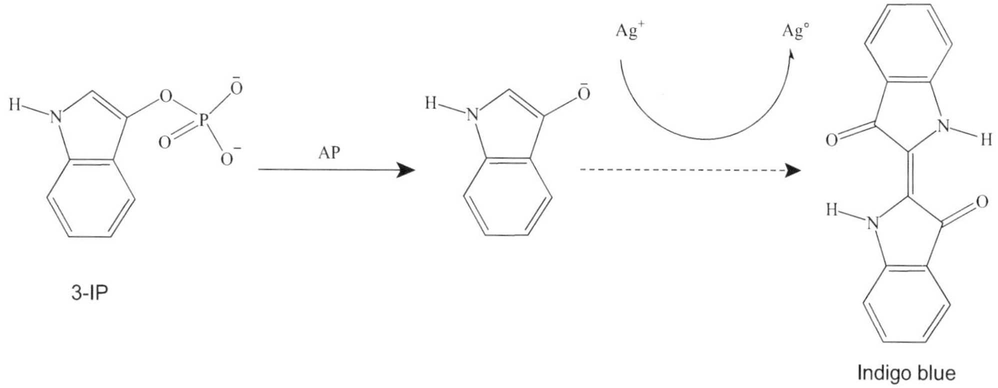

3-IP

Ag$^+$ Ag$^\circ$

Indigo blue

FIGURE 31.2   Scheme of the enzymatic reaction used in the genosensing procedure that generates silver deposits. *3-IP*, 3-indoxyl phosphate; *AP*, alkaline phosphatase.

This kind of enzymatic reaction is highly suitable for multiplex detection, as the substrate deposits over each working electrode without diffusion to the other working electrode. A scheme of the whole hybridization procedure is shown in Fig. 31.1.

This experiment can be carried out in two consecutive laboratory sessions: using the first session (approximately 3 h) to modify the electrodes with AuNPs and streptavidin and prepare the main buffers and solutions needed, while in the second laboratory session (approximately 4 h) the hybridization detection procedure is carried out. This experiment is appropriate for senior-level undergraduate courses in Chemistry and Biotechnology. After the completion of this experiment, students will know how to perform an electrochemical DNA hybridization analysis and simple multiplexed electrochemical analysis.

31.2 Electrochemical cell design

The electrochemical cell consists of d-SPCEs (Fig. 31.3A), which are commercially available together with an edge connector, from DropSens (Llanera, Spain). The electrodes incorporate a four-electrode configuration, printed on ceramic substrates (3.4×1.0 cm), including two working electrodes, one counter electrode and a reference electrode. Both working electrodes (oval-shaped around 4 mm large diameter and 1.5 mm small diameter) and counter electrode are made of carbon inks, whereas the pseudo-reference electrode and electric contacts are made of silver. An insulating layer was printed over the electrode system, leaving uncovered the electric contacts and the electrodes. The area surrounding the four electrodes constitutes the reservoir of the electrochemical cell, with a volume of approximately 70 μL.

31.3 Chemicals and supplies

— *Gold nanostructure source:* Standard gold (III) tetrachloro complex solution in hydrochloric acid (commercially available from Merck).
— *Components of background electrolytes:* Hydrochloric acid, nitric acid, Trizma base, and magnesium nitrate.
— *Enzymatic substrates:* Silver nitrate and 3-IP.
— *Other biological reagents:* Casein from bovine milk, anti-fluorescein (FITC, fluorescein isothiocyanate) antibody labeled with alkaline phosphatase (Ab-AP) and streptavidin (St).
— *Synthetic oligonucleotides:*

  MP (*Mycoplasma pneumoniae*)
     Target: FITC—5′-TTG-GCA-AAG-TTA-TGG-AAA-CAT-AAT-GGA-GGT-TAA-CCG-AGT-G-3′
     Probe: biotin—5′-CAC-TCG-GTT-AAC-CTC-CAT-TAT-GTT-TCC-ATA-ACT-TTG-CCA-A-3′
  SP (*Streptococcus pneumoniae*)
     Target: FITC—5′-CTC-TGA-CCG-CTC-TAG-AGA-TAG-AGT-TTT-CCT-TCG-GGA-CAG-AGG-TG-3′

Probe: biotin—5′-CAC-CTC-TGT—CCC—GAA-GGA-AAA-CTC-TAT-CTC-TAG-AGC-GGT-CAG-AG-3′

— *Analyte:* Synthetic 5′-FITC labeled oligonucleotide sequences matching the DNA sequence of MP or SP.
  Samples: In this experiment, real samples are not studied to simplify the experiment. However, to perform more complex experiments, PCR products can be used as samples. Suitable primers for amplification of both bacteria target sequences are:

Labeled 5′-primer: FITC-5′-TTC-GAA-GCA-ACG-CGA-AGA-A-3′
3′-primer: 5′-TCG-TCA-GCT-CG-3′

— *Electrochemical cell:* Commercially available d-SPCEs (DropSens, Ref. X1110).
— *General materials, apparatus, and instruments:* 1.5 mL centrifuge tubes, general laboratory glassware, weighing scale, timer, pH meter, fridge, freezer ($-20°C$), bipotentiostat, d-SPCEs connector, micropipettes, and corresponding tips.
— Milli-Q water is employed for preparing solutions and washing unless otherwise specified.

31.4 Hazards

Hydrochloric and nitric acids, used for the preparation of background electrolytes, are corrosive, cause severe skin burns and eye damage, and may cause respiratory irritation.

Silver nitrate may intensify fire, is corrosive, causes severe skin burns and eye damage, and is very toxic to aquatic life with long-lasting effects, so a proper disposal of residues is required, as well as maintaining cleanliness in the working space.

Students are required to wear hand gloves, eye protection, lab coat, and other appropriate protective equipment during this experiment.

31.5 Experimental procedure

31.5.1 Preparation of solutions

The following listed solutions and reagents are required in this experiment. The teaching assistant or instructor can prepare in advance some of the solutions or aliquots to save time and minimize the consumption of expensive reagents. Estimate the amount of solutions needed.

— 0.1 M HCl.
— 0.1 mM $AuCl_4^-$ solution prepared in 0.1 M HCl from standard gold (III) tetrachloro complex (1.000 ± 0.002 g of tetrachloroaurate (III) in 500 mL 1 M HCl).
— *Buffer 1:* 0.1 M Tris—HNO_3 pH 7.2 buffer solution.
— *Buffer 2:* 0.1 M Tris—HNO_3 pH 8.0 buffer solution.
— *Buffer 3:* 0.1M Tris—HNO_3 pH 7.2 buffer, containing 2 mM $Mg(NO_3)_2$.
— *Buffer 4:* 0.1 M Tris—HNO_3 pH 9.8 buffer containing 20 mM $Mg(NO_3)_2$.

- Prepare 100 nM biotinylated oligonucleotide probe solutions in buffer 2 for each bacterium: SP and MP. Store them at 4°C until use.
- The mixture solution containing 1 mM 3-IP and 0.4 mM AgNO$_3$ is freshly prepared in buffer 4 and stored in opaque tubes at 4°C until use.
- A 2% (w/v) casein solution is freshly prepared in buffer 1.
- Prepare a 1.25×10^{-7} M St solution in buffer 1 and store it at 4°C until use.
- Prepare a fresh 1:75 dilution from Ab-AP in buffer 3 and store it at 4°C until use.
- Prepare target oligonucleotide solutions with different concentrations by suitable dilution in buffer 2, to perform a calibration curve; e.g., 0, 100, 250, 500, and 750 pM in each bacterium, SP and MP. Some solutions containing only one bacterium are also of interest to check nonspecific adsorptions.
- Prepare an unknown sample for each group of students containing target oligonucleotides at a concentration that fits into the calibration curve and give them to the students to identify the composition of their unknown sample.

31.5.2 Surface modification of dual screen-printed carbon electrodes

Prepare as many electrodes as necessary. Each measurement should be carried out with one electrode and by triplicate. (That is, 18 electrodes in total are required for each group of students; 15 electrodes for the calibration curve and 3 for the measurement of the unknown sample.) To modify the electrochemical cells with AuNPs and streptavidin, proceed as follows:

1. Connect the d-SPCEs to the bipotentiostat and place, covering the four electrodes, a 70 µL drop of the 0.1 mM AuCl$_4^-$ acidic solution. Apply a constant potential of -0.01 V for 2 min. After that, and in the same medium, apply a potential of $+0.1$ V during 2 min to desorb the generated hydrogen. Record the chronoamperogram during the electrodeposition of gold.
2. Rinse the modified d-SPCEs (Au-d-SPCEs) generously with water and let them dry.
3. Cover each working electrode of the Au-d-SPCEs with 4 µL of a 1.25×10^{-7} M St solution and incubate overnight at 4°C.
4. After incubation, rinse gently with buffer 1.

31.5.3 Hybridization reaction procedure

1. Place a 4-µL drop of 100-nM biotinylated oligonucleotide probe in the different working electrodes, each probe to their corresponding working electrode, e.g., MP probe in working electrode 1 (WE1) and SP probe in working electrode 2 (WE2). Incubate for 30 min and rinse with buffer 1.
2. Block the remaining active sites of the electrode surface casting a 70-µL drop of the 2% (w/v) casein solution onto the four-electrode system for 20 min. Then, rinse gently with buffer 3.

(A)

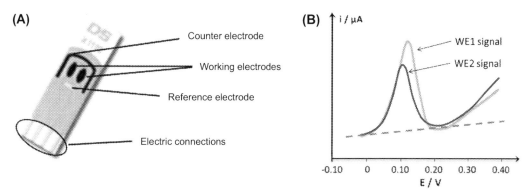

(B)

Counter electrode

Working electrodes

Reference electrode

Electric connections

WE1 signal

WE2 signal

i / μA

-0.10   0   0.10   0.20   0.30   0.40

E / V

FIGURE 31.3   (A) Dual screen-printed carbon electrode employed in this experiment. (B) Scheme of the analytical signals recorded (linear sweep voltammograms).

3. Deposit a 70-μL drop of the fluorescein-labeled target oligonucleotide at different concentrations covering the four electrodes. Incubate for 60 min and rinse gently with buffer 3. Handle the unknown sample in the same way.
4. Place a 70-μL drop of the 1:75 dilution of Ab-AP covering the four electrodes. Incubate for 60 min and rinse gently with buffer 4.
5. Deposit a 70-μL drop of the enzymatic substrate mixture solution (3-IP and $Ag(NO_3)$) covering the electrode system. Incubate for 20 min protected from direct light exposure.
6. Connect the electrode card to the bipotentiostat (using a specific d-SPEs connector or hook clips) and record the linear sweep voltammograms by scanning the potential from −0.02 to +0.40 V at 50 mV/s of scan rate (see Fig. 31.3B).
7. Measure the intensity of the peak current as analytical signal. Express it as an average value with standard deviation.
8. Obtain a calibration curve for both oligonucleotide target sequences using the intensity of the peak current as the signal and obtain a regression equation for each of the target sequences. Keep in mind that the regression equation can be nonlinear, and sometimes, logarithmic transformation is employed for linearization purposes. Precision can be improved by increasing the number of calibration points and replicates. Then, try to find the equation that better fits your data. Estimate also the detection and quantification limits of the method.
9. Identify the composition of your unknown sample, and, if possible, estimate the amount of each oligonucleotide present in your sample.

31.6 Lab report

Once the experiment is finished, write a lab report following the style of a scientific article, including an introduction (discussing the importance of multiplexing in biochemical analysis and the advantages of electrochemical methods role of low-cost mass-produced electrodes as transducers of chemical sensors and the main advantages of the multiplexed electrochemical detection), experimental procedures, results obtained and discussion, finishing with main conclusions. In a more detailed manner, the following points should be considered in the report:

1. Include schematics for explaining the procedure.
2. Indicate the motivation for doing these experiments, and comment the basis of the analytical signal.
3. Include graphical representation of the data obtained along the experiment and remember to include the standard deviation in graphics whenever is possible. The conclusions derived from the experiments should be included.
4. Represent the calibration curves and include the regression equation, sensitivity, limit of detection, and limit of quantification of the sensor for each of the studied bacteria.
5. Identify the different bacteria present in the unknown sample, and if it is possible, estimate the concentration of each bacterium present in the sample, indicating the result with adequate number of significant figures.

31.7 Additional notes

1. Suitable immobilization of streptavidin on electrodes consists on depositing a 4-μL drop of 1.25×10^{-7} M St over each working electrode previously modified with AuNPs according to Ref. [7] (applying −0.01 V for 2 min and +0.1 V for 2 min in a 0.1 mM AuCl$_4^-$ solution, prepared in 0.1 M HCl) and incubating overnight at 4°C. Then, this experiment should be planned as a 2-day experiment, or alternatively the instructor/assistant teacher could have the electrodes prepared in advance. Otherwise, it is possible to acquire commercial screen-printed dual electrodes already modified with AuNPs and St (www.dropsens.com).
2. During washing steps, be careful not to spread liquids over electric contacts of electrodes. In case of spillover, the electric contacts wipe it with a tissue paper.
3. The oligonucleotide sequences have been chosen to detect two specific bacteria after a PCR amplification according to Ref. [7]. However, these oligonucleotides can be substituted by other matching DNA sequences properly conjugated with the selected labels (biotin and fluorescein).
4. Decide at the beginning of the experiment, which probe and target sequence will be analyzed in WE1 and which one in WE2. Keep this decision in mind (note down or mark conveniently the card with permanent markers on the isolating layer) during the immobilization of oligonucleotide probe and analytical signal interpretation.
5. Use one d-SPCE for every measurement.
6. Try to cover the electrodes with a cardboard box during the enzymatic reaction with 3-IP and silver. This step is light-sensitive and a prolonged light exposure can lead to an increase of nonspecific signals.
7. To obtain a representative analytical signal, measurements should be made by triplicate and the analytical signal should be expressed as an average value with standard deviation.

8. The aim of analyzing unknown samples is more qualitative than quantitative, so try to offer different samples to different groups of students. They can contain both, one of them, or even none of the target oligonucleotides (MP or SP targets).
9. To process all the standards and samples in the same batch, try to sequentially process them, i.e., record the measurement of one electrode with 1 min of difference from the previous one, as can be seen in the following timetable:

Standard	1			2			3			4			5			Unknown sample		
Replicate	1	2	3	1	2	3	1	2	3	1	2	3	1	2	3	1	2	3
Biotin-probe	0:00	0:01	0:02	0:03	0:04	0:05	0:06	0:07	0:08	0:09	0:10	0:11	0:12	0:13	0:14	0:15	0:16	0:17
Block	0:30	0:31	0:32	0:33	0:34	0:35	0:36	0:37	0:38	0:39	0:40	0:41	0:42	0:43	0:44	0:45	0:46	0:47
FITC-target	0:50	0:51	0:52	0:53	0:54	0:55	0:56	0:57	0:58	0:59	1:00	1:01	1:02	1:03	1:04	1:05	1:06	1:07
Ab-AP	1:50	1:51	1:52	1:53	1:54	1:55	1:56	1:57	1:58	1:59	2:00	2:01	2:02	2:03	2:04	2:05	2:06	2:07
Enzyme substrate	2:50	2:51	2:52	2:53	2:54	2:55	2:56	2:57	2:58	2:59	3:00	3:01	3:02	3:03	3:04	3:05	3:06	3:07
Readout	3:10	3:11	3:12	3:13	3:14	3:15	3:16	3:17	3:18	3:19	3:20	3:21	3:22	3:23	3:24	3:25	3:26	3:27

10. To evaluate cross-talking between electrodes, or nonspecific signals, a sample containing only one of the bacteria could be evaluated (e.g., MP). Then, it is considered that there is no cross-talking between electrodes or nonspecific signals when the analytical signal for the electrode containing SP probes is near zero. In any case, it could be evaluated.

31.8 Assessment and discussion questions

1. Explain the basis of the analytical signal and the procedure to follow from sample to result.
2. Is it possible to estimate from your data the presence of cross-signal effect (an increment in the analytical signal in one of the working electrodes because of the presence of the second analyte in the sample)? If not, could you propose a way of checking the presence of this effect?
3. According to bibliography, what is the role of electrochemically generated gold in the surface of d-SPCEs?
4. Think about other enzymatic substrates suitable for multiplexed electrochemical analysis.
5. Indicate other possible alternatives different from spatially resolved assays for multiplexing assays.
6. Discuss about possible designs for assaying all the four bacteria: *Mycoplasma pneumoniae, Streptococcus pneumoniae, Legionella pneumophila,* and *Chlamydophila pneumoniae,* in a tetrasensor.

References

[1] T. Saraya, *Mycoplasma pneumoniae* infections: basics, J. Gen. Fam. Med. 18 (2017) 118—125.

[2] C. Cillóniz, A. Torres, M. Niederman, M. van der Eerden, J. Chalmers, T. Welte, F. Blasi, Community-acquired pneumonia related to intracellular pathogens, Inten. Care Med. 42 (2016) 1374—1386.

[3] www.who.int/campaigns/world-antibiotic-awareness-week/en/

[4] C. Dincer, R. Bruch, A. Kling, P.S. Dittrich, G.A. Urban, Multiplexed point-of-care testing — xPOCT, Trends Biotechnol. 35 (2017) 728—742.

[5] S.A. Barghouthi, A universal method for the identification of bacteria based on general PCR primers, Indian J. Microbiol. 51 (2011) 430—444.

[6] T.G. Drummond, M.G. Hill, J.K. Barton, Electrochemical DNA sensors, Nat. Biotecnol. Trends Anal. Chem. 21 (2016) 1192—1199.

[7] G. Martínez-Paredes, M.B. González-García, A. Costa-García, Genosensor for detection of four *pneumoniae* bacteria using gold nanostructured screen-printed carbon electrodes as transducers, Sens. Actuator. B Chem. 149 (2010) 329—335.

[8] R. García-González, A. Costa-García, M.T. Fernández-Abedul, Dual screen-printed electrodes with elliptic working electrodes arranged in parallel or perpendicular to the strip, Sens. Actuator. B Chem. 198 (2014) 302—308.

PART VII

Spectroelectrochemical techniques

Electrochemiluminescence of tris (1,10-phenanthroline) ruthenium(II) complex with multipulsed amperometric detection

Estefanía Núñez-Bajo[1], M. Teresa Fernández Abedul[2]

[1]Department of Bioengineering, Royal School of Mines, Imperial College London, London, United Kingdom; [2]Departamento de Química Física y Analítica, Universidad de Oviedo, Oviedo, Spain

32.1 Background

Electrogenerated chemiluminescence or electrochemiluminescence (ECL) is a phenomenon in which the emission of photons is produced by a high-energy electron transfer reaction between electrogenerated species that is normally accompanied by the regeneration of the emitting species. Although the luminescence generated under electric fields was initially observed in the 1920s [1], it was reported in detail in the 1960s [2]. Since then, this detection technique has been widely studied and applied in many different fields (biosensors, microfluidics, etc.), as can be seen in the many articles that can be found in the literature.

The mechanisms involved in ECL reactions depend on the emitting compound and the species that accompany it. Among others, the ECL of ruthenium complexes and their coreactants has been the most detailed and used in sensors and electron transfer studies [3].

Although modern ECL applications are based exclusively on the use of coreactants, the studies were originally based on ionic ECL annihilation. This mechanism involves the formation of an excited state as a result of an exergonic electron transfer (a process in which a reaction is followed by the emission of photons) between electrochemically generated species. As indicated below, after the emitter R is electrochemically oxidized (Eq. 32.1) and reduced (Eq. 32.2), the cationic ($R^{\bullet+}$) and anionic ($R^{\bullet-}$) radicals are annihilated (Eq. 32.3) forming the excited species R^*, which emits photons when deactivated (Eq. 32.4).

$$R - e \rightarrow R^{\bullet+} \quad \text{(oxidation on the electrode)} \qquad (32.1)$$

$$R + e \rightarrow R^{\bullet-} \quad \text{(reduction on the electrode)} \qquad (32.2)$$

$$R^{\bullet+} + R^{\bullet-} \rightarrow R + R^* \quad \text{(annhilation)} \qquad (32.3)$$

$$R^* \rightarrow R + h\nu \quad \text{(light emission)} \qquad (32.4)$$

When the annihilation reaction is not effective, the use of coreactants increases the intensity of the ECL signal. Unlike the annihilation of electrochemiluminescent ions, in which the electrolytic generation of the oxidized and reduced ECL precursors is required, the ECL with coreactant is usually generated by applying a unidirectional sweep of potentials in a solution containing luminophore species in the presence of an additional agent (coreactant). The redox intermediates of the coreactant are decomposed to produce a very reducing or oxidizing species. Because species with high oxidizing or reducing capacity are formed, ECL reactions are often referred to as "oxidative reduction" ECL (if the intermediate is a reducing agent generated by the electrochemical oxidation of the coreactant) and "reductive oxidation" ECL (if the intermediate is an oxidizing agent, product of an electrochemical reduction). Thus, a coreactant is a species that, after electrochemical oxidation or reduction, immediately undergoes chemical decomposition to form a strongly reducing or oxidizing intermediate. This can react with the oxidized or reduced luminophore, respectively, to generate excited states [4]. Then, the coreactant is consumed but, after ECL, the luminophore is regenerated.

A good coreactant must meet the following conditions, it has: (i) to be soluble in the reaction medium since the ECL intensity is generally proportional to the coreactant concentration, (ii) to produce stable intermediate species in the solution under the electrolytic conditions, (iii) to be easily reduced or oxidized to quickly produce the chemical reaction that leads to the emission, and (iv) to not produce ECL extinction effects or be a luminophore species.

Although there is a wide variety of molecules that exhibit ECL, most publications related to the use of coreactants involve the use of ruthenium organometallic complexes [5] because of their excellent chemical, electrochemical, and photochemical properties even in aqueous media and in the presence of oxygen [6]. In this chapter, $[Ru(phen)_3]^{2+}$ (luminophore) is used in combination with a tertiary amine (coreactant) where the mechanism generally consists of three basic steps: (i) redox reactions at the electrode, (ii) homogeneous chemical reactions, and (iii) formation of excited species. As the mechanisms related to the luminophore/coreactant pair are complex, and the main objective of the experiments is the familiarization with the analytical technique, a detailed description of the mechanism will be obviated.

In this chapter, the dependence of the ECL signal on the luminophore and the medium is studied by cyclic voltammetry (CV) in $[Ru(phen)_3]^{2+}$ solutions. Once the electrochemical and electrochemiluminescent behavior is characterized, the ECL generated by multipulsed amperometric detection (MAD) is evaluated as an alternative to CV. This technique has several advantages: (i) as a pulsed technique, it can increase the sensitivity because the capacitive current decays faster than the faradaic one, increasing the ratio between them (as ECL and electrochemical signals are related, this increase in sensitivity could have a direct effect on the ECL signal), (ii) in case monitorization of the signal with time is desired, application of pulses can be programmed for a long time without performing a manual control, and (iii) once the measurement is finished, file saving and loading is faster because a lower volume of data is obtained when compared with those coming from CV.

Currently, all the commercially available analytical instruments are based on the ECL technology with the use of coreactant in aqueous media. The format and materials of the ECL instrumentation depend on the medium (organic or aqueous), the analyte, the introduction of measurement solutions (in flow or static configurations), and the scope of application (routine or decentralized analysis). Various commercial equipments can be found in the literature [7,8], although numerous studies are carried out in homemade instruments [9]. For the purposes highlighted above, a bipotentiostat/galvanostat and a UV-VIS spectrophotometer combined in a single commercial instrument (μStat-ECL) is used and controlled using specific software [10]. The electrochemiluminescent and electrochemical signals are simultaneously generated and captured (Fig. 32.1), which allows a better understanding of the electrochemical reactions involved in the light emission and the optimization of new analytical methodologies.

The experiment is directed to graduate or Master students of courses related to the development of analytical methodologies (Chemistry, Biotechnology, etc.). It will give them, during a 6-h laboratory session (or two 3-h sessions), an appreciation of the basis of the electrochemiluminescent reactions and the optimization of the electrochemical parameters required as excitation source. They will know about a technique that includes both electrochemical and optical phenomena and is among the most sensitive for bioassays detection. Actually, it is included in many commercial autoanalyzers for high-throughput analysis.

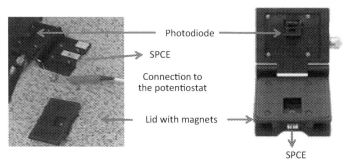

FIGURE 32.1   Different views of the electrochemiluminescence (ECL) cell that is connected to the potentiostat, controlled by a specific software that allows recording simultaneously electrochemical and ECL data. *SPCE*, screen-printed carbon electrode.

32.2 Chemicals and supplies

Solutions

- 10 mM phosphate buffer (PB) pH 8.0.
- 10 mM Tris-HCl buffer (Tris-HCl) pH 8.0.
- 10 mM Tris-1 mM ethylenediaminetetraacetic acid (EDTA) buffer (TE) pH 8.0.
- Dichlorotris(1,10-phenanthroline) ruthenium(II) solution. A 20 mM stock solution of this compound should be prepared in 10 mM Tris-HCl pH 8.0.
- Milli-Q purified water is employed to prepare all buffer solutions.

Instrumentation and materials

- μStat ECL bipotentiostat/galvanostat combined with the ECL cell (Si-photodiode integrated with spectral response range: 340–1100 nm), CAST connector, USB connector, and screen-printed carbon electrodes (SPCEs, DRP-110). All from Metrohm DropSens.
- Micropipettes (1–10, 10–100, and 100–1000 μL) with corresponding tips.
- Microcentrifuge tubes (0.5–1.5 mL), volumetric flasks (10, 25 mL), and other glassware material (beakers, etc.).
- Analytical weighing scale, pH meter, and laboratory spatulas.

32.3 Hazards

EDTA causes serious eye irritation; hence, safety goggles have to be used in every moment. In case of eye irritation, eyes have to be rinsed cautiously with water for several minutes.

The chemical, physical, and toxicological properties of dichlorotris (1,10-phenanthroline) ruthenium (II) have not been thoroughly investigated. It is not considered a hazardous substance according to Regulation (EC) No. 1272/2008 and is not classified as dangerous according to Directive 67/548/EEC.

Students are required to wear gloves, eye protection, lab coat, and other appropriate protective equipment during this experiment.

32.4 Experimental procedure

32.4.1 General setup of the equipment

1. Assure that all the components (potentiostat and ECL cell) are correctly connected as shown in Fig. 32.1.
2. Connect a screen-printed card (SPCE) to the bipotentiostat/galvanostat instrument using the CAST connector.
3. Turn on the μSTAT-ECL instrument by pressing the button in the potentiostat, open the software, and connect the instrument.
4. Open the ECL cell, take out the lid with magnets, and place the SPCE on the slot of the base.

5. Add 10 μL of solution over the three electrodes.
6. Place the lid with magnets carefully, avoiding the splash of the drop and fitting the O-ring on the top of the electrochemical cell (the working electrode should be the only electrode visible through the hole in the lid).
7. Add 40 μL of solution on the top of the electrodes through the hole of the lid with magnets.
8. Close the ECL cell and check that the Si-photodiode fits perfectly.
9. Perform dual electrochemical (EC)/ECL detection.
10. Choose the electrochemical technique (excitation source of ECL) and introduce the corresponding parameters.
11. Run the experiment, record the corresponding curve, and save it following the manual of the software.
12. Take out the clamp, disconnect the CAST connector, and open the ECL cell carefully.
13. Between experiments, take out the lid with magnets, rinse it with water, and dry it.
14. At the end, turn off the equipment, first the software and then the bipotentiostat/galvanostat.

In this chapter, CV and MAD are the methods applied for the electrochemical excitation. The parameters required for each experiment are commented in the corresponding section.

32.4.2 EC and ECL characterization of $[Ru(Phen)_3]^{2+}$ by CV

1. In a 500-μL microcentrifuge tube, prepare a 5 mM $[Ru(Phen)_3]^{2+}$ solution in Tris-HCl buffer pH 8.0 and follow the steps 1–10 described in Section 32.4.1, selecting CV as technique.
2. Introduce the following parameters: $E_{step} = 0.002$ V, $v = 0.1$ V/s, $E_i = +0.3$ V, $E_\lambda = +1.4$ V (Evtx1 in the software), $E_f = +0.3$ V (Evtx2 in the software).
3. Run the experiment and save the voltammogram (I vs. E curve) and ECL emission profile (I_{ECL} vs. E curve). An example is shown in Fig. 32.2.

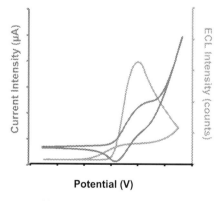

Potential (V)

FIGURE 32.2 Cyclic voltammogram (blue) (dark gray in print version) and electrochemiluminescence (ECL) intensity (green) (gray in print version) versus potential profile recorded simultaneously in a 5 mM $[Ru(phen)_3]^{2+}$ solution in 10 mM Tris-HCl buffer pH 8.0.

4. Repeat the steps 1—3 of this section using two new electrodes.
5. Represent the ECL intensities and current intensities recorded versus the potential.
6. Discuss the redox processes involved. Indicate the average peak current intensity, maximum ECL intensity, and peak potentials.

32.4.3 Effect of the buffer composition on the ECL signal obtained by CV

1. Repeat the procedure described in Section 32.4.2 in different buffer solutions: PB solution, Tris-HCl, and TE pH 8.0 without $[Ru(Phen)_3]^{2+}$ and using one SPCE for each solution. No ECL signal should be observed. If there is any, the solution or the SPCE can contain impurities. Repeat after ensuring everything is clean.
2. In different microcentrifuge tubes of 500 μL, prepare the following solutions:
   - 1 mM $[Ru(Phen)_3]^{2+}$ solution in 10 mM PB pH 8.0
   - 1 mM $[Ru(Phen)_3]^{2+}$ solution in 10 mM Tris-HCl buffer pH 8.0
   - 1 mM $[Ru(Phen)_3]^{2+}$ solution in 10 mM Tris-1mM EDTA (TE) buffer pH 8.0
4. Repeat the procedure described in Section 32.4.2 using three SPCEs for each solution (nine measurements).
5. Represent the ECL emission profiles (I_{ECL} vs. E) recorded in the different ruthenium solutions in a single graph by overlapping the most representative of each solution. Indicate the average ECL peak intensity and peak potential.
6. Considering the introduction section, explain the effect of the buffer on the ECL signal recorded in 1 mM $[Ru(Phen)_3]^{2+}$ solutions observing the voltammograms recorded in different buffer solutions. Choose the appropriate medium for the rest of experiments, as the one that provides the most intense ECL signal.
7. Discuss the excitation potential to be applied to perform MAD.

32.4.4 Optimization of the multipulsed amperometric detection

1. In a 1.5-mL microcentrifuge tube, prepare a 100 μM $[Ru(Phen)_3]^{2+}$ solution in 10 mM TE buffer pH 8.0.
2. Follow the steps 1—10 described in Section 32.4.1 selecting MAD as technique.

The MAD procedure consists of three steps: in the first one the initial potential (E_1) is maintained for a fixed time (t_1), in a second one a pulse is applied, moving the potential to E_2 for a time t_2, and in the final one the potential is moved again to an E_3 value (that can be equal to E_1), where it is maintained for a t_3 time. Then, the cycle is relaxation (t_1 at E_1)/excitation (t_2 at E_2)/relaxation (t_3 at E_3). In the case E_3 is equal to E_1, the procedure can be reduced to two steps: relaxation and excitation. Then, the potentials applied during the excitation and relaxation, as well as the time of application should be optimized. This is made in a one-factor-at-a-time optimization, but an experimental design (see Chapter 35) could also be employed.

32.4.4.1 Optimization of the excitation pulse potential (E_2)
1. Introduce the following parameters in the software: Number of steps: 3, interval: 0.1 s, repetitions: 2, cell 1: 1, E_1: +0.3 V, t_1: 0.1 s, cell 2: 1, E_2: +1.4 V, t_2: 0.5 s, cell 3: 1, E_3: +0.3 V and t_3: 0.5 s.

2. Run the experiment in a 100-μM $[Ru(Phen)_3]^{2+}$ solution in 10 mM TE buffer pH 8.0 and save the recorded ECL emission profiles (I_{ECL} vs. time).
3. Perform the same experiment applying +1.3, +1.35, and +1.45 V of excitation pulse potential, washing the electrode between experiments.
4. Represent the ECL emission profiles (I_{ECL} vs. time) in a single graph and discuss the effect of the relaxation pulse potential on the ECL intensity. Choose the optimal potential for further experiments.

32.4.4.2 Optimization of the excitation pulse width (t_2)
1. Introduce the following parameters in the software: Number of steps: 3, interval: 0.1 s, repetitions: 2, cell 1: 1, E_1: +0.3 V, t_1: 0.1 s, cell 2: 1, E_2: +1.4 V, t_2: 0.5 s, cell 3: 1, E_3: +0.3 V and t_3: 0.5 s.
3. Run the experiment and save the recorded ECL emission profiles (I_{ECL} vs. time). An example is shown in Fig. 32.3. Perform the same experiment applying 0.1, 0.2, 1.0, 5.0, and 10 s as excitation pulse width (t_2) washing the electrode between experiments.
4. Represent the electrochemical excitation signal (potential vs. time) indicating the excitation and relaxation zones in the scheme.
5. Represent the ECL emission profiles in a single graph and discuss the effect of the excitation pulse width on the maximum ECL intensity.

32.4.4.3 Optimization of the relaxation pulse potential (E_1 and E_3)
1. Introduce the following parameters in the software: Number of steps: 3, interval: 0.1 s, repetitions: 2, cell 1: 1, E_1: +0.3 V, t_1: 0.1 s, cell 2: 1, E_2: +1.4 V, t_2: 0.5 s, cell 3: 1, E_3: +0.3 V and t_3: 0.5 s.
2. Run the experiment in 100 μM $[Ru(Phen)_3]^{2+}$ solution in 10 mM TE buffer pH 8.0 and save the recorded ECL emission profiles (I_{ECL} vs. time).
3. Perform the same experiment applying +0.0, +0.1, +0.5, +0.8, and +1.0 V of relaxation pulse potential (E and E, where E = E), washing the electrode between experiments.

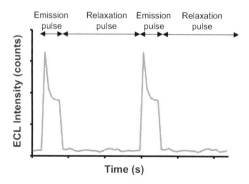

FIGURE 32.3 Electrochemiluminescence (ECL) emission profile recorded in a 100 μM $[Ru(phen)_3]^{2+}$ solution in 10 mM TE buffer pH 8.0. Multipulsed amperometric detection (MAD) signals were recorded applying a MAD program (two repetitions).

4. Represent the ECL emission profiles (I_{ECL} vs. time) in a single graph and discuss the effect of the relaxation pulse potential on the ECL intensity. Choose the optimal potential for further experiments.

32.4.4.4 *Optimization of the relaxation pulse width (t_1 and t_3)*

1. Introduce the following parameters: Number of steps: 3, interval: 0.1 s, repetitions: 2, cell 1: 1, E_1: potential chosen in the previous section, t_1: 0.1 s, cell 2: 1, E_2: +1.4 V, t_2: 0.5 s, cell 3: 1, E_3: potential chosen in the previous section, t_3: 0.5 s.
2. Run the experiment in a 100-µM $[Ru(Phen)_3]^{2+}$ solution in TE buffer pH 8.0 and save the ECL emission profile (I_{ECL} vs. time).
3. Perform the same experiment applying 0.1, 1, 5, and 10 s of relaxation pulse width (changing only t_3).
4. Represent the ECL emission profiles in a single graph and discuss the effect of the relaxation pulse width on the ECL intensity.

32.4.5 Monitoring ECL emission (with multipulsed amperometric detection) with time

1. Considering the results obtained in Section 32.4.4, design a MAD program able to monitor the signal with time (e.g., for 20 min) in a 100-µM $[Ru(Phen)_3]^{2+}$ solution in TE buffer pH 8.0. Introduce the appropriate number of repetitions of a three-step program where the optimized excitation and relaxation parameters (E_1, E_2, E_3, t_1, t_2, t_3) are applied.
2. Run the experiment and save the recorded ECL emission profile (I_{ECL} vs. time). An example is shown in Fig. 32.4.
3. Represent the excitation signal (potential vs. time) and the recorded ECL emission profile (I_{ECL} vs. time). Discuss the stability of the signal with time, the advantages, and a possible analytical application of the method.

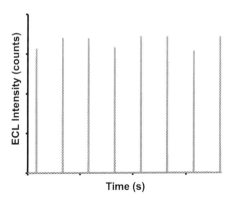

FIGURE 32.4 Electrochemiluminescence (ECL) emission profile recorded in a 100-µM $[Ru(phen)_3]^{2+}$ solution in 10 mM TE buffer pH 8.0, following a multipulsed amperometric detection technology (nine repetitions).

32.5 Lab report

Write a lab report that includes an abstract, a brief introduction explaining the purpose of the experiment, a detailed experimental section, results and discussion, and conclusions. Include tables, graphics, or figures wherever necessary. The following points should be considered:

1. In the introduction, include a revision of reported works with examples of other electro-chemiluminescent species and coreactants. Include also the most common ECL assays where ruthenium complexes are employed and a schematic representation of the components of the equipment. Remark the advantages and drawbacks of ECL with respect to chemiluminescence (CL) and other optical techniques. Indicate also the advantages of the combination of electrochemical and optical principles.
2. In the experimental section, explain the protocols to prepare the solutions, the equipment used, and the electrochemical techniques and parameters employed. Add schematic representations of the excitation signal (potential vs. time).
3. In the results and discussion section, show the most representative curves, including the incidences during the course of the experiment. Add a table comparing the results obtained during the optimization of the excitation and relaxation MAD parameters. Include a figure of the ECL signal recorded by MAD for 20 min and a schematic representation of the electrochemical excitation.
4. In the conclusions, highlight the most representative results and include possible applications of the MAD performed for recording measurements with time (e.g., 20 min).

32.6 Additional notes

1. All reagents in these experiments should be stored protected from light at 4°C.
2. Do not reuse SPCEs for concentrations higher than 1 mM of complex because of the possible poisoning of the electrode.
3. Dirt or strains on the window of the photodiode may cause a drop in light transmittance. Avoid the use of solvents as much as possible for cleaning the photodiode. If such use is unavoidable, use ethanol. Avoid touching the photodiode with bare hands and always use gloves.
4. The most intense signal should be obtained in TE buffer pH 8.0. This could be used to explain the effect of EDTA on the signal.
5. This chapter is thought for two sessions of 3 hours, but if the working time wants to be extended, the effect of the EDTA concentration and pH on the ECL signal could be studied. In addition, MAD experiments can be carried out in solutions with different concentrations of ruthenium complex to represent the calibration plot (I_{ECL} vs. concentration) and know the analytical characteristics of the method. Similarly, comparison with the EC and ECL behaviors of different ruthenium complexes (e.g., $[Ru(bpy)_3]^{2+}$) could be made.

32.7 Assessment and discussion questions

1. Compare the electrochemiluminescent technique with the chemiluminescent and discuss the advantages over other optical detection techniques.
2. Considering the theory explained in the introduction, discuss the effect of the buffer solution, comment the type of mechanism in each case, and explain the effect of EDTA in the signal. Find other examples of coreactants.
3. What are the electrochemical techniques employed here for excitation? Comment the basis of both, deciding critically which one is the best.
4. Considering the stability of the MAD signal with time, comment possible applications of the methodology.
5. A dual detection, optical and electrochemical, can be made. What are the advantages?
6. Find in the bibliography electroactive species, not luminophores, what can affect the ECL signal and explain why.

References

[1] R.T. Dufford, D. Nightingale, L.W. Gaddum, Luminescence of Grignard compounds in electric and magnetic fields, and related electrical phenomena, J. Am. Chem. Soc. 49 (1927) 1858–1864.
[2] D.M. Hercules, Chemiluminescence resulting from electrochemically generated species, Science 145 (1964) 808–809.
[3] A.J. Bard, G.M. Whitesides, Luminescent metal chelate labels and means for detection, U.S. Patent 5 (238) (1993) 808.
[4] H.S. White, A.J. Bard, Electrogenerated chemiluminescence. 41. Electrogenerated chemiluminescence and chemiluminescence of the $Ru(2,21\text{-bpy})_3^{2+}\text{-}S_2O_8^{2-}$ system in acetonitrile-water solutions, J. Am. Chem. Soc. 104 (1982) 6891–6895.
[5] R.M. Wightman, S.P. Forry, R. Maus, D. Badocco, P. Pastore, Rate-determining step in the electrogenerated chemiluminescence from tertiary amines with Tris(2,20-bipyridyl)ruthenium(II), J. Phys. Chem. B 108 (2004) 19119–19125.
[6] A. Juris, V. Balzani, F. Barigelletti, S. Campagna, P. Belser, A. von Zelewsky, Ru(II) polypyridine complexes: photophysics, photochemistry, eletrochemistry, and chemiluminescence, Coord. Chem. Rev. 84 (1988) 85–277.
[7] Meso Scale Discovery: www.mesoscale.com.
[8] Roche Diagnostics Corporation: www.roche.com.
[9] F.R.F. Fan, A.J. Bard, Electrogenerated Chemiluminescence, CRC Press, Boca Ratón, 2004, ISBN 9780824753474.
[10] M.M.P.S. Neves, P. Bobes-Limenes, A. Pérez-Junquera, M.B. González-García, D. Hernández-Santos, P. Fanjul-Bolado, Miniaturized analytical instrumentation for electrochemiluminescence assays: a spectrometer and a photodiode-based device, Anal. Bioanal. Chem. 408 (2016) 7121–7127.

CHAPTER

33

Detection of hydrogen peroxide by flow injection analysis based on electrochemiluminescence resonance energy transfer donor–acceptor strategy

María Begoña González-García, Pablo Fanjul-Bolado

Metrohm Dropsens, Parque Tecnológico de Asturias, Edificio CEEI, Asturias, Spain

33.1 Background

Electrogenerated chemiluminescence (ECL) is a process in which electrochemically generated species combine to undergo electron transfer reactions at an electrode surface to form excited, light-emitting species [1,2]. A growing interest in ECL technique has been reported in the last years (see also Chapter 32), and therefore, it has been implemented in several different sensitive analytical designs [3–5] and combined with other analytical methods [6,7]. Usually, ECL reaction is potentiostatically controlled and photon detection is carried out by high-performance detectors. Different optical sensors are currently available to perform the luminescence capture in ECL reactions. Photomultiplier tubes, charge-coupled device cameras, and photodiodes are the most used [8].

Moreover, the ability to perform multiplex assays of different luminophores is almost mandatory. Potential-resolved ECL, where a potential-controlled switching between emission photons takes place, seems to be the most used alternative for the selective detection of multiple luminophores [9]. However, to distinguish different species that are excited at very similar potentials, the analysis of the wavelength of emission is one of the most interesting strategies. It not only facilitates the identification of multiluminescent species but also results a useful tool for the characterization of new ECL systems. In these cases, the use of spectrometers is necessary, usually in the range of wavelengths comprised between

300 and 900 nm, which allow obtaining instantaneous, synchronized, and real-time collection of multicolor emission spectra.

On the other hand, when a nonradiative energy transfer occurs between an excited donor (as a result of an ECL reaction) and an acceptor, which are in close proximity, an electrochemiluminescence resonance energy transfer (ECL-RET) assay takes place [10,11]. The fact that luminescent signal response of the fluorophore acceptor is controlled by the donor ECL reaction is a significant advantage of this kind of assays because it minimizes inconveniences such as high background signals [11].

Therefore, in this chapter, a proof of concept based on the ECL-RET process triggered by the electrooxidized luminol, in presence of H_2O_2, to the acceptor fluorescein is carried out. The experiments proposed are based on reference [12]. The ECL-RET assay described is developed using a flow injection system (see also Chapters 5, 9 and 28) that uses screen-printed electrodes integrated in one-channel flow cell (TLFCL, thin-layer flow cell). In these electrodes, a transparent slide, which usefully allows the detection of air bubbles inside the cell, is attached over the screen-printed electrodes platform delimiting a flow channel. The injection of sample volume is done through an "in-line luer" injection port, placed closest to the electrochemical cell, where the injection can be easily controlled by the operator through a precise syringe. This configuration brings important advantages because it simplifies operability and effectiveness of working in flow injection analysis (FIA) systems. By combining the bipotentiostat/galvanostat and a UV-VIS spectrometer, both available in one-piece spectroelectrochemistry instrument (SPELEC device), the ECL signal is simultaneously electrochemically generated and captured (with the aid of a reflection probe that is connected to the spectrometer). For that purpose, a specific support suitable to insert the disposable TLFCL electrode in an FIA system is used. The electrode, perfectly hold, is connected to the potentiostat with a specific cable. On the other hand, the connection to the spectrometer is established by employing a reflection optical fiber that faces the TLFCL working electrode in one end and connects the spectrometer by the other.

Fig. 33.1 shows the FIA system used in this experiment coupled to a compact SPELEC equipment, controlled by a specific software that allows recording simultaneously electrochemical and ECL data.

The experiment is directed to graduate or Master students of courses related to the development of analytical methodologies (Chemistry, Biotechnology, etc.). It will give them, during a 3-h laboratory session, an appreciation of the chemical basis of the electrochemiluminescence reactions and the particular case of this kind of assays where resonance energy transfer is used.

33.2 Chemicals and supplies

Reagents:

— 0.1 M phosphate buffer (PB) pH 8.0.
— Luminol sodium salt. A 10 mM solution of this compound should be prepared in 0.1 M PB pH 8.0. It should be sonicated for obtaining a proper solution (5 min).
— Hydrogen peroxide. Dilutions of this compound should be done in 0.1 M PB pH 8.0.

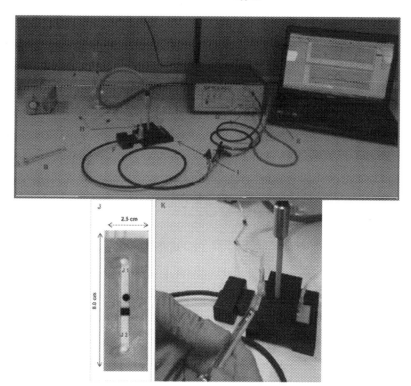

FIGURE 33.1 Image of the miniaturized flow injection analysis system employed and spectroelectrochemistry (SPELEC) instrument: (A) Miniature peristaltic pump with thin-layer flow cell (TLFCL) flow fitting, (B) Hamilton syringe, (C) SPELEC instrument, (D) spectrometer, (E) bipotentiostat/galvanostat, (F) TLFCL connector (CAC-TLFCL), (G) reflection probe (RPROBE), (H) TLFCL in-line port, and (I) TLFCL support (REFLECELL) holding TLFCL110-CIR electrode. (J) TLFCL110-CIR with (J1) TLFCL in-line port insertion hole and (J2) TLFCL-flow fitting for waste disposal. (K) shows a zoom with details of (F), (H), and (I).

- Fluorescein sodium salt. A 10 mM solution of this compound should be prepared in 0.1 M PB pH 8.0. It should be sonicated for obtaining a proper solution (5 min).
- Luminol, hydrogen peroxide, and fluorescein solutions should be protected from light.
- Milli-Q purified water is employed to prepare all buffer solutions.

Instrumentation and materials:

- SPELEC UV-VIS instrument, reflection probe VIS-UV (DRP-RPROBE-VIS-UV), spectroelectrochemical reflection flow cell for thin-layer flow cell screen-printed electrodes (DRP-TLFCL-REFLECELL), thin-layer flow cell graphite screen-printed electrodes with circular working electrode (DRP-TLFCL110-CIR), connector for thin-layer flow cell screen-printed electrodes (DRP-CAC-TLFCL), flow fittings and in-line luer injection port for TLFCL screen-printed electrodes (DRP-TLFCL-FLOWFITTING and DRP-TLFCL-INLINEPORT), all from Metrohm DropSens.
- Peristaltic pump.
- Hamilton syringe 10–100 μL.

- Micropipettes (10–100 and 100–1000 µL) with tips.
- Micro–test tubes (1.5–2 mL).
- Volumetric flasks (10, 25 mL).
- Opaque vials.
- Bath sonicator and magnetic stirrer.
- Analytical balance and laboratory spatulas.
- Beakers and wash bottles.

33.3 Hazards

Luminol and fluorescein can cause skin and respiration irritation and also serious eye irritation. Inflammation of the eye is characterized by redness, watering, and itching. In the case of fluorescein, it may cause adverse reproductive effects (maternal) based on animal studies. It may affect genetic material and cause cancer (tumorigenic) based on animal studies. Hydrogen peroxide is a strong oxidant. It is not flammable itself, but it can cause spontaneous combustion of flammable materials and continue support of the combustion because it liberates oxygen as it decomposes. Hydrogen peroxide mixed with magnesium and a trace of magnesium dioxide will ignite immediately. Keep container tightly closed and in a cool and well-ventilated area. Separate from acids, alkalis, reducing agents, and combustibles.

Avoid entering all compounds into the drain, so proper disposal of residues is required, as well as maintaining clean and duty the working space.

Students are required to wear hand gloves, eye protection, lab coat, and other appropriate protective equipment during this experiment.

33.4 Experimental procedure

33.4.1 Preparation of the flow system to obtain ECL signals

1. First of all, looking at Fig. 33.1, assure that all components (pump, flow tubes, in-line port, flow fittings, etc.) of the flow system are correctly connected.
2. Put an electrode TLFCL 110-CIR into the TLFCL-REFLECELL. For this, open the upper cover of the holder and place the electrode on the base of the holder.
3. Place the in-line luer injection port in the inlet channel (the hole closer to the circular working electrode) and the other flow fitting in the outlet hole of the electrode.
4. Put in the flow carrier container 0.1 M PB pH 8.0 solution.
5. Turn on the pump and fix a flow rate of 1 mL/min.
6. Check the system (there are no leaks, no bubbles on the electrode surface). If all is right, stop the pump and place the upper cover of the holder and turn on again the pump.
7. Connect the electrode to the bipotentiostat/galvanostat of the SPELEC instrument using the appropriate cable connector (see *Instrumentation and materials*).

8. Place the reflection probe into the hole in the upper cover. This hole is over the circular working electrode. Connect one of the ends of the reflection probe to the spectrometer of the SPELEC instrument (the wider end). (CAUTION: The reflection probe has two ends. In spectroelectrochemical experiments, one of the ends is usually connected to the light sources and the other one to the spectrometer. In this case, the light sources are not going to be used in these experiments. Please keep closed the other end during the experiments).

9. Turn on the SPELEC instrument, open the SPELEC software, and connect the instrument.

10. Load the procedure for the amperometric detection technique.

11. In the electrochemical tab, introduce the following parameters:
    - Conditioning potential (E_{cond}) = 0 V
    - Conditioning time (t_{cond}) = 0 s
    - Deposition potential (E_{dep}) = 0 V
    - Deposition time (t_{dep}) = 0 s
    - Equilibration time (t_{eq}) = 3 s
    - Detection potential (E) = +0.6 V
    - Interval time (interval) = 0.1 s
    - Measurement time (t) = 1000 s

12. Introduce the followed optical parameters in the optical tab:
    - Integration time = 1000 ms
    - Scan to average = 1
    - Boxcar = 1

13. Set dark and reference spectrum (in this case is the same).

14. Optical information can be shown in intensity (counts) or intensity minus dark. Select the last one.

15. Run the experiment. The amperogram is then recorded in the electrochemical window, and simultaneously, in the optical window, one spectrum is recorded per second.

16. Wait until the amperogram reaches a stable background line before making the first injection.

17. Make injections of 60 μL of buffer solution or mixtures of luminol, hydrogen peroxide, and fluorescein. For injections, use the Hamilton syringe to load 60 μL and introduce the needle of the syringe into the septum of the in-line luer injection port and push carefully the plunger of the syringe.

18. When the experiment finishes, save it.

19. Execute the function spectra versus time from the "Experiments" (see software manual), select the wavelength of the maximum emission of the luminol or/and fluorescein, and send the graph to the secondary axis of the workspace (a new window will open in the screen). After this, select EC curve of the tree of experiments and send it to the principal axis of the workspace.

Fig. 33.2 shows an example of the ECL and EC signals that can be obtained.

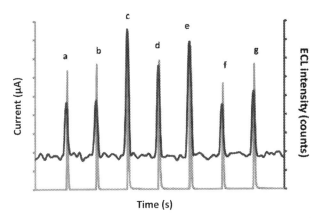

FIGURE 33.2 Luminol-H_2O_2 electrogenerated chemiluminescence (ECL) assay: synchronized electrochemical response (EC, blue color) (black color in print version) and ECL intensity peaks (ECL, red color) (dark gray in print version) versus time, for different H_2O_2 concentrations (a, b, f, g: 2.5 mM; c, e: 7.5 mM; d: 5 mM). Experimental conditions: [luminol] = 1 mM. Integration time: 1000 ms. Amperometric detection was performed at +0.6 V (vs. Ag pseudo-reference electrode). Sample volume injection: 60 μL. Flow carrier: 0.1 M phosphate buffer pH 8.0 with 1 mL/min of flow rate. Selected wavelength for spectra versus time function (red curve): 425 nm.

33.4.2 Injections of different mixtures of luminol, hydrogen peroxide, and fluorescein

1. In six tubes of 1.5 mL (protected from light or opaque tubes), labeled with letters A−F, prepare the following solutions:
   - Tube A: 2 mM of luminol in 0.1 M PB pH 8.0 solution.
   - Tube B: 50 mM of H_2O_2 in 0.1 M PB pH 8.0 solution.
   - Tube C: 0.1 mM fluorescein in 0.1 M PB pH 8.0 solution.
   - Tube D: Mixture of luminol (2 mM) and H_2O_2 (50 mM) in 0.1 M PB pH 8.0 solution.
   - Tube E: Mixture of fluorescein (0.1 mM) and H_2O_2 (50 mM) in 0.1 M PB pH 8.0 solution.
   - Tube F: Mixture of luminol (2 mM), fluorescein (0.1 mM), and H_2O_2 (50 mM) in 0.1 M PB pH 8.0 solution.
2. Make consecutive injections of the six tubes (duplicate injections for each tube) in the flow system, starting from tube A and finishing with tube F. Follow the steps described in the previous section and observe what happens after each injection in both electrochemical and optical windows. It is possible to visualize again the experiment after it finishes using the tool named "experiment film" in the menu experiments (see software manual for more details).
3. When the experiment finishes, save it and later represent the spectra versus time in the workspace (see previous section) for two wavelengths: 425 and 535 nm.
4. Send to the workspace the EC curve as described in previous section.
5. Save the workspace or export the curves to excel (see software manual).
6. Explain what happens for each injection and the reason of the electrochemical and optical behaviors.

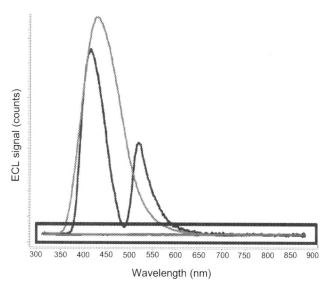

FIGURE 33.3 UV-Vis spectra obtained for injections of each tube. Tubes D (blue color) (black color in print version) and F (red color) (dark gray in print version). The rest of the tubes are in the square. No bands are observed for the rest of injections.

7. Obtain a graph ECL counts versus wavelength for each tube. For this, follow the next steps:
   - Note in the lab notebook the time values of the peak currents obtained in the amperometric curve for each injection (in the case of injections of mixture without luminol, no peak is recorded so the time values will be estimated).
   - Clean the workspace (see manual for details)
   - In the "experiments menu," select the function named "Visible spectrum at EC point," put the cursor on the amperogram of the electrochemical window and click. Only the spectrum that corresponds to this value of time will appear. Send it to the workspace (principal axis).
   - Repeat the previous step for all the values of time (all tubes).

   An example of this graph is shown in Fig. 33.3.

33.4.3 Optimization of concentrations of luminol and fluorescein

33.4.3.1 Optimization of fluorescein concentration

1. First of all, prepare in 1.5-mL tubes mixtures of luminol, H_2O_2, and fluorescein keeping constant the concentrations of luminol (2 mM) and H_2O_2 (50 mM) and varying the concentration of fluorescein from 0 to 5 mM.
2. Make injections of each tube following the procedure explained in Sections 33.4.1 and 33.4.2.
3. Represent in the workspace only the spectra versus time graph for the wavelength 535 nm (maximum emission wavelength of fluorescein) and measure the peak height for each injection.

4. Plot the peak height versus fluorescein concentration and select the best concentration. Explain the results.
5. Obtain the spectrum for each injection and send it to the workspace (before, clean the workspace). Plot in the same workspace all the spectra obtained for each concentration of fluorescein. Explain the behavior of the emission bands of luminol and fluorescein.

33.4.3.2 *Optimization of luminol concentration*
1. Prepare in tubes of 1.5 mL mixtures of luminol, H_2O_2, and fluorescein keeping constant the concentration of fluorescein (selected in the last study) and H_2O_2 (50 mM) and varying the concentration of luminol from 0 to 5 mM.
2. Make injections of each tube following the procedure explained in Sections 33.4.1 and 33.4.2.
3. Represent in the workspace only the spectra versus time graph for the wavelength 535 nm (maximum emission wavelength of fluorescein) and measure the peak height for each injection.
4. Graph the peak height versus luminol concentration and select the best concentration of luminol. Explain the results.
5. Obtain the spectrum for each injection and send to the workspace (before, clean the workspace). Graph in the same workspace all the spectra obtained for each concentration of luminol.

33.4.4 Calibration plot for hydrogen peroxide

1. Follow the procedure explained in previous sections (Section 33.4.1 and 33.4.2) and, using the same parameters for the amperometric detection and optical spectra, make consecutive injections of increasing concentrations of H_2O_2 comprised between 0 and 50 mM. Repeat three times each concentration.
2. After running and saving the experiments, send the spectra versus time graph to the workspace for wavelengths 425 and 535 nm.
3. Plot the peak intensity obtained for each wavelength versus the concentration of hydrogen peroxide. Also, include the deviation for each measurement using error bars. From this graph, estimate the linear range, the sensitivity (as the slope of the calibration curve), the limit of detection (as the concentration corresponding to a signal that is 3 times the standard deviation of the intercept [or of the estimate] in the lower range of concentrations), and the limit of quantification (as the concentration corresponding to a signal that is 10 times the standard deviation of the intercept or estimate).

An example of calibration plot is shown in Fig. 33.4:

33.5 Lab report

Write a lab report following the typical style of a scientific article. It should include an abstract, a brief introduction explaining the purpose of the experiment, a detailed experimental

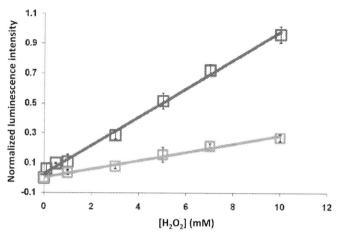

FIGURE 33.4 Calibration plots for hydrogen peroxide at 425 nm (blue color, emission of luminol) (gray in print version) and at 535 nm (orange color, emission of fluorescein) (dark gray in print version).

section, results and discussion, and conclusions. Include tables, graphics, or figures wherever necessary. The following points should be taken into account:

1. In the introduction, include a revision of reported works with other examples of ECL-RET. Include also other kind of assays where resonance energy transfer is used, as for example chemiluminescence with resonance energy transfer (CL-RET). Discuss if it would be possible to convert some of these CL-RET samples in an ECL-RET system. Remark the advantages and drawbacks of ECL-RET with respect to CL-RET assays.
2. In the experimental section, explain the protocols to prepare the solutions in detail and the equipment used.
3. In the results and discussion section, include figures with representative raw data and results presented in tables and graphs (e.g., spectra vs. time or calibration curves), paying special attention to the significant figures in each case.
4. Discuss the values obtained considering the expected results and the incidences during the course of the experiment.

33.6 Additional notes

1. All reagents in these experiments should be stored protected from light.
2. The measurement time (running time) recommended here is 1000 s because 1000 ms of integration time means that a total of 1000 spectra will be recorded in each run. However, in the spectra tree, not all spectra are visible. In case all the spectra want to be seen, check the software manual.
3. This chapter is thought for sessions of 3 hours, but if the working time wants to be extended, parameters such as integration time, flow rate, detection potential, or injection volume can be studied.

4. The software of this instrument has available, among others, an interesting tool named "experiment film" that allows to see a film of the whole experiment after it finishes. For using it, check the software manual.
5. In all the sections, the recording of the amperogram and the optical spectrum should be well synchronized. The instrument and the software guarantee this synchronization.

33.7 Assessment and discussion questions

1. Discuss the reactions involved in the light emission of luminol and the mechanism that explains the light emission fluorescein.
2. Luminol presents anodic electrochemiluminescence. Give other examples of anodic ECL. Include also some examples of cathodic electrochemiluminescence processes.
3. In the optimization of fluorescein concentration, it is expected that the intensity of the peak increases with increasing concentration of fluorescein. However, from a certain high fluorescein concentration, the intensity of the emission band decreases. Explain the reason of this behavior.

References

[1] M.M. Richter, Electrochemiluminescence (ECL), Chem. Rev. 104 (2004) 3003–3036.
[2] L. Hu, X. Guobao, Applications and trends in electrochemiluminescence, Chem. Soc. Rev. 39 (2010) 3275–3304.
[3] J. Ballesta Claver, M.C. Valencia Mirón, L.F. Capitán-Vallvey, Disposable electrochemiluminescent biosensor for lactate determination in saliva, Analyst 134 (2009) 1423–1432.
[4] K. Muzyka, Current trends in the development of the electrochemiluminescent immunosensors, Biosens. Bioelectron. 54 (2014) 393–407.
[5] A.M. Spehar-Délèze, R. Gransee, S. Martínez-Montequín, D. Bejarano-Nosas, S. Dulay, S. Julich, H. Tomaso, C.K. O'Sullivan, Electrochemiluminescence DNA sensor array for multiplex detection of biowarfare agents, Anal. Bioanal. Chem. 407 (2015) 6657–6667.
[6] Y. Chi, J. Duan, S. Lin, G. Chen, Flow injection analysis system equipped with a newly designed electrochemiluminescent detector and its application for detection of 2-thiouracil, Anal. Chem. 78 (2006) 1568–1573.
[7] J. Wang, Z. Yang, X. Wang, N. Yang, Capillary electrophoresis with gold nanoparticles enhanced electrochemiluminescence for the detection of roxithromycin, Talanta 76 (2008) 85–90.
[8] M.A. Carvajal, J. Ballesta-Claver, D.P. Morales, A.J. Palma, M.C. Valencia-Mirón, L.F. Capitán-Vallvey, Portable reconfigurable instrument for analytical determinations using disposable electrochemiluminescent screen-printed electrodes, Sensor. Actuator B Chem. 169 (2012) 46–53.
[9] E.H. Doeven, G.J. Barbante, C.F. Hogan, P.S. Francis, Frontispiece: potential-resolved electrogenerated chemiluminescence for the selective detection of multiple luminophores, ChemPlusChem 80 (2015) 456–470.
[10] A. Roda, M. Guardigli, E. Michelini, M. Mirasoli, Nanobioanalytical luminescence: Förster-type energy transfer methods, Anal. Bioanal. Chem. 393 (2009) 109–123.
[11] M. Mirasoli, M. Guardigli, E. Michelini, A. Roda, Recent advancements in chemical luminescence-based lab-on-chip and microfluidic platforms for bioanalysis, J. Pharm. Biomed. Anal. 87 (2014) 36–52.
[12] M.M.P.S. Neves, M.B. González-García, D. Hernández-Santos, P. Fanjul-Bolado, A miniaturized flow injection analysis aystem for electrogenerated chemiluminescence-based assays, ChemElectroChem 4 (2017) 1686–1689.

Determination of tris(bipyridine) ruthenium(II) based on electrochemical surface-enhanced raman scattering

María Begoña González-García, Alejandro Pérez-Junquera, Pablo Fanjul-Bolado

Metrohm Dropsens, Parque Tecnológico de Asturias, Edificio CEEI, Asturias, Spain

34.1 Background

Spectroelectrochemical techniques have attracted great interest in the last years because they allow obtaining simultaneous information of both electrochemical and optical characters, multiplying the possibilities of studying different chemical systems. Raman spectroelectrochemistry provides information about the vibrational states of molecules and, therefore, about their functional groups and structure, so that is one of the most useful spectroelectrochemical techniques. However, because of the inherent low sensitivity of the normal Raman technique, there are certain technical variations that could significantly increase the signals and, thus, provide a greater capacity to be used in solution for spectroelectrochemical studies such as mechanistic studies or analytical determinations. One of these variations is the surface-enhanced Raman scattering (SERS). It was first introduced by Fleishman et al. in 1974 from their studies using electrochemically roughened silver surfaces and pyridine as a model Raman probe [1]. In this experiment they evidenced the spectacular increment of the Raman signal in a roughened silver electrode. Henceforth, the use of such SERS substrates was generalized and extended to other metals such as gold and copper [2]. Since the first discovery, SERS substrates have been mainly used for qualitative analysis [3]. However, the quantitative approach has not been widely exploited yet, motivated by some problems regarding the sensitivity and/or the reproducibility, among others. In this sense, the recent use of metal nanoparticles (NPs) or in situ electrochemical activation of metal substrates has shown the usefulness of these hybrid techniques for quantitative analysis with good sensitivity and reproducibility [4,5]. In spite of the current generalized use of metal NPs, in many quantitative electrochemical surface-enhanced Raman scattering (EC-SERS) experiments, the employment of roughened metal electrodes presents some advantages over substrates based on chemically synthesized NPs because of the easier, faster, and simpler preparation that could offer in situ activation.

Therefore, the use of an electrochemical pretreatment to achieve a roughened SERS substrate represents a benefit in this sense. However, in most of the cases, because of the difficulty of coupling electrochemical techniques with Raman spectroscopy, the substrate preparation is carried out ex situ and then the species are left to adsorb on the surface for several minutes to obtain an appropriate optical signal. Although this approach is usually useful, a possible time-dependent decrease of the surface and plasmon properties of the substrate could occur (by adsorption of other species, changes in the surface features, etc.), which would lead to optical signals lower than those obtained by using in situ activation. Because of these unavoidable changes, many quantitative experiments carried out so far can suffer from lack of reproducibility, making the technique less reliable. In situ preparation of the SERS substrate overcomes these drawbacks.

On the other hand, screen-printed electrodes (SPEs) have been recently described as substrates for SERS and there are only few examples where they are used as EC-SERS tools [6,7]. Among others, metal SPEs have the advantage of being disposable, cheap, and easy-to-use electrodes. They also can be reproduced in shape and composition with a reasonable precision. Besides, all the three electrodes are integrated in the same device, namely, position of the reference and counter electrode is absolutely reproducible, which could confer more repeatability in the protocol of preparation. Finally, SPEs could be coupled to a specific cell that provides a reproducible methodology in the experimental protocol. All these features make metal SPEs suitable candidates to be used in Raman spectroelectrochemical analytical applications.

This chapter is based on reference [5]. A real-time in situ activation of readily available silver SPEs with excellent time/potential resolution (one spectrum acquired each 50 ms) is proposed to obtain information about the processes leading to a SERS-active substrate. This effect is compared with the same substrate without any electrochemical activation and with a graphite substrate, material where no SERS effect is observed. On the other hand, $[Ru(bpy)_3]^{2+}$ is a very interesting molecule because it is widely employed as a model for spectroelectrochemical systems [5,8,9] and it is the most common detection label in electrochemiluminescence assays [10] because of its great performance and sensitivity.

This experiment is directed to Master students of courses related to the development of analytical methodologies (Chemistry, Biotechnology, etc.). It will give them, during a 3-h laboratory session, an appreciation of the usefulness of Raman spectroelectrochemistry and the SERS effect.

34.2 Spectroelectrochemical setup

In Fig. 34.1, a fully integrated Raman spectroelectrochemistry equipment and a setup for SPEs are shown.

The equipment for Raman spectroelectrochemistry consists on a compact and integrated instrument, which contains a laser source of 785 nm, a bipotentiostat/galvanostat, and a Raman spectrometer. This instrument is connected to a bifurcated reflection probe and a

FIGURE 34.1 Fully integrated instrument for Raman spectroelectrochemistry and setup for screen-printed electrodes.

specific cell for SPEs. The laser spot size is about 200 μm. The equipment is controlled by DropView SPELEC software, which allows performing simultaneous and real-time spectroelectrochemical experiments, with totally synchronized data acquisition.

34.3 Chemicals and supplies

Reagents:

- Tris(2,2′-bipyridyl)dichlororuthenium(II) hexahydrate ($[Ru(bpy)_3]^{2+}$). Solutions of this reagent should be prepared in 0.1 M KCl and protected from light.
- 0.1 M potassium chloride solution (0.1 M KCl).
- Milli-Q purified water is employed to prepare all solutions.

Instrumentation and materials:

- SPELEC RAMAN instrument, Raman probe (DRP-RAMANPROBE), Raman spectroelectrochemical cell (DRP-RAMANCELL), silver SPEs (DRP-C013), graphite SPEs (DRP-110), connector for SPEs (DRP-CAST) from Metrohm DropSens
- Micropipettes (10−100 and 100−1000 μL) with tips
- Microtest tubes (1.5−2 mL)
- Volumetric flasks (10, 25 mL)
- Opaque vials
- Analytical balance and laboratory spatulas
- Beakers, wash bottles, magnetic stirrer

34.4 Hazards

Tris(2,2′-bipyridyl) dichlororuthenium(II) hexahydrate can cause skin and respiration irritation and can cause serious eye irritation. Avoid entering into the drain, so proper disposal of residues is required, as well as maintaining clean and duty the working space.

Students are required to wear hand gloves, eye protection, lab coat, and other appropriate protective equipment during this experiment. For use of laser (785 nm), students should wear special Raman laser safety glasses for this excitation wavelength.

34.5 Experimental procedure

Several experiments will be carried out with and without electrochemical activation. The objective is to show the differences between the Raman spectra obtained without any SERS effect (graphite electrodes), with SERS effect (silver electrodes), and Electrochemical (EC)-SERS effect (activated silver electrodes).

34.5.1 Raman spectra of $[Ru(bpy)_3]^{2+}$ on graphite and silver electrodes without electrochemical activation

1. Open SPELEC software and connect SPELEC Raman instrument.
2. Disconnect the potentiostat (see manual for more details).
3. Introduce the followed optical parameters in Optical tab:
    - Integration time = 50 ms
    - Scan to average = 1
    - Boxcar = 1
    - Laser power = 0.5 V
    - Experimental time = 2 s
4. Place the graphite SPE into the Raman cell and add 60 μL of 1 mM $[Ru(bpy)_3]^{2+}$ solution with the micropipette (ensure a proper contact with the three electrodes).
5. Take the solution with the micropipette from the electrode surface and close carefully the Raman cell. Place again the solution on the electrode by the cell aperture.
6. Introduce the Raman probe.
7. Set dark spectrum.
8. Run the experiment.
9. Change the x-axis from wavelength (nm) to Raman shift (cm^{-1}). Raman spectra are usually expressed in Raman shift. Take into account that the laser wavelength is 785 nm.
10. Optical information can be shown in intensity (counts) or intensity minus dark. Select the last one.
11. Save the experiment.
12. Select spectrum 1 from the spectra collection and send to the workspace (secondary axis). A new window will be opened in the screen. This window keeps opening during all experiments.
13. Remove the electrode Raman probe and the electrode from the Raman cell, clean the cell, and place a silver SPE.
14. Repeat the steps 4–12.

Repeat the experiments with graphite and silver electrodes using a higher concentration of $[Ru(bpy)_3]^{2+}$ (10 mM) and another integration time (1000 ms) and compare the

(A)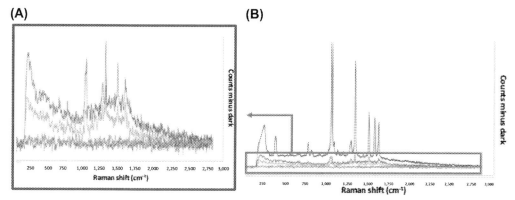

(B)

FIGURE 34.2 (A) Raman spectra obtained without any electrochemical activation for 1 mM [Ru(bpy)$_3$]$^{2+}$ with an integration time of 50 ms (light blue [gray in print version] and green spectra [light gray in print version]) and for 10 mM of [Ru(bpy)$_3$]$^{2+}$ with an integration time of 1000 ms (brown [dark gray in print version] and light brown [light gray in print version]) using graphite (green [gray in print version] and light brown spectra [light gray in print version]) and silver (light blue [gray in print version] and brown spectra [dark gray in print version]) electrodes. (B) Raman spectrum (blue color) (gray in print version) obtained with electrochemical activation using silver electrodes for 1 mM [Ru(bpy)$_3$]$^{2+}$ and an integration time of 50 ms. The inset shows the spectra obtained in part (A).

results obtained. Fig. 34.2A shows examples of Raman spectra obtained following this procedure.

34.5.2 Raman spectra of [Ru(bpy)$_3$]$^{2+}$ on silver electrodes with electrochemical activation

After finishing the experiments carried out in the previous section, connect again the potentiostat (see manual for more details) and follow the next steps:

1. Load the procedure for cyclic voltammetry technique.
2. In the electrochemical tab, introduce the following parameters:
   - Conditioning potential (E_{cond}) = 0 V
   - Conditioning time (t_{cond}) = 0 s
   - Deposition potential (E_{dep}) = 0 V
   - Deposition time (t_{dep}) = 0 s
   - Equilibration time (t_{eq}) = 3 s
   - Initial potential (E_{begin}) = +0.3 V
   - Vertex 1 potential (E_{vtx1}) = −0.4 V
   - Vertex 2 potential (E_{vxt2}) = +0.3 V
   - Step potential (E_{step}) = 0.002 V
   - Scan rate (S_{rate}) = 0.050 V/s
   - Number of scans (n_{scans}) = 1
3. Introduce the same optical parameters than those used in the previous section (see Section 34.4.1) in Optical tab. Now the experimental time is fixed by the electrochemical technique (28 s).

4. Place the silver electrode into the Raman cell and add 60 μL of 1 mM [Ru(bpy)$_3$]$^{2+}$ solution with the micropipette (ensure a proper contact with the three electrodes).

5. Take the solution with the micropipette from the electrode surface and close carefully the Raman cell. Place again the solution on the electrode by the cell aperture.

6. Introduce the Raman probe.

7. Set dark spectrum.

8. Run the experiment.

9. Change the x-axis from wavelength (nm) to Raman shift (cm^{-1}). Raman spectra are usually expressed in Raman shift. Take into account that the laser wavelength is 785 nm.

10. Optical information can be shown in intensity (counts) or intensity minus dark. Select the last one.

11. Save the experiment.

12. Select spectrum with higher peaks (approximately spectrum 309 for this integration time) from the spectra collection and send to the workspace (secondary axis).

Fig. 34.2B shows an example of Raman spectrum obtained with silver electrodes electrochemically activated (EC-SERS). Compare all spectra and explain the results.

Graph the spectra versus EC curve for the last spectrum of [Ru(bpy)$_3$]$^{2+}$ obtained with EC-SERS effect. For this, the software of the equipment has a tool available that allows us to obtain this graph (see software manual). Execute this function for a Raman shift of 1040 cm^{-1} (this Raman shift can be a little higher or lower) and send the graph to the secondary axis of workspace (the workspace should be cleaned previously). After that, select the EC curve of the tree of experiments and send it to the principal axis of the workspace. Fig. 34.3 shows an example of the results that should be obtained. Compare both curves and explain the relationship between them.

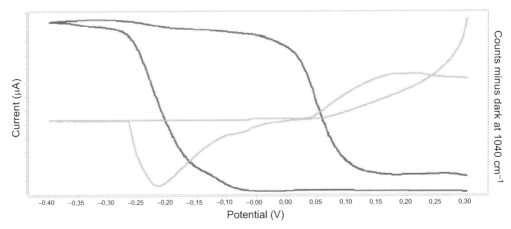

FIGURE 34.3   Spectra versus EC (red curve) (dark gray in print version) and cyclic voltammogram (blue curve) (black color in print version) at 1040 cm^{-1} for 1 mM [Ru(bpy)$_3$]$^{2+}$, obtained with activated silver electrodes.

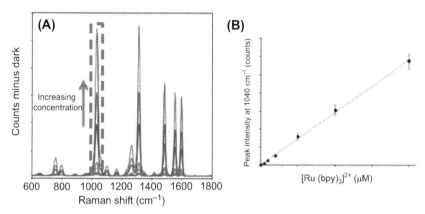

FIGURE 34.4 (A) Raman spectra obtained for different concentrations of $[Ru(bpy)_3]^{2+}$ increasing from 0.05 to 5 µM, and (B) corresponding calibration plot.

34.5.3 Calibration plot for $[Ru(bpy)_3]^{2+}$ through EC-SERS effect

Following the procedure explained in the previous section (Section 34.5.2) and using the same parameters for CV and Raman spectra (except the integration time that should be changed to 2000 ms), make consecutive records for increasing concentrations of $[Ru(bpy)_3]^{2+}$, comprised between 0.05 and 5 µM using a new silver electrode for each assay. Repeat three times each concentration. After running and saving each experiment, send the Raman spectrum with higher peaks to the workspace.

Plot the peak intensity of the band at $1040 \, \text{cm}^{-1}$ versus the concentration of ruthenium complex. Include also the deviation for each measurement using error bars. From this graph, estimate the linear range, the sensitivity (as the slope of the calibration curve), the limit of detection (as the concentration corresponding to a signal that is 3 times the standard deviation of the intercept [or of the estimate] in the lower range of concentrations), and the limit of quantification (as the concentration corresponding to a signal that is 10 times the standard deviation of the intercept or estimate).

An example of the analytical signals obtained and a calibration plot are shown in Fig. 34.4.

34.6 Lab report

Write a lab report following the typical style of a scientific article. It should include an abstract, a brief introduction explaining the purpose of the experiment, a detailed experimental section, results and discussion, and conclusions. Include tables, graphics, or figures wherever necessary. The following points should be taken into account:

1. In the introduction, include a revision of the published works where SERS and EC-SERS are applied, focusing in the main differences between them. Include also a list of the main substrates that are usually employed in Raman experiments based on the SERS effect (with or without electrochemical activation).

2. In the experimental section, explain the protocols to prepare solutions in detail and the equipment used.
3. In the results and discussion section, include figures with representative raw data and results presented in tables and graphs (e.g., spectra vs. EC, calibration curve), paying special attention to the significant figures in each case.
4. Discuss the results obtained considering those expected and the incidences during the course of the experiment.

34.7 Additional notes

1. Ruthenium complex solutions should be stored protected from light.
2. A volume of 60 µL is recommended to cover the three electrodes and at the same time to avoid the Raman probe immerse into solution.
3. In Raman experiments, the distance from the probe to the sample is very important. The focal distance of the Raman probe recommended in this chapter is 7.5 mm and the Raman cell is designed considering this value. If another setup is used, especial attention should be paid to this distance because it affects considerably the intensity of the Raman bands.
4. The values of integration time and the power of laser can vary from one instrument to another, so the values indicated in these experiments can be changed. Then, it is recommended to check these values.
5. Anyway, it is very interesting to check the effect of the integration time and laser power in the Raman spectrum.
6. The software of this instrument has available an interesting tool named "experiment film" that allows to see a film of the whole experiment after it finishes. To use it, check the software manual.
7. In Sections 34.4.2 and 34.4.3, records of cyclic voltammograms and Raman spectra should be well synchronized. The instrument and the software guarantee this synchronization.

34.8 Assessment and discussion questions

1. Discuss the main theories that explain the SERS and the main differences between SERS and EC-SERS.
2. Search in the bibliography the electrochemical behavior of $[Ru(bpy)_3]^{2+}$ and answer these questions:
   (a) During the EC-SERS experiments proposed in this chapter, does the ruthenium complex suffer any oxidation/reduction process?
   (b) Is the experiment proposed in Sections 34.3 and 34.4 a Raman spectroelectrochemical assay? Explain the response.
   (c) Search in the bibliography examples of SERS or EC-SERS with other substrates. Are they Raman spectroelectrochemical assays?

References

[1] M. Fleischmann, P.J. Hendra, A.J. McQuillan, Raman spectra of pyridine adsorbed at a silver electrode, Chem. Phys. Lett. 26 (1974) 163–166.

[2] B. Sharma, R.R. Frontiera, A. Henry, E. Ringe, R.P. Van Duyne, SERS: materials, applications, and the future surface enhanced Raman spectroscopy (SERS) is a powerful vibrational, Mater. Today 15 (2012) 16–25.

[3] S.E.J. Bell, N.M.S. Sirimuthu, Quantitative surface-enhanced Raman spectroscopy, Chem. Soc. Rev. 37 (2008) 1012–1024.

[4] J.F. Betz, W.W. Yu, Y. Cheng, I.M. White, G.W. Rubloff, Simple SERS substrates: powerful, portable, and full of potential, Phys. Chem. Chem. Phys. 16 (2014) 2224–2239.

[5] D. Martín-Yerga, A. Pérez-Junquera, M.B. González-García, J.V. Perales-Rondon, A. Heras, A. Colina, D. Hernández-Santos, P. Fanjul-Bolado, Quantitative Raman spectroelectrochemistry using silver screen-printed electrodes, Electrochim. Acta 264 (2018) 183–190.

[6] D. Li, D.-W. Li, J.S. Fossey, Y.-T. Long, Portable surface-enhanced Raman scattering sensor for rapid detection of aniline and phenol derivatives by onsite electrostatic preconcentration, Anal. Chem. 82 (2010) 9299–9305.

[7] L. Zhao, J. Blackburn, C.L. Brosseau, Quantitative detection of uric acid by electrochemical-surface enhanced Raman spectroscopy using a multilayered Au/Ag substrate, Anal. Chem. 87 (2015) 441–447.

[8] C.A. Schroll, S. Chatterjee, W.R. Heineman, S.A. Bryan, Thin-layer spectroelectrochemistry on an aqueous micro-drop, Electroanalysis 24 (2012) 1065–1070.

[9] D. Martín-Yerga, A. Pérez-Junquera, D. Hernández-Santos, P. Fanjul-Bolado, Electroluminescence of $[Ru(bpy)_3]^{2+}$ at gold and silver screen-printed electrodes followed by real-time spectroelectrochemistry, Phys. Chem. Chem. Phys. 19 (2017) 22633–22637.

[10] J. Li, E. Wang, Applications of tris(2,20-bipyridyl)ruthenium(II) in electrochemiluminescence, Chem. Rec. 12 (2012) 177–187.

PART VIII

General considerations

Design of experiments at electroanalysis. Application to the optimization of nanostructured electrodes for sensor development

Pablo García-Manrique[1], María Begoña González-García[2], María Carmen Blanco-López[1]

[1]Departamento de Química Física y Analítica, Universidad de Oviedo, Oviedo, Spain;
[2]Metrohm Dropsens, Parque Tecnológico de Asturias, Edificio CEEI, Asturias, Spain

35.1 Background

Experimental results are extremely important in Analytical Chemistry. During the development of a method of analysis, they are used to take conclusions and optimize the response. Many researchers use one-factor-at-a-time experiments: they fix the value of one factor while they record the response as they change another one. This would be fine in the cases where the variables are independent. But very often the variables interact, and the signal to optimize in an experiment depends on the level chosen for another factor. At those cases, only a multivariate approach could lead to a true optimum signal.

Design of experiments (DoE) is a set of statistical methods used to find out the relationship between several factors and quantify their effect over the response (the analytical signal in instrumental analysis). One of the main objectives of this approach is to obtain the maximum information at minimum cost, while providing a strong support to the conclusions obtained [1,2]. Statistical analysis is used to separate and evaluate the effects of the factors involved.

In this chapter, we show how to introduce this approach for the in situ synthesis of gold nanoparticles (AuNPs) at the surface of screen-printed carbon electrodes (SPCEs). These nanostructured surfaces will be subsequently used as transducers for mercury determination in tap water. The experiment could be carried out too with carbon nanotube (CNT)—modified SPCEs.

Mercury is a very toxic metal, which causes serious environmental problems. It is released to the environment mainly as a component of pesticides or industrial wastes. It passes then to the water cycle, where the most abundant species is Hg(II). It can be accumulated in the liver, heart, brain, and bones, causing then nervous disorders, intellectual damage, or even death. Most of the regulations have fixed 1 ng/mL as the maximum concentration allowed in drinking water.

The normalized protocols for mercury analysis in the European Union are based on atomic absorption spectroscopy [3] and atomic fluorescence spectroscopy [4]. These and other methods such as cold vapor atomic absorption spectroscopy [5] or inductively coupled plasma mass spectrometry [6] are laborious and time-consuming, requiring bulky instrumentation and specialized analysts. Electroanalytical methods, in contrast, use inexpensive and portable devices. The US Environmental Protection Agency (EPA) has recommended the use of stripping voltammetry with gold electrodes for heavy metal analysis, including mercury [7]. Several electrochemical procedures for Hg(II) can be found in the literature [8,9].

When an electrode is modified with nanoparticles, it acquires the properties and advantages given by them, such as more active surface or improvement of the signal-to-noise ratio. AuNPs offer the possibility to work with gold surfaces, but at lower cost. They have also catalytic activity, and they lower the overpotential required to improve reaction kinetics. As consequence, processes are often more reversible than those observed with a conventional electrode made of the same material.

Mercury can be electrochemically deposited below its reduction potential (under potential deposition, [UPD]) because of a strong attraction between the metal and a gold surface (see Chapters 4 and 23 on the use of the UPDs of lead and cadmium for their determination). This results into a thin alloy layer when the metal is reduced. Based on this effect, a method for Hg(II) determination in water samples with AuNPs-SPCE combined with liquid—liquid microextraction was developed [10]. Some other times a hybrid structure AuNPs-CNTs was used [11]. Fig. 35.1 shows the cyclic voltammograms obtained for a 500 μg/mL of Hg(II) at (A) SPCEs, (B) screen-printed gold electrodes, and (C) AuNPs-CNT-nanostructured electrodes. As it can be observed, at the SPCE only the total deposition of mercury takes place, whereas at graphs B and C there are more cathodic processes.

The peaks labeled as C1 (at +0.42 V approximately) and C2 (+0.25 V) correspond to the first and second UPDs. They form two redox pairs with their corresponding anodic peaks A1 and A2. AuNP-modified electrodes provide more sensitive measurements at lower cost. The first UPD will be used for selective mercury determinations.

In this experiment, an electrochemical sensor for the determination of Hg(II) will be developed based on the use of AuNPs synthesized at SPCEs. DoE will be used for optimization of the nanostructure generation. As a simple example, three variables were chosen: gold precursor concentration, intensity of the current applied, and time. The analytical signal is the peak intensity for Hg oxidation by anodic stripping voltammetry (by square wave voltammetry, SWV), following a preconcentration step at the positive potential corresponding to the first UPD. The principles for SWV have been explained in Chapter 3 and the principle for anodic stripping voltammetry in Chapter 4.

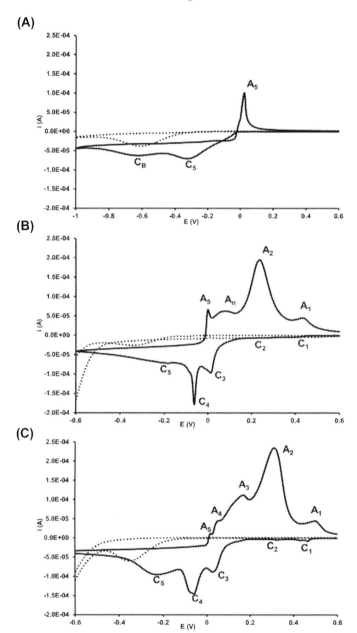

FIGURE 35.1 Mercury redox behavior studied by cyclic voltammetry in a 500 µg/mL Hg solution at (A) screen-printed carbon electrode, (B) screen-printed gold electrode, and (C) gold nanoparticles carbon nabotube–nanostructured electrodes. Anodic (A_{1-5}) and cathodic (C_{1-5}) peaks are shown. Start potential: +0.6 V. Scan rate: 100 mV/s. Dashed lines correspond to the background electrolyte (0.1 M HCl). *Adapted from Sensors and Actuators B 165, 2012, 143–150.*

The DoE approach will allow to:

— Obtain the maximum information about the procedure with the minimum number of experiments.
— Identify the significant effects (i.e., those that cause a response different from the experimental error).
— Evaluate their effect and interactions among the variables used.
— Find the combination of factors that yield the optimum analytical signal.

35.2 Electrochemical cell

The electrochemical cell consists of SPCEs, as in Chapter 23.

35.3 Chemicals and supplies

Reagents and solutions

— *Gold nanostructure source:* Standard $HAuCl_4$ solution (1.000 ± 0.002 g in 500 mL of 1 M HCl) commercially available from Merck. From this solution, aliquots will be taken to prepare the necessary dilutions in 0.1 M HCl for nanostructuration.
— *Background electrolyte:* 0.1 M hydrochloric acid.
— *Stock solution of mercury(II) acetate* (0.5 g/L in 0.1 M HCl). Standards solutions will be prepared by dilution with 0.1 M HCl.
— Milli-Q water is employed for preparing solutions and washing.

Materials and instruments

— Potentiostat/galvanostat.
— SPCEs (DropSens DRP-110).
— DSC connector (DropSens) working as the interface between the SPCEs and the potentiostat.
— General materials and apparatus: 1.5-mL centrifuge tubes, analytical balance micropipettes, and corresponding tips.
— Computer with MiniTab software or an equivalent statistical software package.

35.4 Hazards

This experiment involves the use of a strong acid (HCl). The use of gloves, lab coat and safety glasses is mandatory through the whole experience. On the other hand, because of elevated toxicity, all waste solutions containing mercury should be collected in an appropriate labeled container, and correctly processed according to institution waste policy management.

35.5 Experimental procedure

35.5.1 Preparation of the nanostructured electrodes

An aliquot of 40 mL of $AuCl_4^-$ solution at the concentration chosen is deposited on the SPCEs, ensuring coverage of the three electrodes at the screen-printed card (with reference, working, and auxiliary electrodes). The AuNPs will be electrochemically generated. For the optimization of the nanostructure, three factors will be studied: the effect of Au concentration (0.1 mM or 1 mM), intensity of the current applied ($-5\,\mu A$ or $-100\,\mu A$), and time (60 s or 300 s). According to the literature, those are the most important factors for the formation of gold nanostructures. This study will be carried out following a factorial design (see below). The response function will be the oxidation peak of Hg. The optimum nanostructure will be that giving the maximum intensity.

A different electrode will be used for each step of the optimization. Once all the electrodes are ready, they are rinsed with distilled water, and an aliquot of 40 μL of a 50 ng/mL Hg(II) solution is deposited on them. A potential of $+0.30$ V is applied during 200 s and then the potential is scanned between $+0.30$ and $+0.55$ V. SWV is used at this step (frequency: 40 Hz; amplitude: 20 mV; step potential: 8 mV). The peak obtained for Hg anodic striping would be measured.

35.5.2 Design of experiments for in situ generation of gold nanoparticles

The DoE experimental procedure will follow three main steps:

1. Identification of the parameters involved at the generation of the AuNPs.
2. Randomization of the set of experiments to minimize systematic errors.
3. Use of statistical analysis to separate and evaluate the effect and interaction among variables.

The simplest factorial design requires two levels for each factor. In this case, we have a 2^3 factorial design, which stands for three factors at two levels. The upper and lower values chosen for the factorial design are shown at Table 35.1.

The values of the three factors should be changed simultaneously, covering all the possible combinations. For the 2^3 design, eight combinations are possible ($2\times 2 \times 2$). Each combination will be called "run." Several replicates are necessary to estimate random error. If the order of experiments is randomized, it is possible to distinguish experimental errors from significant changes at the response because of the change of the level of the factors. For this experiment, the optimization of the nanostructures will be carried out by using duplicate measurements (16 runs).

TABLE 35.1 Upper and lower levels for the three factors chosen. The code used for the upper level is (1) and that for the lower level is (-1).

	$[AuCl_4^-]$/mM	$i/\mu A$	t/s
Low level (-1)	0.1	-5	60
High level (1)	1	-100	300

TABLE 35.2   Randomized set of experiments for the optimization of the nanostructured electrodes using a 2^3 full factorial design. The last column is ready to be filled as explained in Section 35.5.3.

Run order	[Au]	i	t	[Au]/mM	I/μA	t/s	Signal (μA)
1	1	1	1	1	−100	300	
2	−1	1	1	0.1	−100	300	
3	−1	−1	−1	0.1	−5	60	
4	−1	1	1	0.1	−100	300	
5	1	1	−1	1	−100	60	
6	−1	−1	1	0.1	−5	300	
7	−1	1	−1	0.1	−100	60	
8	1	−1	1	1	−5	300	
9	1	−1	−1	1	−5	60	
10	1	1	1	1	−100	300	
11	−1	−1	−1	0.1	−5	60	
12	1	−1	1	1	−5	300	
13	1	−1	−1	1	−5	60	
14	−1	−1	1	0.1	−5	300	
15	1	1	−1	1	−100	60	
16	−1	1	−1	0.1	−100	60	

The statistical software packages allow designing randomized combinations. An example of the resulting spreadsheet is shown at Table 35.2. This table could be distributed by the instructor at the beginning of the experiment. The students will complete then the column corresponding to the analytical signal (intensity of the oxidation peak for Hg). Then, data collected will be analyzed with the aim of identifying the principal effects and interactions among factors (Section 35.5.3).

35.5.3 Optimization of the nanostructured electrodes

Once all the electrodes have been nanostructured, they are washed with distilled water, and a 40 μL aliquot of Hg(II) solution (50 ng/mL) is deposited on them. A potential of +0.30 V is applied during 300 s. Then, the potential is swept from +0.30 to +0.55 V. For this step, SWV is advised (frequency: 40 Hz; pulse amplitude: 20 mV; step potential: 8 mV). The anodic stripping voltammetry peak for Hg is measured, and the last column at Table 35.2 is completed with the peak current intensity value corresponding to each run. This variable is the so-called *response* at DoE software (i.e., the output for what models are created).

Once all runs are completed and results are listed into Table 35.2, those values can be introduced into a statistical software program. ANOVA analysis allows identifying the significance level of the factors and distinguishing their effect from the experimental error (random). A deep explanation of the predictive model is beyond the scope of this work. It can be assisted by references 1 and 2. Moreover, a glossary of the specific terms used is included at the end, in Section 35.7. However, for a practical interpretation, the following steps could be followed, as a complement of a chemometrics course:

1. Residual analysis. Validity of the model.

    The results are valid only if the random error is not affected by the level of the factor (homogeneity of variance) and the residues follow a normal distribution. These properties can be easily checked at Figs. 35.2A—D.

    There are some considerations to be taken into account:
    - If the experimental data follow a normal distribution, the accumulated frequency represented versus the data should follow an "S"-shaped curve. This could be shown as a line if the y-axis follows a logarithmic scale (Fig. 35.2A). Experimental points should fit approximately on the normal probability plot.
    - Residuals should be randomly distributed above and below the corresponding fitted values (Fig. 35.2B). In the case that individual high residuals were observed, they would indicate outliers and should be explained.

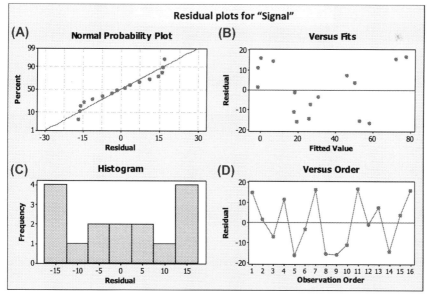

FIGURE 35.2   Examples of different residuals plot: (A) normal probability plot, (B) residuals versus fitted value, (C) frequency of residuals, and (D) residuals versus observation order. These graphs help to verify randomness, normality, and homogeneity of variance. *Screenshots from Minitab software (v.14).*

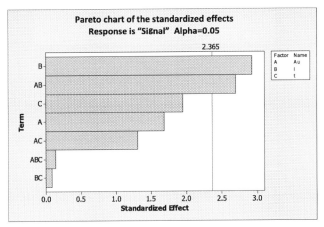

FIGURE 35.3   Pareto chart of the standardized effects related to electrode nanostructuration for electrochemical detection of mercury. Red line limits statistical significance. *Screenshots from Minitab software (v.14) analysis for real experimental results.*

- Frequency on the residuals should not show a preferable trend (generally, small residuals are more frequent than large residuals) (Fig. 35.2C). They could also indicate different dispersion depending on the level of the factor.
- A constant increase or decrease of the residuals following the run order should not be observed (example of Fig. 35.2D).

2. Identification of the most significant factors and interactions: Pareto chart

In a second step, those factors with higher influence over the response can be identified by the Pareto chart (Fig. 35.3). This is based on the Pareto principle that assumes that the majority of the variability observed is due only to a few of all the possible effects. Factors or interactions (see Section 35.7) are listed from the most important (upper) to the least (lower), and their magnitude is represented by a bar graph (histogram). The chart includes a reference line to separate the effects that are statistically significant. Very frequently, this threshold is defined by $\alpha < 0.05$ (or $P = 0.05$). This means that there is only a 5% probability that the change observed for the "response" (analytical signal) is due to random error. Therefore, in the experiment described here, we can conclude with a 95% of probability that the response changes are mainly due to the different level of the factor "gold precursor concentration" (A in the Pareto chart) and the interaction effect among that one and "intensity" (AB in the Pareto chart).

3. Effects and interactions plots

Complementary to this graph, statistic software creates very useful graphs to understand the effect of factors. This way, we can easily check when factors and response are positively or negatively correlated (i.e., an increase at the level of the factor is related with an increase in the response or the opposite). Those graphs are the main effects and interaction plots for the response (Fig. 35.4). Interaction and principal effects are defined at Section 35.7.

(A)

(B)

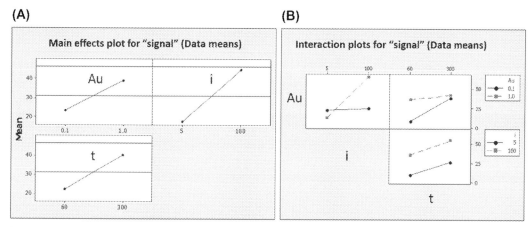

FIGURE 35.4 Graphical representation of main effects (A) and interactions (B) from the factors studied in the electrode nanostructuration for electrochemical detection of mercury. *Screenshots from Minitab software (v.14) analysis for real experimental results.*

In this experiment, we can observe that response increases at the upper level of all the factors. Parallel lines at the interactions plot indicate lack of interaction among those factors, whereas secant lines confirm the interaction. This is the case of the factors "gold precursor concentration" and "intensity," which show a strong interaction. There is interaction too between "gold precursor concentration" and "time," but according to Fig. 35.3, that one is not significant.

Finally, with all the information, the response can be optimized through the selection of the most intense signal obtained with its specific combination of factor's levels. Some software includes optimization tools for this purpose and allows a complete exploration of the experimental region for this purpose. Moreover, more than one response can be co-optimized, including responses where a conflict of interest can arise (such as cost-effective fabrication of devices or maximum sensitivity).

As the analytical signal displayed the maximum values at the upper level of the factors studied, we have selected the following values for the continuation of this experiment:

- Concentration of the gold precursor solution: 1 mM.
- Intensity of the current to be applied: -100 μA.
- Time to apply the intensity current: 300 s.

Higher values were not explored looking for a compromise among sensitivity and time and cost of the experiment.

35.5.4 Determination of mercury in tap water

The determination of Hg(II) will be carried out with the most favorable combination of factors according to the ANOVA analysis of the factorial design. The SPCE will be nanostructured following the optimized procedure: an aliquot of 40 μL of the 1 mM $AuCl_4^-$

solution will be deposited on the SPCE. Then, an intensity current of −100 μA will be applied for 300 s. After rinsing with Milli-Q water, the electrode is ready for Hg(II) determination. A volume of 40 mL of standard (or sample) solution should be then deposited on the nanostructured electrode. Dilutions should always be made in 0.1 M HCl. A potential of −0.5 V should be applied during 120 s. Then, SWV will be performed, sweeping the potential from +0.30 to +0.55 V and using 50 Hz as frequency, 25 mV as amplitude, and 4 mV of step potential. The electrode could be reused after a rinsing step with Milli-Q water.

The interelectrode reproducibility could be evaluated by using a 40 ng/mL standard on five different nanostructured electrodes and calculating the relative standard deviation.

The determination will be performed using an external calibration curve that could be obtained by plotting the signal of Hg standards at the range 2−100 ng/mL. Three replicates of every calibration point are advised. If a constant shift to higher values is observed within the replicates, a different electrode should be used for every measurement. The concentration of mercury in spiked tap water samples can be determined by using the external standard calibration or by the standard addition procedure. Dilution of the spiked sample should be made with 0.1 M HCl. Differences between both slopes (external calibration and standard addition) would indicate the presence of matrix effects.

35.6 Lab report

The report can be done individually or by groups. In any case, this document should contain the following aspects:

— Detailed information about the workflow of the practice.
— Study of the residuals.
— An interpretation of ANOVA results.
— Interpretation of Pareto chart and main/interaction effects graphs.
— Interelectrode variation.
— Calibration plot and quantification of real/spiked samples.

35.7 Additional notes

Glossary of terms

Factorial design: Experimental design used to study the individual effect and interaction among factors involved at a given response. The levels of all factors are changed simultaneously. This is opposed to the univariate design that changes only the level of one factor while keeping all the others constant. In this chapter, *response* is the analytical signal.

Factor: Each variable involved at a change at the response value.

Principal effect: Difference between the average response at the upper level of the factor and the average response of the lower level.

Interaction effect: Two factors interact significantly over the response when the effect of one of them depends on the level of the other factor.

Residuals: Difference between the experimental value for response and the fitted value calculated by the model.

ANOVA: Analysis of the total variation observed, separated into every source contributing to that total variance value.

35.8 Assessment and discussion questions

The student should be able to explain the lab results and procedures. The following activities are suggested as optional upgrading activities:

1. Introduce complementary techniques to fully characterize nanostructured electrodes based on bibliography.
2. Describe how DoE could be used to optimize a liquid—liquid extraction procedure, based on reference 10.

References

[1] J.N. Miller, J. Miller, Statistics and Chemometrics for Analytical Chemistry, fourth ed., Pearson Educational Limited, 2000 (chapter 7).
[2] D. Montgomery, Design and Analysis of Experiments, ninth ed., John Wiley & Sons, 2017.
[3] ISO 12846:2012. https://www.iso.org/standard/51964.html.
[4] ISO 17852:2006. https://www.iso.org/standard/38502.html.
[5] J. Murphy, P. Jones, S.J. Hill, Determination of total mercury in environmental and biological samples by flow injection cold vapor atomic absorption spectrometry, Spectrochim. Acta B: Atom Spectrosc 51 (1996) 1867—1873.
[6] J.S. dos Santos, M. de la Guardia, A. Pastor, M.L.P. dos Santos, Determination of organic and inorganic mercury species in water and sediment samples by HPLC on-line coupled with ICP-MS, Talanta 80 (2009) 207—211.
[7] EPA "Mercury in Aqueous Samples and Extracts by Anodic Stripping Voltammetry" Technical Report December, 1996.
[8] C. Gao, X.-J. Huang, Voltammetric determination of Hg (II), Trends Anal. Chem. 51 (2013) 1—12.
[9] D. Martín-Yerga, M.B. González-García, A. Costa-García, Electrochemical determination of mercury: a Review, Talanta 116 (2013) 1091—1104.
[10] E. Fernández, L. Vidal, D. Martín-Yerga, M.C. Blanco, A. Canals, A. Costa-García, Screen-printed electrode based electrochemical detector coupled with ionic liquid dispersive liquid—liquid microextraction and microvolume back-extraction for determination of mercury in water samples, Talanta 135 (2015) 34—40.
[11] D. Martín-Yerga, M.B. González-García, A. Costa-García, Use of nanohybrid materials as electrochemical transducers for mercury sensors, Sens. Actuator. B 165 (2012) 143—150.

Bibliographic resources in electroanalysis

M. Jesús Lobo-Castañón, M. Teresa Fernández Abedul

Departamento de Química Física y Analítica, Universidad de Oviedo, Oviedo, Spain

This is an era where information can be obtained from very different sources and in an extraordinary amount. However, it is becoming of paramount importance to obtain accurate and reliable information. In this chapter, we will comment some possible sources of information that can be useful for entering the field of electroanalysis and for increasing the knowledge in specific issues of the area.

The two main bibliographic resources are books and journal articles. They provide different types of contents and both are required to either enter or deepen into the world of electroanalysis. As commented in Table 36.1, the objectives and content are very different for both resources. Most of the books deal with fundamental knowledge; however, journals include more or less specialized content. One approaches a new discipline usually looking for a book with a wide and comprehensive view. Thus, they are extremely important now that multidisciplinarity becomes almost a requirement. They are a way to connect disciplines, understanding the basis of the different areas. However, when one is working in a discipline, growing occurs, in part, thanks to the specialized work that is published in "specialized" journals. The focus (scope) of the journals is narrower, but issues are dealt with extreme depth. A book is considered as a learning tool

TABLE 36.1 Main diffferences between books and journals.

Books	Journals
• Fundamental knowledge	• Specialized knowledge
• Wide angle	• Narrow focus
• Comprehensive view	• Extreme depth
• Learning tool	• Latest research/new results
• New topics	• Applied techniques
• Connecting disciplines	• Growing disciplines

Laboratory Methods in Dynamic Electroanalysis
https://doi.org/10.1016/B978-0-12-815932-3.00036-X

for understanding new topics, while journals are usually approached to look for latest research or new results. In any case, both are required because scientists are continuously deepening into their disciplines and interacting with new ones.

36.1 Books and monographs

There is a vast amount of books dealing with fundamentals of electrochemistry and electroanalytical techniques. This section introduces a selection of general manuals explaining the principles of the electroanalytical techniques and selected monographs presenting their applications in specific areas of analysis:

1. A.J. Bard, G. Inzelt, F. Scholz (Eds.), Electrochemical Dictionary, Springer-Verlag, Berlin, 2008.
   A reference manual with short explanations of the scientific terms in the field of electro chemistry. It also includes brief biographies of eminent electrochemists.

2. A.J. Bard, L.F. Faulkner, Electrochemical Methods: Fundamentals and Applications, second ed., John Wiley & Sons, New Jersey, 2001.
   A classical book with a rigorous treatment for the fundamentals of the electrochemical techniques. It includes problems and practice examples. Recommended as reference book or as a textbook for an advanced course in electrochemistry.

3. H.H. Girault, Analytical and Physical Electrochemistry, Marcel Dekker, New York, 2004.
   A book that presents the physical bases of the electroanalytical techniques in a rigorous manner.

4. V.S. Bagotsky, Fundamentals of Electrochemistry, second ed., John Wiley & Sons, New Jersey, 2006.
   A textbook that includes basic concepts, useful for undergraduate students in electro-chemistry, and more advanced topics for postgraduate level. It is organized in four parts devoted to basic concepts, kinetics of electrochemical reactions, applied aspects of electrochemistry and selected topics, which includes electrocatalysis, bioelectrochemistry or nanoelectrochemistry.

5. P.F. Lefrou, J.C. Poignet, Electrochemistry: The Basics, with Examples, Springer-Verlag, Berlin, 2012.
   A textbook presenting the basic electrochemical concepts, its principles and main applications.

6. P.T. Kissinger, W.R. Heineman, Laboratory Techniques in Electroanalytical Chemistry, second ed., Marcel Dekker, New York, 1996.
   Reference book that includes basic concepts on main electroanalytical techniques as well as other more applied issues such as carbon electrodes, film electrodes, solvent and sup-porting electrolytes, or specific topics such as electroorganic synthesis or electrochemical detection in liquid chromatography and capillary electrophoresis.

7. P.M.S. Monk, Fundamentals of Electroanalytical Chemistry, John Wiley & Sons, New York, 2001.
A book on the main techniques in electroanalytical chemistry with some experimental issues such as electrode preparation also included. The book is intended for those wanting to learn at a distance or in absence of a tutor. Accordingly, the approach taken is a series of tutorial questions and worked examples, interspersed with question for students to attempt in their own time.

8. J. Wang, Analytical Electrochemistry, third ed., John Wiley & Sons, New Jersey, 2006.
A textbook covering modern electroanalytical techniques, with clear and reader-friendly explanations of the fundamentals of these techniques. The last chapter is devoted to electrochemical sensors.

9. F. Scholz, Electroanalytical Methods. Guide to Experiments and Applications, second ed., Springer-Verlag, Berlin, 2010.
A textbook that provides the theoretical foundations of electrochemistry and describes in depth the most frequently used electroanalytical techniques.

10. R.G. Compton, C.E. Banks, Understanding Voltammetry, third ed., World Scientific, New Jersey, 2018.
A textbook providing the fundamentals of the various forms of voltammetry, with insight into the design of real experiments. It can be complemented with:
   – R.G. Comptom, C. Batchelor-McAuley, E.J.K. Dickinson. Understanding voltammetry: problems and solutions. Imperial College Press. London, 2012 and,
   – R.G. Comptom, E. Laborda, C.R. Ward. Understanding voltammetry: Simulation of electrode processes. Imperial College Press. London, 2014.

11. S. Alegret, A. Merkoçi (Eds.), Electrochemical Sensor Analysis, Comprehensive collective volume dedicated to electrochemical sensors, Elsevier, Amsterdam, 2007.

12. A. Escarpa, M.C. González, M.A. López (Eds.), Agricultural and Food Electroanalysis, John Wiley & Sons, 2015. A book covering the use of electroanalytical techniques in agricultural and food analysis.

36.2 Journals

There is a category in Journal Science Reports entitled Electrochemistry, which includes 28 journals devoted to covering all the basic and applied aspects in the field. Among them, those listed below, ranked by their impact factor in 2018, include in their scope the publication of original work and reviews in Analytical Electrochemistry.

1. Biosensors and Bioelectronics, Elsevier
2. Sensors and Actuators B: Chemical, Elsevier
3. Electrochimica Acta, Elsevier
4. Bioelectrochemistry, Elsevier
5. Electrochemistry Communications, Elsevier
6. ChemElectroChem, Wiley VCH

7. Journal of Electroanalytical Chemistry, Elsevier
8. Journal of the Electrochemical Society, ECS
9. Sensors, MDPI
10. Electroanalysis, Wiley VCH
11. Journal of Applied Electrochemistry, Springer

In addition, journals profiled on general Chemistry (Analytical or Multidisciplinary) publish frequently works on electroanalytical techniques and electrochemical sensors. Some of these journals, ranked by the impact factor (2018), are listed below:

1. Chemical Reviews, ACS
2. Advanced Materials, Wiley
3. Advanced Functional Materials, Wiley
4. Journal of the American Chemical Society, ACS
5. Angewandte Chemie International Edition, Wiley
6. TrAC, Trends in Analytical Chemistry, Elsevier
7. ACS Sensors, ACS
8. Lab on a Chip, Royal Society of Chemistry (RSC)
9. Analytical Chemistry, American Chemical Society (ACS)
10. Microchimica Acta, Springer
11. Analytica Chimica Acta, Elsevier
12. Talanta, Elsevier
13. Critical Reviews in Analytical Chemistry, Taylor & Francis
14. Analyst, RSC
15. Analytical and Bioanalytical Chemistry, Springer
16. Microchemical Journal, Elsevier
17. RSC Advances, RSC
18. Analytical Biochemistry, Academic Press Elsevier
19. Analytical Methods, RSC
20. Analytical Letters, Taylor & Francis

Journals of the categories of Nanoscience and Nanotechnology, and Materials Science, Multidisciplinary, include in many cases work related to electroanalysis, especially in the field of (bio)sensors and for the purpose of characterization:

1. ACS Nano, ACS
2. Nano Letters, ACS
3. Small, Wiley
4. Nanoscale, RSC
5. Advanced Electronic Materials, Wiley
6. Advanced Healthcare Materials, Wiley

Apart from these, specific applications can be covered by more specialized journals in different categories:

1. Clinical Chemistry, American Association of Clinical Chemistry
2. Journal of Pharmaceutical and Biomedical Analysis, Elsevier
3. Food Chemistry, Elsevier

4. Journal of Agricultural and Food Chemistry, ACS
5. Environmental Science and Technology, ACS

When considering the publication in journals, open access should be considered. This issue is included in Responsible Research and Innovation (RRI) European policies and in many other similar. This is an alternative to the traditional "pay-to-read" publication system because the increasing demand for information has forced people to new approaches. Although economical issues have still to be optimised, it is widely recognized that making research results more accessible contributes to better and more efficient science, as well as to innovation in public and private sectors. In this way, traditional journals are introducing open access possibilities and, on the other hand, new open access journals are continuously arising, Such as "Biosensors and Bioelectronics: X", "Sensing and Bio-sensing Research" or "Sensors and Actuators Reports" from Elsevier or the journals of the editorial MDPI, publisher of Open Access journals (e.g., "Sensors").

36.3 Web resources

The amount of information related to electroanalysis that can be found in different web pages is enormous. This is certainly an advantage, but here it is very important to find accurate and reliable information. Another characteristic of this information, which has to be taken into account, is that it is very "volatile" information: access to pages is not always available and many of them are not updated. As commented below, different agents can deliver this information: scientific societies, universities, providers of equipment and instrumentation, etc.

1. Electrochemistry Societies
   — (ISE) International Society of Electrochemistry (www.ise-online.org)
     It was founded in 1949 by leading European and American electrochemists. The web page of this society contains updated information regarding new books in electrochemistry and forthcoming congresses and meetings organized or sponsored by ISE.
   — The Electrochemical Society (www.electrochem.org), former American Electrochemical Society. It was founded in 1902 with the mission of advancing theory and practice at the forefront of electrochemical and solid-state science and technology. It offers open access to, e.g., the "Journal of the Electrochemical Society" or "ECS Electrochemistry Letters."
   — Societies of Electrochemistry of different countries or groups: e.g., European Society of Electroanalytical Chemistry (ESEAC), Indian Society of Electroanalytical Chemistry (ISEAC: www.iseac.org.in), European Society of Electroanalytical Chemistry (ESEAC), and the Electrochemical Society of Japan (www.electrochem.jp)
2. Electronic resources
   There are some specific places where it is possible to find information on electroanalysis. This is the case of the Analytical Sciences Digital Library (ASDL) (www.asdlib.org, last access date February 2019), a free access peer-reviewed website that catalogs various digital resources related to chemical measurements and instrumentation. The ASDL website is one of the several collections initially funded by the National Science Digital Library program of the National Science Foundation, and it is currently supported by the Division of Analytical Chemistry of the American Chemical Society.

The site http://community.asdlib.org/blog/category/Techniques/electrochemistry-Techniques/, February, 2019, provides multiple links to electrochemistry resources, including laboratory manuals, videos, and lecturer notes.

3. Resources from providers

A compilation of videos of electrochemistry applications can be found in this web page: https://electrochemistryresources.com, February, 2019. It is a noncommercialized site maintained by Gamry Instruments and contains contributions from the electrochemists on staff.

It is also possible to find some interesting webinars in other web pages of commercial sites (as is the case of, e.g., https://metrohm.com).

Index

Note: 'Page numbers followed by "f" indicate figures and "t" indicates tables'.

Printed in the United States
By Bookmasters